AF355988

Hydraulic Engineering and Modelling of Water Flow by Use of Computational Fluid Dynamics (CFD) and Modern Hydraulic Analysis Methods

Hydraulic Engineering and Modelling of Water Flow by Use of Computational Fluid Dynamics (CFD) and Modern Hydraulic Analysis Methods

Guest Editor

Charles R. Ortloff

Basel • Beijing • Wuhan • Barcelona • Belgrade • Novi Sad • Cluj • Manchester

Guest Editor
Charles R. Ortloff
University of Chicago
Chicago, IL
USA

Editorial Office
MDPI AG
Grosspeteranlage 5
4052 Basel, Switzerland

This is a reprint of the Special Issue, published open access by the journal *Water* (ISSN 2073-4441), freely accessible at: https://www.mdpi.com/journal/water/special_issues/hydraulic_engineering_CFD.

For citation purposes, cite each article independently as indicated on the article page online and as indicated below:

Lastname, A.A.; Lastname, B.B. Article Title. *Journal Name* **Year**, *Volume Number*, Page Range.

ISBN 978-3-7258-2653-7 (Hbk)
ISBN 978-3-7258-2654-4 (PDF)
https://doi.org/10.3390/books978-3-7258-2654-4

Contents

Preface

One aim of the chapters to follow is to provide examples of the historical continuity of hydraulic engineering practice from ancient to modern times. Although much information exists in the archeological literature in descriptive terms related to ancient water conveyance systems, the underlying water engineering behind the design and function of water conveyance structures used by ancient water engineers can now be brought forward using Computational Fluid Dynamics (CFD) analysis. This is accomplished with computer models of the geometry of existing ancient canals, aqueducts, reservoirs, and channels using the CFD-calculated water flow patterns through them to extract the design intent of ancient water engineers and, thus, their knowledge of water flow principles. Although CFD results use modern hydraulic engineering terminology, there likely exists an equivalent ancient knowledge base of water flow behavior used and understood by ancient water engineers, albeit in undiscovered texts. Much can be learned about ancient water engineers from water flows observed in nature, which is the ancient equivalent of learning modern water engineering principles from hydraulic flume tests. The investigations presented in the leading chapter give examples of water engineering discoveries obtained from CFD analysis at archeological sites in South America, Europe, Asia, and the Middle East—the examples presented introduce a new water engineering field, Paleohydraulics, to add to traditional anthropological and archeological studies based on socio-political, socio-religious and socio-economic studies. Based on water technologies of several World Heritage Old and New World societies investigated in my field studies, an adjoining manuscript related to Complexity Theory shows how advances in water engineering underlie and influence advances in societal development. Historical climate records tell us that many ancient societies ceased to exist or went through water system modifications to limit climate damage from flood and drought anomalies. The effects of climate variations on water collection and transfer structures, as they affect the agricultural base of a society and its continuity, can be determined through the use of the Complexity Theory chapter; this follows from insights into the water engineering knowledge base used for water conveyance structures in ancient societies as obtained from CFD studies. Thus, we can use CFD analysis coupled with Complexity Theory analysis to obtain estimations and insight into the advance or decline of a society given the degree of sophistication of its water supply and distribution systems' water engineering together with climate change effects. To a degree, many of the Old and New World's water systems are in the early stages of discovery and analysis; several of these systems are discussed in the chapters to follow to give readers a current view of the state of Paleohydraulic studies.

One goal of all societies, ancient and modern, is to have a stable agricultural base to support population growth. This concern has led to many different agricultural field system configurations for ancient and modern societies that are appropriate for different geophysical ecologies and their sources of irrigation water. In many cases, the search for an optimally productive agricultural system stems from trial-and-error field system designs given the variability of crop types, weather and climate fluctuations, and different geophysical soil and aquifer types, as well as the inventiveness of different societies. One manuscript in this collection indicates that research on this subject continues in present times to devise field system designs with optimal productivity. Further, currently, green energy production to limit future destructive climate change effects is a prime concern involving the reduction of CO_2 and methane from the atmosphere. To address this issue, there is a focus on efficiency refinements related to dam systems for energy production in several manuscripts. While energy production refinements related to turbine designs and generator efficiency increases continue, the totality of dam energy-efficiency technologies includes much in the way of water spillage control, energy extraction from post-turbine water flows, and downstream flow conditioning to limit stream bed erosion. Additional manuscripts relate to accessory pipeline ducting to control ground erosion from discharged dam water, spillway design for smooth water discharge, and unique agricultural field system designs necessary to obtain maximum agricultural production. Similar attention is paid to water channel bed erosion concerns for agricultural irrigation water supply systems and maximizing water supplies for cities in ancient and in modern times given the many manuscripts on ancient and modern aqueduct systems presented in the open literature and in this collection.

Studies of ancient water systems in this and previous journal and book presentations from water engineering scholars indicate, in several example cases, sophisticated ancient water engineering technologies in use that predate their official European discovery by centuries and even millennia. The present manuscript collection is intended to present analyses of several ancient water engineering systems followed by analyses of modern-day water control and distribution systems. There is clearly a long history of water engineering development and discovery from ancient to modern times that can be brought forward using modern analysis methodologies as discussed in the presented manuscripts. The manuscripts in this book are meant to detail both modern and ancient hydrodynamics engineering research studies intended to foster further studies using modern analysis methodologies from the water engineering community. The new field of Paleohydraulics, with examples of its use given in the leading manuscript, have the potential to bring forward many discoveries of ancient water engineers that are presently unknown to the history of water engineering and to give credit to ancient water engineers' accomplishments for the first time.

Charles R. Ortloff
Guest Editor

Editorial

Engineering and Modeling of Water Flow via Computational Fluid Dynamics (CFD) and Modern Hydraulic Analysis Methods

Charles R. Ortloff [1,2]

[1] Department of Anthropology, University of Chicago, 1126 East 59th Street, Chicago, IL 60637, USA; ortloff5@aol.com

[2] CFD Consultants International, Ltd., 18310 Southview Avenue, Los Gatos, CA 95033, USA

Citation: Ortloff, C.R. Engineering and Modeling of Water Flow via Computational Fluid Dynamics (CFD) and Modern Hydraulic Analysis Methods. *Water* **2024**, *16*, 3086. https://doi.org/10.3390/w16213086

Received: 15 October 2024
Revised: 18 October 2024
Accepted: 21 October 2024
Published: 28 October 2024

The manuscripts presented in this Special Issue will both present engineering analyses of problems of contemporary importance in fluid mechanics and hydrodynamics and describe historical water systems and the technologies used to design and operate them by ancient Old and New World water engineers. This approach will give readers the opportunity to understand the historic development of water engineering from ancient to modern times. While modern water engineering structures depend upon knowledge of known engineering technologies for their design and function, surviving water structures from centuries past leave gaps in our understanding of the water engineering technologies used for their construction and use. This knowledge gap regarding ancient engineering practices used to design and build complex water systems is addressed by several of the manuscripts focused on providing insights and answers regarding the historical development of water engineering.

To fill this gap in our understanding of ancient water technologies, Computational Fluid Dynamics (CFD) analysis of models of ancient canals and other hydraulic features are made to visualize water flow patterns, outlining the design intent of and usage of these systems by ancient water engineers. As CFD analysis of ancient water engineering technologies is conducted in terms of modern hydraulic terminology, it implies that a comparable knowledge base and understanding of water engineering principles existed in ancient times but was described in ancient technical nomenclature not preserved in the archaeological record. The introductory manuscript [1] gives nine examples of CFD use applied to examine prominent Middle Eastern, European, Asian and South American archaeological World Heritage Sites' water systems dating from 2500 BC to 1400 AD—this approach will identify the ancient water engineering technologies used in their design and operation. Further manuscripts [2,3] provide hydraulic engineering analysis results for 17th-century aqueducts and water storage hydraulic structures in Europe and North America to increase our knowledge of the hydraulic engineering knowledge bases used in later Old and New World sites' design and construction. One gap in the archaeological record concerns elaborate complex water engineering structures used for urban and agricultural use and their connection to and influence on the development and advancement of the social structure of a society. This gap is addressed via Complexity Theory [4] to determine the formal relationship between water technologies used by several ancient societies and the social structure evolution of these ancient societies, which existed in the 500–1450 AD time period. Complexity Theory increases our understanding of the effects of climate change in the form of extended drought and excessive rainfall periods on societal extinctions noted from archaeological investigations—this is now a topic of great importance given present-day extreme climate excursions equal to, or well beyond, those reported in earlier historical periods in terms of intensity. Descriptions of ancient water technologies given in [1–4] will close the gap in our knowledge of the water engineering structures of the ancient world identified through CFD use and Complexity Theory, determining the evolution, growth, and continuance or extinction of ancient societies based on their hydraulic engineering technologies when subject to extreme climate anomalies. Although this identification of ancient water engineering history comes with the downside of not knowing the mathematics and physics technologies available to past structures' water engineers when designing and operating complex water systems, the use of CFD analysis to frame in modern terms the hydraulic engineering technologies of several ancient societies gives a preliminary version of the water engineering knowledge base used in the design, construction and use of their water engineering structures. As an alternate source of water engineering knowledge beyond the equation-based technologies in current use, the observation and recording of water flow characteristics in natural settings by ancient water engineers could be the source of their water technology knowledge comparable to what modern age equations governing water flows duplicate and predict; this observation likely would serve as a knowledge base used by ancient water engineers in the design and construction of their water systems. This observation of ancient methods used to gain knowledge of hydraulic phenomena is comparable to the use of modern hydraulic test facilities to first observe hydraulic phenomena and then devise mathematical equivalencies to describe what has been observed.

Complex water system analysis problems and solutions based upon current mathematical and water physics knowledge are discussed in the remaining six manuscripts. In modern times, one important use of water engineering knowledge concerns societies' need for green energy production. This issue is related to major climate change effects attributed to carbon dioxide released into the atmosphere through organic fuel use, which is the source and cause of catastrophic flooding and drought effects now observed in extreme world weather patterns. Thus, a new emphasis on the importance of green energy production through the use of dams incorporating turbines connected to multiple electricity generators has become a major focus, and this subjects is the focus of the following six manuscripts. Modern green energy production research now places further emphasis on water control and distribution irrigation systems necessary to provide ample food supply for growing world populations. To examine current research on these areas of major concern, six manuscripts present advanced research on the engineering aspects of water energy dissipation regarding dam spillways [5], the structural integrity of hydropower discharge systems [6], the water supply pipeline sourcing of water-related power generating systems [7], and the use of CFD simulations of rotors at dam pipeline outlets as further sources of green energy electrical generation [8]. Additionally, water conservation measures related to the CFD modeling of Ringlet water reservoirs [9] together with CFD investigations of a Tyrollian lake's capacity to accommodate discharged spillway water used for off-season water supply for crops are the subjects of further investigations [10]. Of interest regarding the study of energy dissipation in water flowing in spillways of the 17th-century French Lugdunum aqueduct system [2] is a similar 21st-century study of spillway water patterns and their erosional effects on downstream soil channels [5]—this issue indicates that water transfer effects in channel erosion and energy recovery still constitute major concerns that have remained present over centuries of research. In the same vein, many of the nine examples of ancient water system technologies discussed in [1] still have similar modern counterparts, particularly regarding technologies used for efficient food production increase—this has been a topic of major concern for ancient and modern societies throughout history.

This Special Issue addresses water engineering issues, many of which originate from antiquity, that still generate scholarly concern and interest millennia later insofar as they relate to the survival, continuity and development and progress of advanced societies (as reference [4] illustrates), as well as the need for stable food resources together and clean water resources for cities. This universal need, addressed in ancient and modern times for the development and advancement of cities, continues to be the subject of research studies in the present day. The manuscripts presented in this book provide research results relevant to efficient water use for societal development, a topic that has retained relevance from the earliest days of mankind's existence to the present day. Considering that the improvement and understanding of water technologies applied for urban and agricultural use remain key issues, the totality and range of this Special Issue's manuscripts will advance research on both ancient and modern versions of water technologies.

Conflicts of Interest: Author C.R. Ortloff was employed by the company *CFD Consultants International, Ltd.* The author declares that the research was conducted in the absence of any commercial or financial relationships that could be construed as a potential conflict of interest. The rest authors declare no conflict of interest.

References

1. Ortloff, C.R. CFD Investigations of Water Supply and Distribution Systems of Ancient Old and New World Archaeological Sites to Recover Ancient Water Engineering Technologies. *Water* **2023**, *15*, 1363. [CrossRef]
2. Kessener, P. The Aqueducts of Lugdunum. *Water* **2024**, *16*, 2117. [CrossRef]
3. Wright, K. A Summary of 25 Years of Research on Water Supplies of the Ancestral Pueblo People. *Water* **2024**, *16*, 2462. [CrossRef]
4. Ortloff, C.R. Paleohydraulics and Complexity Theory: Perspectives on Self Organization of Ancient Societies. *Water* **2023**, *15*, 2071. [CrossRef]
5. Mikalsen, L.; Thorsen, K.; Bor, A.; Lia, L. Investigation of Improved Energy Dissipation in Stepped Spillways Applying Bubble Image Velocimetry. *Water* **2024**, *16*, 2432. [CrossRef]
6. Kolahdouzan, F.; Afzalimehr, H.; Siadatmousavi, S.; Jourabloo, A.; Ahmed, S. The Effect of Pipeline Arrangement on Velocity Field and Scouring Process. *Water* **2023**, *15*, 1321. [CrossRef]
7. Radzi, M.; Zawawi, M.; Abas, M.; Mazian, A. Structure Integrity Analysis using Fluid-Structure Interaction at Hydropower Bottom Outlet Discharge. *Water* **2023**, *15*, 1039. [CrossRef]
8. Farouk, M.; Kriaa, K.; Elgamal, M. CFD Simulation of a Submersible Passive Rotor at a Pipe Outlet under Time-Varying Water Jet Flux. *Water* **2022**, *14*, 2822. [CrossRef]
9. Mat Desa, S.; Jamal, M.H.; Mohd, M.S.F.; Samion, M.K.H.; Rahim, N.S.; Muda, R.S.; Sa'ari, R.; Kasiman, E.H.; Mustaffar, M.; Ishak, D.S.M.; et al. Numerical Modelling on Physical Model of Ringlet Reservoir, Cameron Highland, Malaysia: How Flow Conditions Affect the Hydrodynamics. *Water* **2023**, *15*, 1883. [CrossRef]
10. Bor, A.; Szabo-Meszaros, M.; Vereide, K.; Lia, L. Application of Three-Dimensional CFD Model to Determination of the Capacity of Existing Tyrolean Intake. *Water* **2024**, *16*, 737. [CrossRef]

Article

CFD Investigations of Water Supply and Distribution Systems of Ancient Old and New World Archaeological Sites to Recover Ancient Water Engineering Technologies

Charles R. Ortloff [1,2]

1   CFD Consultants International, Ltd., 18310 Southview Avenue, Los Gatos, CA 95033, USA; ortloff5@aol.com
2   Research Associate in Anthropology, University of Chicago, Chicago, IL 60637, USA

Abstract: New to archaeological studies is the field of paleohydrology characterized by the use of modern hydraulic engineering and computational fluid dynamics (CFD) analysis of ancient urban and agricultural water supply and distribution systems. Examples are presented in nine chapters of CFD investigations of old and new world archaeological sites (several of which are World Heritage sites) using FLOW-3D CFD software to bring forward new discoveries revealing the depth of ancient water engineers' knowledge and creativity not previously noted in the archaeological literature. As modern analysis methods reveal technical details of ancient water systems, equivalent ancient technologies exist that were used in the design and operation of ancient water systems, albeit in formats, texts, and origins yet to be discovered. The nine chapters to follow present brief summaries of ancient sites' water systems and the use of CFD and modern hydraulic engineering methods to discover the water engineering knowledge base used by ancient water engineers. Results are new revelations unknown in the current literature of ancient sites. Paleohydrology studies presented serve to add a further dimension to the history of ancient new and old world archaeological sites by bringing forward added details of water engineering projects accomplished by ancient engineers.

Keywords: archaeology; paleohydrology; CFD; ancient hydraulic engineering; ancient world heritage sites; new and old world archaeological sites; recovery of lost technologies

Citation: Ortloff, C.R. CFD Investigations of Water Supply and Distribution Systems of Ancient Old and New World Archaeological Sites to Recover Ancient Water Engineering Technologies. *Water* 2023, *15*, 1363. https://doi.org/10.3390/w15071363

Academic Editor: Giuseppe Pezzinga

Received: 11 February 2023
Revised: 25 March 2023
Accepted: 25 March 2023
Published: 1 April 2023

1. Introduction

Water supply and distribution systems in the ancient world reflect hydraulic engineering technologies that in many ways are comparable to modern practice. One way to uncover the hydraulic engineering technologies used by ancients is to create computational fluid dynamics (CFD) models of their urban and agricultural water transfer and distribution systems and, together with use of modern hydraulic engineering analysis methods, uncover hydraulic engineering technologies used by the ancients in their water system designs. Use of CFD analysis in the new field of paleohydrology provides details of the water engineering used to address the many problems ancient water engineers faced to supply their agricultural and urban water requirements. CFD models together with analysis of ancient fluid mechanics structures provide visualization and understanding of water flow patterns within these structures that reveal the design intent and function created by ancient water engineers. By observation of the computed water flow patterns and the hydraulic engineering designs that produce these patterns, insight into the technology base available to ancient engineers is recovered by CFD analysis to gain insight into ancient engineering practice. To illustrate the use of paleohydrology to understand ancient water engineering, examples of ancient water transport and control features used in canals, channels, aqueducts, reservoirs, and pipelines serve as examples from the archaeological record for sites in ancient South America and the Middle East. For the nine sites examined in the present paper, technical details of FLOW-3D CFD usage (3D mesh size, number of computational cells, turbulence model, solution convergence criteria, fluid properties,

CFD model relation to Google Earth and onsite field photographs, CFD model description details) are noted in the specific references listed each chapter. It is noted that hundreds, if not thousands, of years of destruction by nature's forces as well as man's destructive interventions render many ancient water system remains a fraction of their original form and obscure original design intent. This constraint is typical of archaeological remains where limited data combined with unlimited imagination provide probable answers to the problems and solutions of ancient water engineers. As few documents on water system engineering have survived the centuries and those that do have technical terms that cannot be associated with modern hydraulic engineering technology usage, CFD analysis provides a start to uncover what ancient water engineers had available in their knowledge base to solve water supply problems. In chapters to follow, CFD usage applied to several sites illustrates the use of paleohydrology to uncover the hydraulic knowledge base of ancient societies, and serves as an introduction to this new field.

2. The Water Supply and Distribution System of (300 BCE–CE 106) Petra in Jordan

Aspects of the water supply and distribution system of Petra in Jordan are detailed in the literature ([1], pp. 244–264; [2], pp. 137–169; [3], pp. 137–169; [4–8]). Of interest as a paleohydraulic CFD application is a hilltop water feature located in Figure 1, B-2, lower left side corner (Al Habis), associated with part of the lower Farasa housing district and a contributor to the Great Temple's (Figure 2) water supply system. From Figure 1, branches of the springs at Ain Braq and Ain Ammon serve the outlying areas near the Treasury (El Kazneh located in Figure 1, 1-C) and provide a partial water source for the Roman Soldier's Tomb complex (16, Figure 1), the Renaissance Tomb area sites (17), and a channel to serve the southern canal branch of the Siq water system. Water supply to the Wadi Farasa hilltop area (Figure 1, lower lefthand corner of B-2) may originate from an Ain Braq canal branch and/or the northern branch Siq pipeline to serve the downhill residence quarter with further continuance on to the Great Temple area and other sites on Colonnade Street (25, Figure 1) that include the marketplace areas and the recently excavated temple-pool complex [9–11] on the northern part of Colonnade Street.

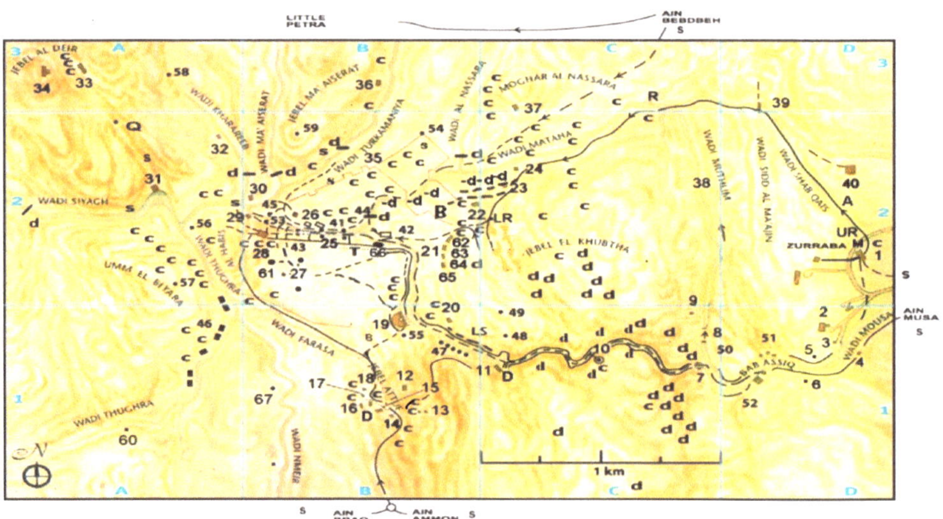

Figure 1. Petra with numbered site names given in Appendix A.1. Figure from [2], p. 266.

Figure 2. The Great Temple (28, Figure 1). Figure from [2], p. 293.

The Great Temple's water supply originates partially from extension of the Siq pipeline past the Theater (B-2, 19), the Marketplace (B-2, 25), the adjacent Paradeisos Water Basin, and the Temenos Gate Al Habis hilltop to the south of the Great Temple. From notes available from the early (1812) Petra explorer Johann Burckhardt, bridges crossing the Wadi Mataha from a castellum (Figure 1, C-2) likely transferred additional pipeline water to the Great Temple and Paradeisos basin areas from springs in that area. Most pipelines are of Nabataean origin, dating from ~300 BCE to CE 106 with later Roman pipelines primarily in the marketplace area, the Great Temple area, and the Roman–Byzantine bath complex [12,13] located in Figure 2 (the top left structure) dating from late Roman occupation of the city. This bath complex had an earlier Greek precedent structure with a novel water supply system later modified or rebuilt in later Roman–Byzantine times, as Section 8 details.

From the Al Habis hilltop junction (lower left corner of B-2, Figure 1) higher in height and to the south of the Great Temple, a pipeline component assemblage exists (Figure 3), likely sourced from the Ain Braq pipeline (B-1) and/or a Siq pipeline extension. A further pipeline branch on the hillside western slope above the Wadi Farasa streambed (B-1 to the southwest corner of B-2) originating from Ain Braq spring source exists but is yet to be confirmed by excavation. Dual downslope pipelines emanate from the Figure 3 assemblage, with one pipeline leading to a downslope housing district and the other to the Great Temple. Remnants of this pipeline's presence were noted by conversations with excavators from the Brown University Joukowsky excavation team active in Great Temple excavations in the 1990s.

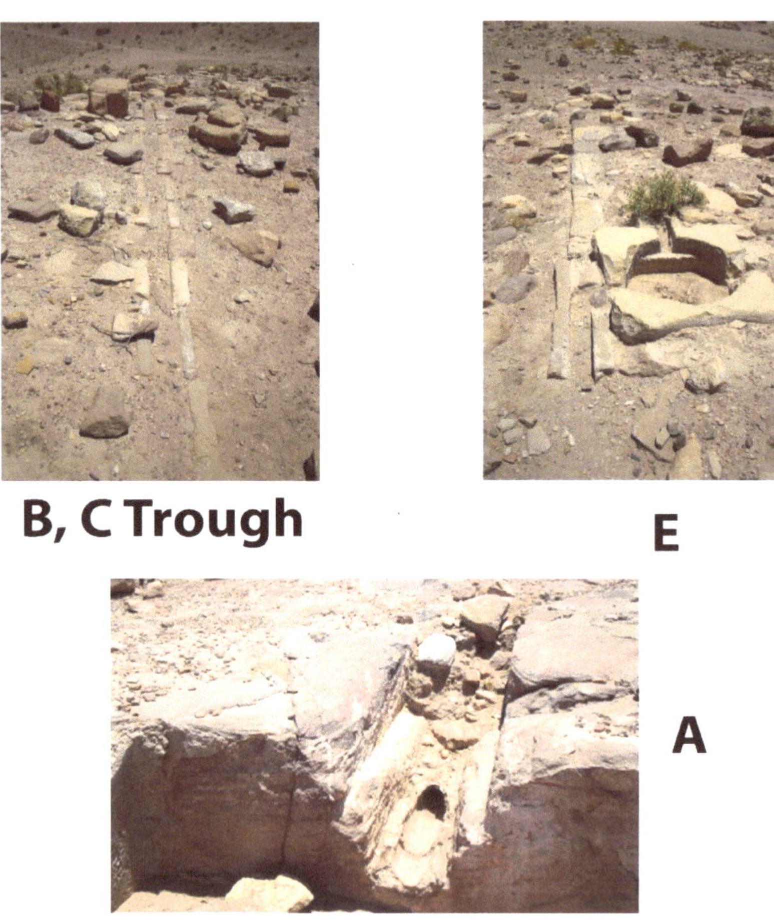

Figure 3. Assemblage components A–C and E of Figure 1, lower left hand B-2 corner of the El Habis hilltop pipeline water supply system. Photo by author on site in Jordan.

It is noted that a Crusader outpost exists on the Al Habis hilltop, indicating a water source from a pipeline emanating from the Ain Braq spring source. Excavated pipeline traces outside of the southernmost back wall of the Great Temple exist ([1], p. 266), further indicating a pipeline connection (Figure 4D,C) derived from an Ain Braq pipeline source to the Figure 3 assemblage. Also located in the Great Temple area above its inner semicircular *Theatron* (Figure 2) is a rear wall, outside of which was a trough (no longer present) that captured pipeline water from the Figure 4D,C pipelines and channeled it to an underground reservoir in the paved area west of the *Theatron*. This reservoir had a top channel to lower reaches of the Great Temple, as excavations from the Joukowsky excavation team indicated. As Great Temple excavation soil was deposited on Al Habis slopes over this deeply buried pipeline, positive determination of its existence awaits excavation. In summary, the Great Temple had multiple water sources. One pipeline from the Figure 3 assemblage was sourced by the Ain Braq spring and Siq pipeline extensions (Figure 1, B-1). A further water source likely exists from a further branch of the Ain Braq pipeline into the Great Temple area, yet to be confirmed by excavation. The Ain Braq pipeline shows multiple branching to the near El Kaznah Treasury area, the Roman Warrior's Tomb complex, and now to the Wadi Farasa area to provide supplemental water supplies to the elite housing area and further on to the Great Temple area and the Paradeisos pool complex located on the main thoroughfare of Petra [9–11]. Only further excavations of the totality of the Ain Braq water supply system can reveal its true extent.

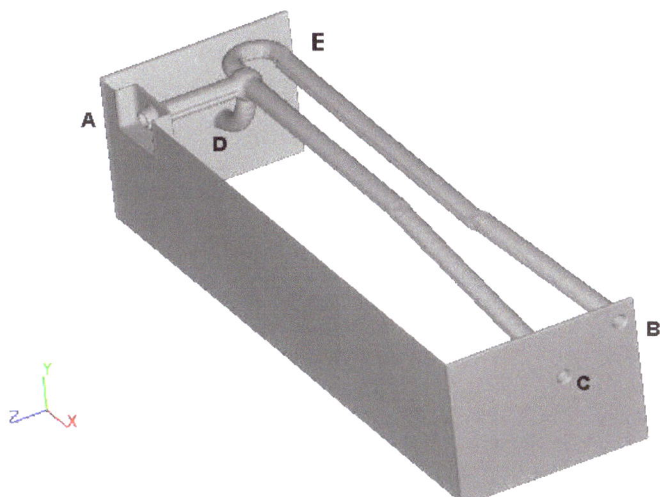

Figure 4. FLOW-3D CFD model of the pipeline connections constructed from Figure 3. Pipeline inner diameter ~15 cm. Figure from [2], p. 291.

Combining excavated connection details of the components shown in Figure 3, the CFD model shown in Figure 4 originated from in-place field observations of A–C and E components.

Assuming an Ain Braq and/or Siq extension supply branch to deliver water to pipeline D, the pipeline component assemblage (Figure 4) consists of basin A with a pipeline segment to E shown in Figure 3. From junction E, pipeline E–B supplied water to the downhill Farasa housing district ([9–14], p. 136) [15], while pipeline branch E–C provided water to the Great Temple and to the nearby Paradeisos water basin area through additional piping from the Great Temple area. Given water supply to the Great Temple, water was transferred to internal Great Temple reservoirs and directly to lower portions of the Great Temple that had drainage to the Wadi Mataha stream bed (Figure 1, path C-3, C-2, B-2, A-2) with drainage to the Wadi Siyagh, A-2. The FLOW-3D computer model (Figure 4) utilizes an input water velocity for full flow input conditions in a level portion of the pipeline at D before the dual pipeline branches E–B, E–C continue to steeper terrain to supply downhill destinations.

The A basin served to manage transient flow excursions, as Ain Braq spring output can vary depending on spring recharge history from seasonal rainfall duration and intensity as well as water diversions into other pipeline branches upstream of the main distribution center with its dual pipeline branches.

The next step is to determine the hydraulic function of the basin A connection to the distribution junction E. For one of several hydraulic functions, the basin pipeline A served to divert flow away from the dual downhill branches if these were blocked during repair or if the distribution center was blocked to divert water to other pipeline branches of the Ain Braq system. The blockage then diverts water to basin A shown in Figure 3. The flow into the basin continues until a hydrostatic pressure balance of water accumulated in basin A balances with the hydrostatic pressure in water supply pipeline D. Thus, an automatic means for system shutoff exists once the lower exit ports of B and C (Figure 3) are blocked.

A further, more sophisticated hydraulic engineering use of the Figure 3 pipeline assemblage configuration originates from the observation that subcritical full flow conditions exist in the supply pipeline D due to the low slope lead-in pipelines to D and pipeline flow resistance from inner-wall pipeline roughness. Due to the hydraulically steep declination angles for dual pipelines E–B and E–C, input full flow past the E junction area is transforms to supercritical partial flow in steep pipeline branches E–B and E–C. This flow transition from full to partial flow induces partial vacuum regions above the partial flow water surface in E–B and E–C pipelines. As the subterranean branch pipelines exits are submerged into reservoirs at their destination locations, air is drawn into pipeline exit openings toward the partial vacuum regions. Transient water/air bubble streams then proceed upstream in dual steep declination pipelines branches to counter partial vacuum regions, leading to flow intermittency and an oscillatory hydraulic jump within the pipeline branches converting pipeline supercritical flow to subcritical flow downstream of the hydraulic jump. This unsteady hydraulic phenomenon affects flow stability of the entire pipeline system, leading to pulsating discharges into the B, C downstream reservoirs as well as affecting the stability of input flow at D.

Additionally, pulsating flows cause internal pressure level changes that can flex pipeline joint connections, causing leakage. A cure to transient unstable conditions in the steep declination dual E–B and E–C pipelines was anticipated by Nabataean water engineers by air supplied through the basin pipeline A (Figure 3) into the dual pipelines at junction E to counter and eliminate partial vacuum regions to promote stable flow conditions.

Air led into partial vacuum regions cancels the vacuum regions and subsequent flow instabilities in the dual subterranean pipeline branches. For situations for which hydraulic instabilities cause a backflow to propagate upstream in the D supply in line, and/or water surges in the E–D supply line, Figure 5 indicates a water deposit in the A basin to help damp flow instabilities.

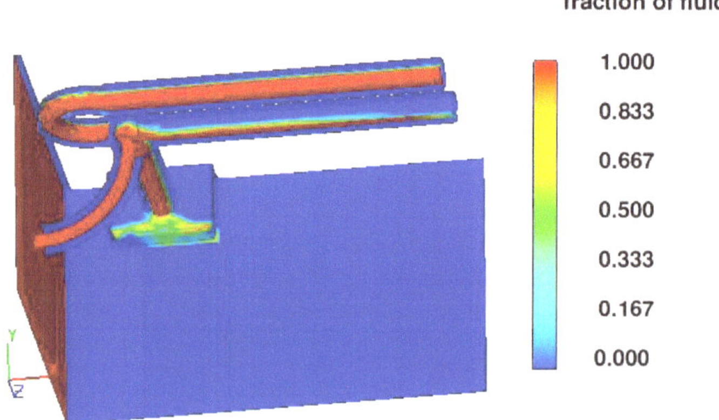

Figure 5. Backflow into the A basin caused by surges in the E–D and E–C water supply line. Figure from [2], pp. 291–292.

In the figures to follow, fluid fraction ff = 1 denotes water; ff = 0 denotes air; and intermediate ff values denote water droplet/air mixtures. Figure 5 indicates full flow at E with the development of partial, supercritical flow in pipeline branches E–B and E–C due to the large declination slopes (~15–20°) of the E–B and E–C pipelines.

Figure 5 indicates the transfer of air into the E–B and E–C pipelines through the A air pipeline to prevent flow transient instabilities from developing as the partial vacuum regions leading to flow instabilities are cancelled. Figure 6 represents the steady state establishment of flow to downstream housing areas and to the Great Temple. Water that has been deposited into the A basin during a transient event together with full subcritical flow from D supplies the E location where partial flow originates in the E–B and E–C pipelines due to their large declination slopes. Figure 6 represents steady-state conditions.

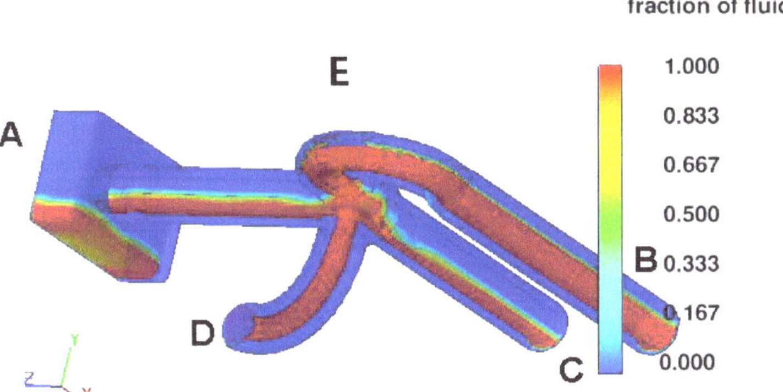

Figure 6. Establishment of steady-state flow conditions in pipelines. Figure from [2], pp. 291–292.

In Figures 5–7, blue areas denote the presence of air in pipelines, red areas denote water, and intermediate colors denote water–air mixtures caused by splashing and/or local turns in the piping system that cause transient water instabilities leading to a localized surface air–water mixing event. For Figure 6, the steady-state design flow rate is achieved; for Figure 5, backflow into basin A has occurred due to an overdesign flow rate occurrence.

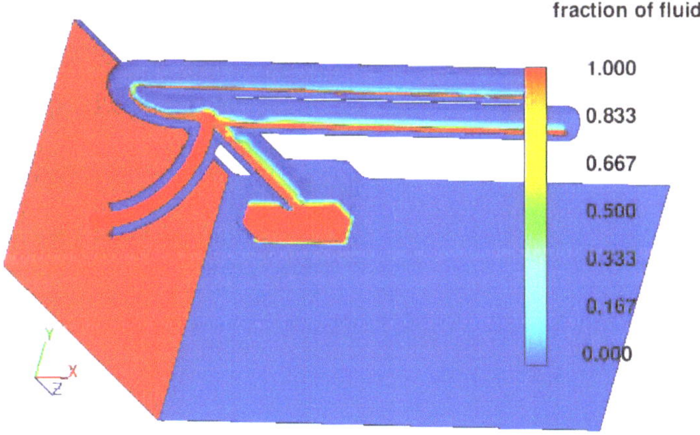

Figure 7. Automatic termination of flow by balancing input D hydrostatic pressure with back pressure from water accumulation in basin A. Figure from [2], pp. 291–292.

Figure 7 represents a situation where an excessive flow transient occurs in supply pipeline D. Due to flow resistance of downstream piping, water accumulates in basin A as the design system flow rate is exceeded. When hydrostatic pressures match from basin A and input pipeline D, then flow terminates as now pipelines E–C and E–B contain mostly air. Before this situation, the water transfer to basin A helps to damp transient water, and pressure surges as may occur upon system start up to promotes rapid development of steady-state flow conditions. The basin A air pipeline then serves to (1) minimize transient water surges originating from air pulses entering partial vacuum regions by providing an air passageway to E; (2) facilitate damping of transient water instabilities in the downstream high declination slope dual pipelines E–B and E–C by transferring backup water caused by transient hydraulic jump motion to the basin and flow rate surges from supply pipeline D, thus promoting steady-state flow stability. One further hydraulic function of basin pipeline A is to divert flow away from the dual downhill branches if these were blocked during repair or if the main water distribution center was blocked to divert water to other pipeline branches of the Ain Braq water supply system. Thus, it is an automatic way to maintain B and C system shut-off through a pressure balance using A and D features.

In summary, the hydraulic engineering use of the pipeline configuration shown in Figure 3 originates from the observation that subcritical full flow conditions exist in the lengthy supply pipeline to D due to its low slope and the inner pipeline surface roughness pipeline conditions resulting from the many pipeline joint connections. Given half-meter individual pipeline sections, and the long length of the supply pipeline, many thousands of joint rough connections ensure full flow conditions. Due to the hydraulically steep declination angle in dual pipelines E, input full flow past the junction area is converted to supercritical partial flow in steep pipeline branches E–B, E–C. This flow transition induces partial vacuum regions above the partial flow water surface. As the subterranean branch pipelines exits are submerged into reservoirs at their destination locations, air is drawn into pipeline exit openings toward the partial vacuum region. The intermittent air bubble stream proceeds upstream in dual steep declination pipelines branches to counter the vacuum regions, leading to flow intermittency and an oscillatory hydraulic jump in the dual pipeline branches. This hydraulic phenomenon affects the flow stability of the entire system, leading to pulsating discharges into the E–B, E–C downstream reservoirs as well as affecting the stability of the input flow at D. Additionally, pulsating flows cause internal pressure changes that can affect pipeline joint connections causing leakage. The cure to this unstable condition was anticipated by Nabataean water engineers when air is supplied by the basin pipeline A. Air led into the partial vacuum regions cancels the vacuum regions and subsequent flow instabilities in the dual subterranean pipeline branches. As pipeline air access is shown to have substantial benefits, use of this technology was evident in other contemporary sites. For example, top holes in stone block piping are evident to alleviate partial vacuum creation as flow transitions from full to partial flow on a steep slope ([1], p. 312) for a Laodicean water supply system and are evident in much of the surface pipelines at Ephesus where partial, open channel flow exists in pipelines [15]. Given access to a wide range of water technologies as a trade center, Petra could incorporate many learned water control technologies from trade route partners from different continents.

3. The Water Control System for Moche-Chimú (CE 400–1400) Pacatnamu in the Jequetepeque Valley of North Coast Peru

The ceremonial site of Pacatnamu in the Jequetepeque Valley of north coast Peru originates from Mochica origins in the Early Intermediate Period (CE 250–750) with later occupation by Chimú conquest in the Late Intermediate Period (CE 1000–1400). The Late Intermediate Period and earlier Middle Horizon Sicán influence plays a significant role in valley aqueduct design, as evidenced by coastal far north water systems in their homeland areas. Later occupation of the Jequetepeque Valley by Inka conquest in Late Horizon times (CE 1400–1532) indicates Pacatnamu site abandonment and Inka efforts to collaborate with marginal Chimú occupation of the valley area centered at the site of Farfan to provide water to Inka administrative buildings vital to control valley resources [16]. Vital to Pacatnamu existence during Moche and Chimú occupation was a reliable water supply for urban and agricultural use provided by the nearby Jequetepeque River during multiple occupations by different societies. Details of the site's long history and water system development are provided in ([1], pp. 111–122), with one area of special interest that demonstrates an application of advanced water engineering knowledge.

Figure 8 indicates a plan view of a main canal originating from the Jequetepeque River source that has multiple earth-fill aqueducts crossing multiple F6 shallow erosion gullies formed from drainage of the eastern valley hills from rainfall and El Niño events. As the canal approached the deep Hoya Hondada Quebrada, valley water engineers decided on a different aqueduct design that lies at a low level (about four meters) above the Hoya

Hondada quebrada bed supplied by a steep declination canal that runs from the elevated embankment northern edge to the low-level aqueduct (Figure 9). The canal proceeds across this low aqueduct to supply water to Moche and Chimú farming settlements as well as continuing to agricultural field systems south of Pacatnamu. The main canal has multiple off-take canals that supply irrigation water to field systems west of the main aqueduct, as indicated in Figure 8. Of interest is that field system agriculture on irrigated flat surfaces is supplemented by fields located on sloped embankments to the F6 quebradas; this is carried out for crops that need only partial sunlight for their growth and is a testament to the need to use any available land suitable for agriculture.

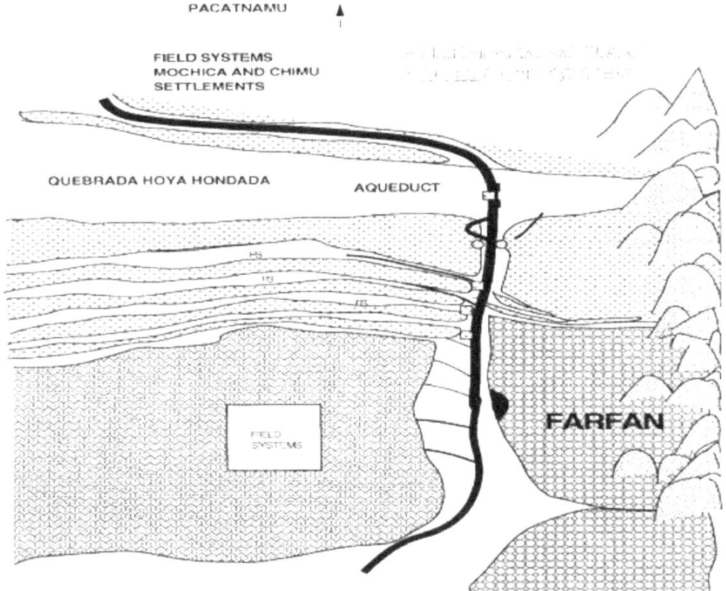

Figure 8. Details of the Farfan–Pacatnamu aqueduct system. Figure from [1], p. 114.

Figure 9 shows the low-level aqueduct crossing the bottom of the wide Hoya Hondada aqueduct. As indicated from Figure 8, the canal continues westward on to the shallow slope of the northern Hoya Hondada embankment to provide water to Moche and Chimú farming settlements serving the main southern field systems along its route toward Pacatnamu. The low aqueduct is an earth-fill structure without canal bed lining other than layers of silt deposit from extensive use. No notable culverts built through the bottom of the low aqueduct to drain accumulated rainfall runoff behind the aqueduct are apparent from field observation, nor are any other culverts found at multiple aqueducts crossing F6 quebradas, as indicated in Figure 8. As five steep short aqueducts supply the lower Figure 8 field systems, high-speed water flowing down these steep channels encounters dual stone "chokes" that create a hydraulic jump before the choke to lower the water velocity entering the flat field systems. The choke opening widths at the bottom of the steep aqueducts are carefully spaced to make the passage flow at the maximum critical flow with excess flow as the source of the hydraulic jump. This design was made to limit the entry water speed to level field systems to limit soil transfer from field system internal channels.

Figure 9. The low-height Hoya Hondada Aqueduct viewed from the northern embankment. The aqueduct channel proceeds downhill along the southern bank of the quebrada to a plateau with Moche and Chimú settlements. Figure from [1], p. 115. Photo by author at site in Peru.

If the low aqueduct height were constructed at the same level as the ~15 m high embankment, it would require construction of a massive ~5 m high earth-fill dam structure to support the canal at the same slope as the approach canal length. During a massive El Niño flood event, water accumulating behind the ~15 m high dam would have hydrostatic pressure forces sufficient to wash the dam away, as it is composed of unconsolidated soils; this failure disaster was anticipated by ancient water engineers by using the low earth-fill dam of ~4 m height that would survive low hydrostatic pressures derived from water accumulated behind the low height dam. However, this design choice introduced a new problem. A steep canal descent was required from the high embankment to the low aqueduct height. As the low aqueduct channel was unlined, water descending the steep canal from the high embankment to the low aqueduct accelerates to high velocity due to gravity and would erode the unlined canal at the impact junction location. The solution to this erosion impact failure problem by water engineers is considered by a FLOW-3D CFD canal model of the canal geometry on the southern bank (Figure 10) just before the steep canal descent to the low Figure 9 aqueduct.

The Figure 10 canal segment has several features: (1) the canal width contracts (location A) as it approaches the descent canal continuance to the lower aqueduct; (2) an elevated left-side canal wall opening before the dual opposing obstacles exists, directing water flow into a side drainage channel (location C); (3) two opposing stone obstacles contract the flow in the contracted canal section (B); (4) further downstream, a right wall opening (location D) exists connected to the main canal. The canal segment shown in Figure 9 is on the order of 20 m length; canal bottom widths range from ~2.5 m in the canal wide (A) area to ~1.3 m in the downstream contracted (D) area. Canal bottom depth from the adjacent ground surface is on the order of ~1 m. While over 1000 years separates the construction of the canal by Chimú water engineers to the form observed from current field observations, excavation clearing of this canal segment provided a best estimate of its key features for the CFD model used to explore its design function to regulate water flow. Upstream of this canal segment, the canal maintains the same near-rectangular cross-section shape with a bottom dimension of ~2 m.

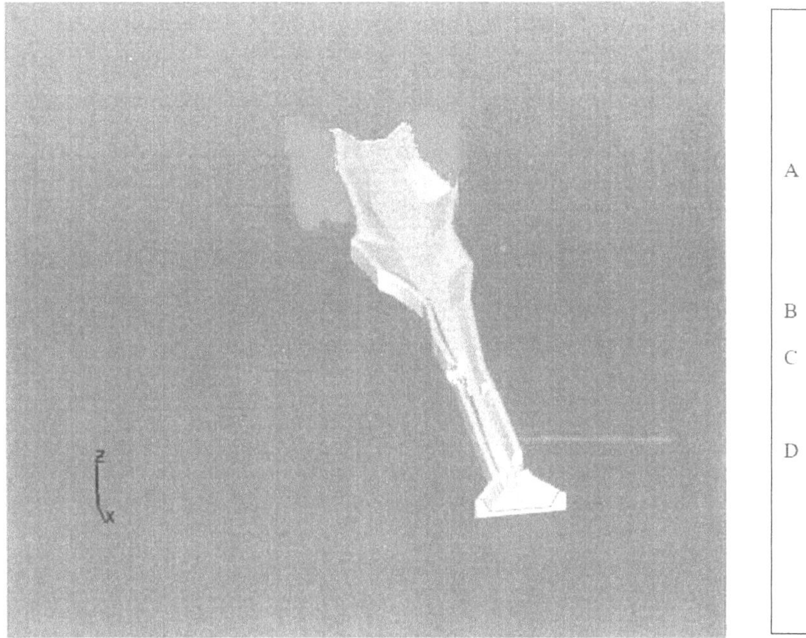

Figure 10. Northern embankment canal geometry ahead of the steep descent canal to the lower aqueduct. FLOW-3D model [1], p.118 modified by author.

Figure 10 indicates a further channel constructed from sidewall opening (Figure 10, location D) that carries a fraction of the canal water to intersect at point E, Figure 11, with the main body of water proceeding down the steep incline canal to the lower aqueduct. Given the low slope of the Figure 11 channel, the subtracted flow is at a low subcritical velocity value. This slow flow now intersects the supercritical flow in the steep declination channel at point E, Figure 11, causing a hydraulic jump to occur in the declination channel elevating the post-jump flow height and decreasing its water velocity. This solution lowers the impact velocity of water flowing on to the lower aqueduct from the steep descent channel and limits the erosion of the lower earth-fill aqueduct at the impact location. Water from both sources then continues across the low aqueduct to agricultural field systems.

FLOW-3D CFD analysis [17] reveals the design intent of the Chimú engineering solution. Again, a major design consideration is to limit damage to the low Hoya Hondada earth-fill canal system from El Niño flood events that can destroy the entire system. Subcritical velocity water proceeding down the Figure 10 sloped canal segment increases in velocity and height as the canal contraction zone (A,B) is approached. As the dual side wall (choke) obstacles (C) are encountered, critical flow matching the design canal flow rate is established that establishes the safe design flow rate through the area between side wall obstacles. When the canal flow rate exceeds this design flow rate from El Niño flood water washing into the canal along its length, water height before the choke elevates and proceeds through the elevated sidewall opening (Figure 10, location C) with the excess flow directed to a lower drainage area. To this point, the canal design flow rate (or flow rates lower than the design flow rate) is achieved to limit canal damage from excessive flood conditions. Flow at the design flow rate proceeds to location D (Figure 10), where a fraction of the flow rate is subtracted in the contour channel (Figure 11) to cure the water impact canal erosion problem on the low-height Hoya Hondada aqueduct impact location. The

Chimú water engineers thus solve multiple design problems to diminish the water impact destruction of the lower earth-fill aqueduct at the impact location on the lower aqueduct as well as threats from El Niño flood wash-in events. The aqueduct impact problem is largely mitigated by rerouting a portion of the canal flow, as Figure 11 describes, where a subcritical subtracted flow from the main canal intersects a supercritical flow in the steep declination canal to induce, through a hydraulic jump, a lower velocity combined flow, limiting erosion of the lower Hoya Hondada aqueduct.

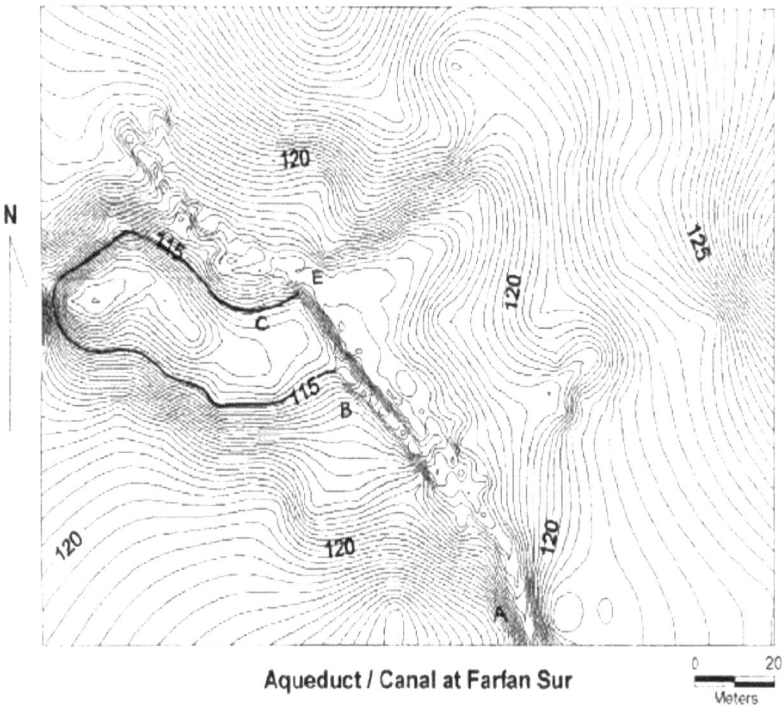

Figure 11. Low slope channel path from Figure 10, D channel wall opening to an intersection location midway on the steep decline canal path to the low aqueduct. Figure from [1], pp. 121–122.

Further details of the hydraulic engineering of the Chimú aqueduct are found in ([1], pp. 111–124). As precise surveying is required for the success of this engineering design, previous investigation of surveying accuracies ([2], p. 89) achieved by ancient Andean societies indicates a $\sim 10^{-3}$ degree accuracy achievement by the Late Intermediate Period Chimú society in Moche Valley canals adjacent to their capital city of Chan Chan. Later Late Horizon Inka canals achieve $\sim 10^{-3}$ degree accuracies, while an earlier Middle Horizon Tiwanaku canal pair indicates an accuracy of $\pm 10^{-4}$ degrees in a dual canal system characterized by two nearby canal water flows in opposite directions. By comparison, Roman surveying accuracies achieve similar accuracies as measured from onsite measurements at the Pont du Gard aqueduct.

Figure 12 illustrates the flow subtraction from Figure 10, D location at the design flow rate from FLOW-3D CFD analysis of the water system to counter aqueduct damage. The combined set of design adjustments indicated by Figures 10 and 11, B and C flow subtractions serve to preserve the canal from flood wash-in erosion damage and indicate the hydraulic engineering capabilities of Chimú water engineers to anticipate problems and provide innovative solutions to the problems.

Figure 12. FLOW-3D CFD computed exit flow subtraction from the Figure 10, D location at the design flow rate. Figure from [1], p. 120.

The flow innovations represent a degree of hydraulic knowledge that only CFD analysis can bring forward to demonstrate aspects of the state of hydraulic engineering knowledge achieved by South American pre-Columbian societies.

4. CFD Analysis of a Self-Cleaning Water Drainage Channel at the Mid-Eastern Greek Site (300 BCE–CE 100) of Priene in Coastal Western Turkey

The site of Priene, located ~15 km from the coastal area of western Turkey, is unique as a Greek colony city without later Roman architectural influence. Given the tolerance of Greek civilization to multiple religions, beyond temples to Greek and Egyptian gods found on site (Figure 13A,B,D,J,M,N), an early Hebrew synagogue was recently found by archaeologists on site. Originally an ally of Athens, Priene was conquered by the Persians in the 6th century BCE. Later, as a member of Greek cities in western Turkey participating in the Ionian revolt against Persian occupation under Darius, the city was relocated in 350 BCE to its present inland site. The city was visited by Alexander the Great in 334 BCE, where his stay featured the planning of the conquest of the Persian Empire; a house reputed to be his residence during his stay in the city was given his name and served as my base of operations during many site visits. Later occupation by Seleucids and the Pergammon administration followed, with the city later incorporated into the Roman Empire. From Constantinople as the seat of the eastern Roman Empire, there followed later century occupation by Byzantine forces. Later in the city's history, a Byzantine church (P, Figure 13) indicates Christian occupation until later Muslim conquest of sites by the Ottoman Empire (and Constantinople) itself in the 15th century CE, when the Priene was finally abandoned.

Details on the history and water systems of the city are given in ([1], pp. 338–339; [15,18]). City water supplies largely originated from karstic spring sources in the city northern highlands that, through an extensive pipeline network, distributed water to many areas of the city, including the marketplace agora, the theater, the stadium, the multiple gymnasiums, and the city's council chambers (Figure 13).

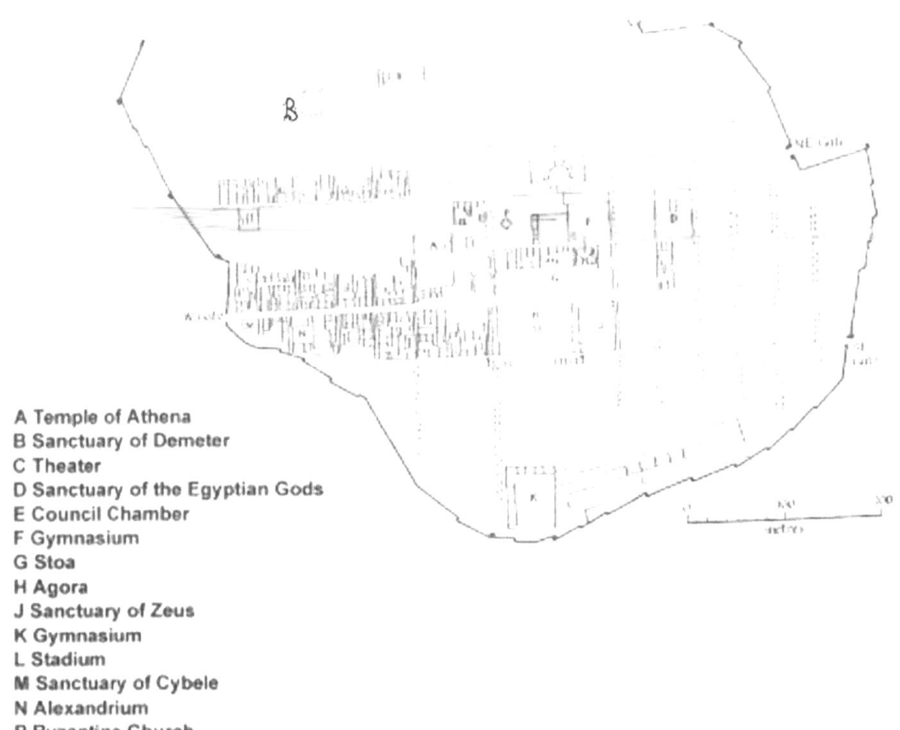

Figure 13. Detailed map of walled urban Priene and inner city main structures. Figure from [1], p. 338.

Of interest to the present study is the water drainage system at the West Gate area of the city. The city (Figure 13) is composed of urban living quarters centered around the main city street leading to the West Gate. To supplement the pipeline water distribution system, large underground cisterns were built in the central housing district to collect channeled rainfall as well as "used" water from city residences supplied from the continuously flowing pipeline system. Drainage water from all occupied areas, and particularly from the busy marketplace agora site, was largely delivered to an open channel (Figure 14) on the main thoroughfare for passage through a bottom opening in the city wall. Water for ceremonial temples (listings given in Figure 13) required a high degree of purification for rituals; this was achieved by multiple silt collection basins inserted into pipelines that had access ports for frequent cleaning. Of interest then are water purification systems and water structure designs that have an "automatic" method of self-cleaning built into their design.

At Priene's West Gate (Figure 13), debris-laden drainage water from housing, the agora marketplace (H) and government buildings, temples, and the theater was served by springs north of the main street pass through the street channel (Figure 14) into the West Gate drainage opening (Figure 15) through the city outer wall. The street approaching the drainage structure has a steep channel that accelerates drainage water to supercritical (Fr > 1) conditions with the high-velocity water height approaching normal depth as it proceeds toward the drainage opening [19–21]. While a straight channel exit continuing the geometry of the Figure 14 inlet channel would normally be expected, the complex geometry drainage structure indicated in the Figure 15 CFD model reconstruction is in place from field measurements. Rapidly flowing, debris-laden water enters this structure from the positive y direction (Figure 15) and interacts with the complex internal shaping before exiting into a canyon outside of the city wall. The channel water height approaching the Figure 15 drainage structure is at the lowered height, high-velocity, normal depth and is contained within the lead-in channel without spillage by design. A stone baffle plate exists to cover the top front part of the Figure 15 drainage structure that contains the input flow to avoid spillage and flooding of housing structures close to the city wall (Figure 13). The Figure 14 rectangular cross-section channel has a width of ~1.5 m, a depth of ~ 0.5 m, and a declination slope of ~20°. FLOW-3D CFD calculations (Figure 16) indicate complex streamline vortex structures accompanied by high levels of turbulence induced by flow passage through the complex Figure 15 structure. The intent of the Figure 15 design is now revealed: severe agitation and entrainment of the debris-laden flow ensures that debris settling does not occur as would be the case if a straight-through, lead-in continuance channel design were emplaced. The debris water load comprises suspended lighter particles and heavier mud and debris particles settled on the channel bottom that slowly move from water shear effects toward the Figure 15 drainage structure. The drainage system is "self-cleaning", as agitation of the water flow from CFD calculation (Figure 16) provides resuspension of heavier and lighter debris particles to clear the channel of suspended and settled waste particles.

Figure 14. Steep channel along the center of the Priene main east–west throughfare, leading water drainage from housing and the agora to the complex West Gate drainage structure shown in Figure 15. Approximate mean channel width ~ 1.2 m; approximate mean channel depth ~ 0.25 m. Arrow denotes flow direction. Photo by author at Priene in Turkey.

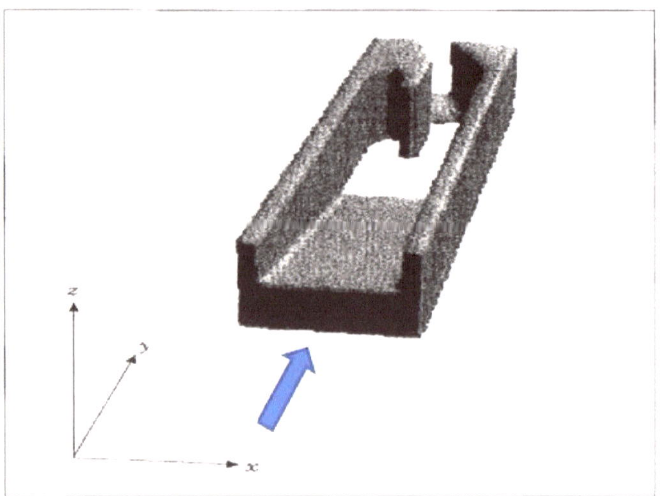

Figure 15. Drainage structure passing through the base of the West Wall, accepting drainage water from the Figure 14 channel. The arrow denotes the water flow direction. From [1], p. 341; FLOW-3D CFD model by author.

Figure 16. X plane CFD computed streamline patterns within the Figure 15 structure, indicating complex vortex patterns and high turbulence levels. Arrow denotes flow direction.

In summary, Greek water engineers provide a sophisticated self-cleaning channel exit design to facilitate the city's high hygiene standards observed by many water purification devices. It is noted that the pressurized pipeline supplying ceremonial water basins in the Sanctuary of Demeter (Figure 13B) proceed underground lower than the basins then rise upward to the basins; this design ensures that debris settles in lower pipelines to provide purified water to the basins.

5. Water Supply and Distribution System of 800–1400 CE Cambodian Angkor West Baray

The history and cultural development of Angkor in Cambodia is well known. Of interest are the two major 9th century CE barays (water reservoirs) shown in Figure 17, known as the West Baray and East Baray. For the scale of the site, the dimensions of the West Baray are 8.3 km length, 2.4 km width, and a depth of ~10 m from the top of the built-up containing mound structures bounding the West Baray. The two barays were designed to capture heavy rainfall from two monsoon seasons as well as water from northern rivers. Within the city itself, the north-to-south flowing Siem Reap river provided water to inner city settlements, temples, agricultural zones, internal city reservoirs, and local population settlement areas through many surface channels. As suspended soil particles are carried by rivers from rainfall runoff, silt is continually delivered into the West and East Barays, causing deep layers of sediments to deposit in the barays over centuries since initial construction in the ninth century CE. The original purpose for the barays is somewhat obscure, as various scholars have attributed their construction to ceremonial purposes or city water needs through seasonal changes in rainfall availability. Various scholars have interpreted the purpose of the West Baray as the irrigation source for rice fields in the dry season; others propose a religious function based upon the discovery of an upper body fragment of a 10 m high bronze statue of Vishnu at the Mebon structure central to the West Baray. Of interest are the sites of Ta Prom, Pre Rup south of the West Baray, the sites of Ta Nei and Ta Keo to the west of the baray, and Angkor Wat to the far southeast of the West Baray. These sites have as their water source a diversion of the Siem Reap river that was designed to replenish the West Baray in the dry season. Over the centuries, the southern reaches of the West Baray have been modified to provide channeled surface water irrigation to downslope rice farmlands.

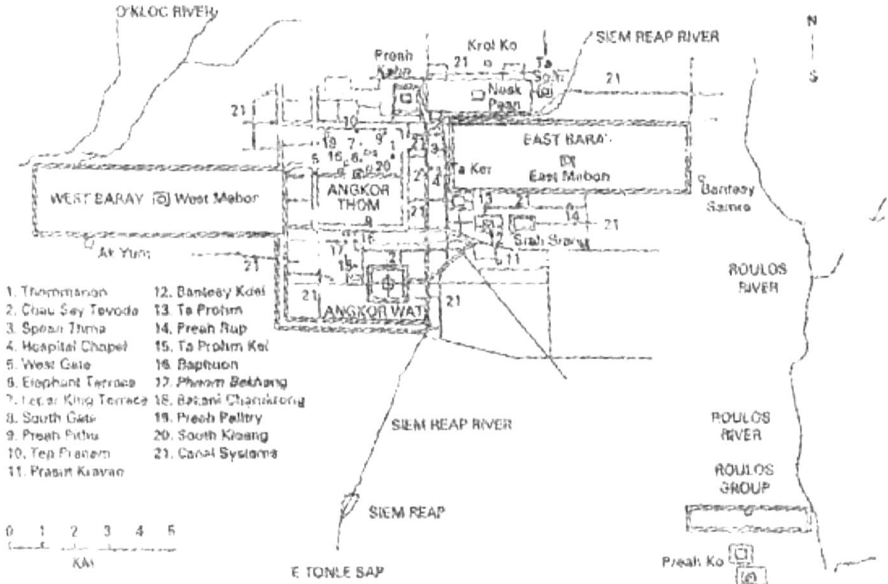

Figure 17. Plan view of urban Angkor and surroundings showing East and West Barays. Site names in Appendix A.2. Figure from [1], p. 359.

The present chapter Flow-3D CFD study focuses on the use of the West Baray in off-season water supply conditions to supplement rice farming areas south of the baray. Given the large population of Angkor, multiple-season rice farm cropping continuity was vital to sustain the city during the dry season. Efforts to maintain a high West Baray water level by rainfall collection and channeled river water were vital to provide the southern farming area sufficient groundwater to sustain continuous rice field agriculture.

In the figures to follow, t (time) represents months after the dual monsoon season ends and the dry season initiates; z denotes depth in meters below the original interior baray bottom surface. The ff fraction of fluid scale represents aquifer water concentration: ff = 1 represents water; ff = 0 represents air in porous aquifer soil; intermediate ff values represent air/water mixture states within the porous aquifer. Rainy season water seepage from the West Baray into the southern, downward-sloped agricultural field system aquifer south of the baray is maintained from rainfall and river sources providing water to the baray. The high hydrostatic pressure from West Baray water depth contained by the ~10 m high encircling embankment (when the baray is full) provides sufficient hydrostatic pressure to extend the water penetration distance and depth into the southern aquifer for several kilometers, sufficient to provide water to root systems for several crop types as well as to excavated pool areas for rice cultivation.

During the dry season, with limited input river flow, high evaporation rates, and limited seepage from areas north of the baray, the West Baray's water level drops to a lower level with reduced seepage into the southern field system area to maintain height groundwater height. For rice farming, a stable saturated aquifer underlying water pool bottoms is necessary to continuously maintain the rice farming conditions.

Given different varieties of rice cultivated in the southern area field systems, various field system configurations necessary for their growth can be implemented given different water pit levels dug into the saturated aquifer that sustain different crops moisture levels. The original West Baray bottom surface is approximately at the same level of the farming area just outside the ~10 m high, mountainous earth-fill barrier berm that encompasses the baray interior.

For the CFD study, the dry season West Baray water level is assumed low due to extended drought with a relatively dry southern field system aquifer. Figures 18–21 detail the computed CFD progression of water seepage from the West Baray into the southern reaches of the farm area aquifer during initiation of wet season rainfall. With the end of the extended drought period, monsoon rains return, filling the West Baray water level through rainfall accumulation, water input from the nearby river, and groundwater seepage from the northern, higher-level recharged aquifer. The CFD study then informs of the reactivation of the West Baray's main purpose as the monsoon season initiates and starts water transfer to the southern farm area aquifer to reinstate rice farming. The farm area saturation level maintained under wet season conditions is now of importance as the dry season initiates to determine the ability to continue rice farming as baray water levels decrease. Figures 18 and 19 show the aquifer fluid fraction (ff) concentration 3.75 and 4.08 m below the southern agricultural area surface 2.53 months into the dry season, resulting from a limited wet season rainfall level. The red zone is the low West Baray water level. Results indicate that water transfer into the aquifer initially proceeds slowly due to the low hydrostatic pressure at the baray bottom. Near-surface aquifer penetration is low, with remnant aquifer water fluid fractions remaining at depth from earlier monsoon season water deposits.

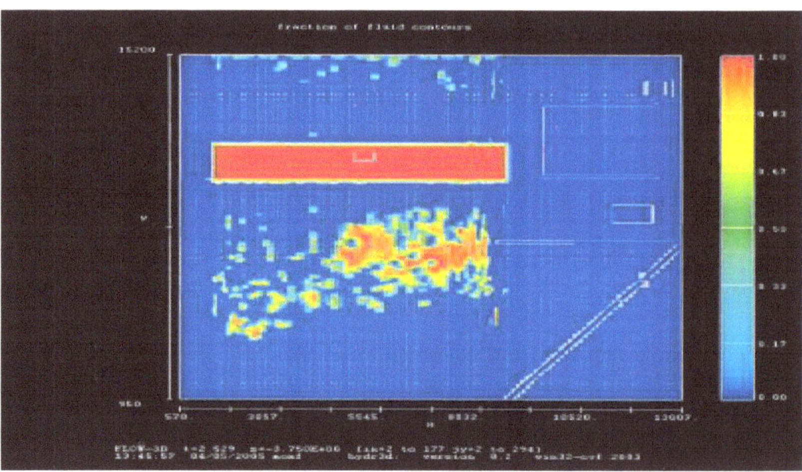

Figure 18. Aquifer fluid fraction (ff) concentration at 3.75 m below the southern farm area surface 2.53 months after the extended drought ends. Color version added to [1], p. 367.

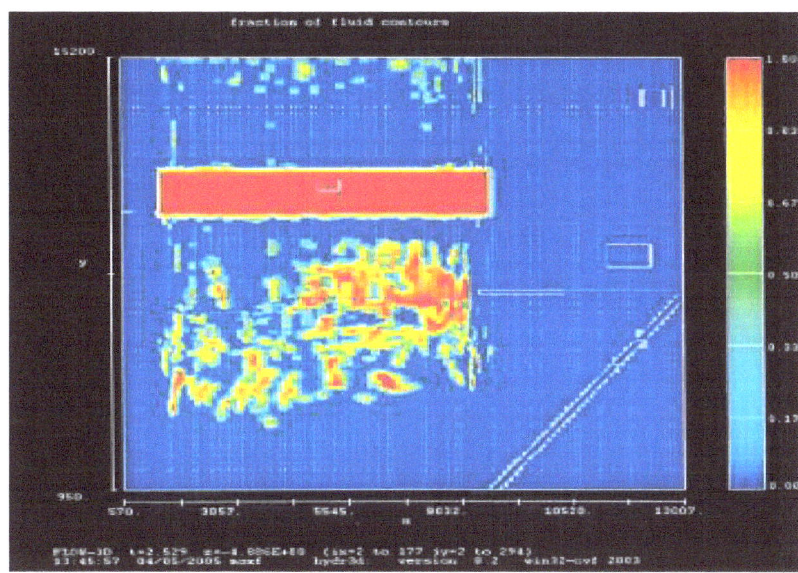

Figure 19. Aquifer fraction of fluid ff concentration at 4.08 m below the southern farm area surface 2.53 months after the monsoon season initiates. Aquifer water levels remain high at depth from earlier rainy season deposits. Color version added to [1], p. 367.

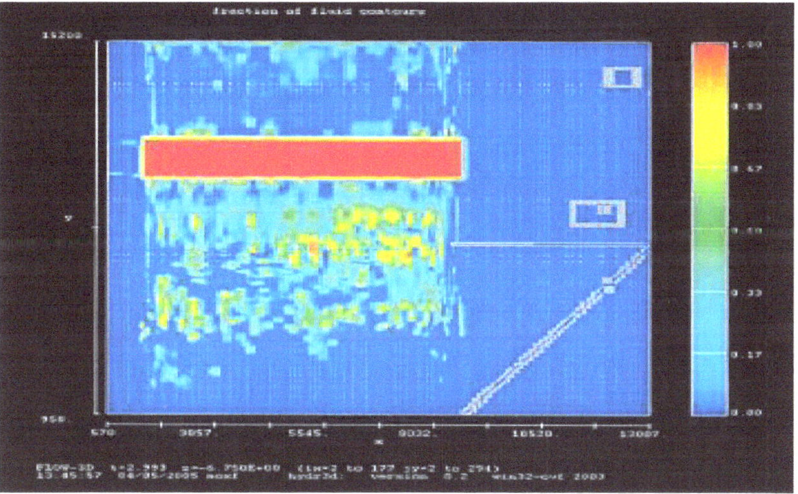

Figure 20. Aquifer fraction of fluid ff concentration at 6.75 m below the southern farm area several months into the monsoon season. Color version added to [1], p. 367.

As the rainy season initiates months later with baray infilling, near-surface aquifer infilling initiates (Figure 20)—this process is aided by the saturated aquifer at depth left over in the dry season, as Figures 17 and 18 indicate.

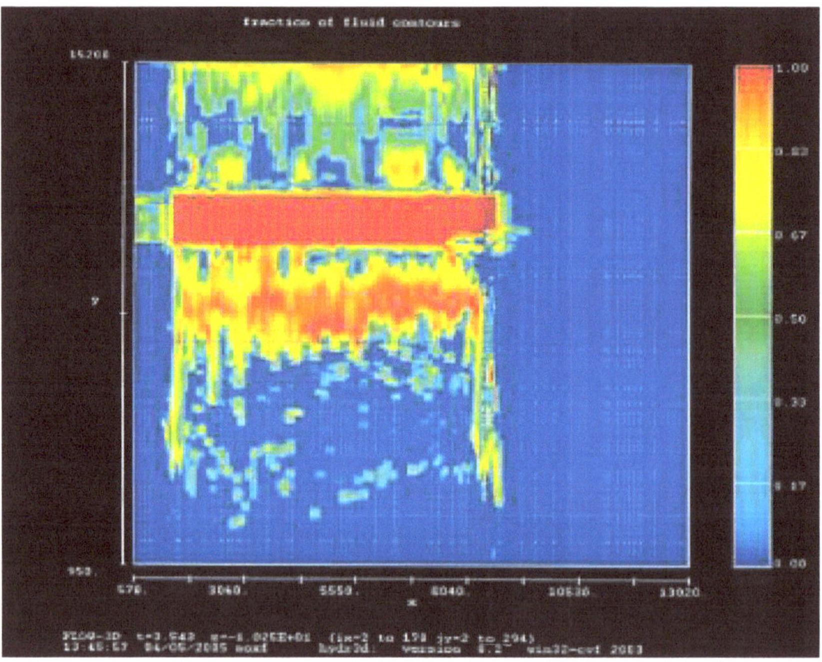

Figure 21. Aquifer fluid fraction ff concentration at 1.07 m below the southern farm area surface 3.54 months after the monsoon season ends. Note the later groundwater supply from ingested rainfall from the northern mountain area contributing to the baray water supply. Color version added to [1], p. 367.

Figure 20 indicates that as the West Baray continues to fill to higher water levels as the monsoon season initiates, the aquifer is now experiencing saturation to a large depth relative to the ground surface as the lower part of the aquifer is already saturated and is unable to accept further water input. This observation is confirmed by Figure 21 which indicates that at a later time, with continued rainy season baray filling, at a depth of 1.07 m below the ground surface, the southern agricultural area is saturated to a large depth ~ 1.5 km from the baray. This is the intended purpose of the West Baray, as now-excavated rice farming basins excavated into the deeply saturated ground surface fill with water suitable for rice cultivation. Under normal climate circumstances where infrequent drought periods exist, the West Baray retains a high water level from higher-elevation northern area aquifer seepage and river water input. Figure 21 is then the standard condition for rice farming through both wet and dry seasons, as (1) in the dry season, saturated soil conditions are maintained at depth from the previous wet season deposits and accessible for rice farming by pits dug to lower depths or for other crops with deep root systems; (2) in the wet season, a saturated aquifer at depth is available for rice farming (or other crops). With continued availability for farming during wet and dry seasons, the large population of Angkor can be well supplied with food resources.

At present times, the barays are no longer functional due to a large depth of silt accumulation that occurred after site abandonment at ~1430 CE. Several channels exiting from the central south side and westernmost corner of the West Baray are present, indicating water transfer to surface field systems to the west and south of the West Baray. These high-level through-berm water supply channels may have originated later in the site's life when West Baray high silt accumulation limited previously used aquifer-provided agricultural resources to southern field systems.

CFD results proceed from specified aquifer properties [22] (porosity, hydraulic conductivity, particle size, specific surface) and water properties (density, viscosity). Given that fertile soil deposition from rain erosion of northern mountain range soils had occurred over millennia, the southern farming areas of the West Baray were well positioned to provide high rice yields. As this area is nearby to urban population centers, great interest in developing this area for rice farming was a priority for Angkor water engineers. CFD results presented indicate that Figures 18–21 represent the basis for sustained agricultural production throughout yearly seasonal changes, given the two monsoon events in June–September and October–November. Of note is that silt washing into and depositing into the barays over the many years of Angkor's existence slowly reduced their main function, as silt deposits presented a barrier for water transfer into the southern aquifer and reduced baray water storage capacity. At present, after centuries of site abandonment, many meters of silt cover the original bottom of the barays, cancelling their former practical use.

6. Landscape Change, Beach Ridge Formation, and Agricultural Field System Change from El Niño Flood Events at Preceramic (2500–1600 BCE) Caral on North Coast Peru

The preceramic site of (2500–1600 BCE) Caral is in the Supe Valley on the north coast of Peru, approximately 40 km inland from the Pacific Ocean coastline; Caral is adjacent the south side of the Supe River, as indicated by 13, Figure 22.

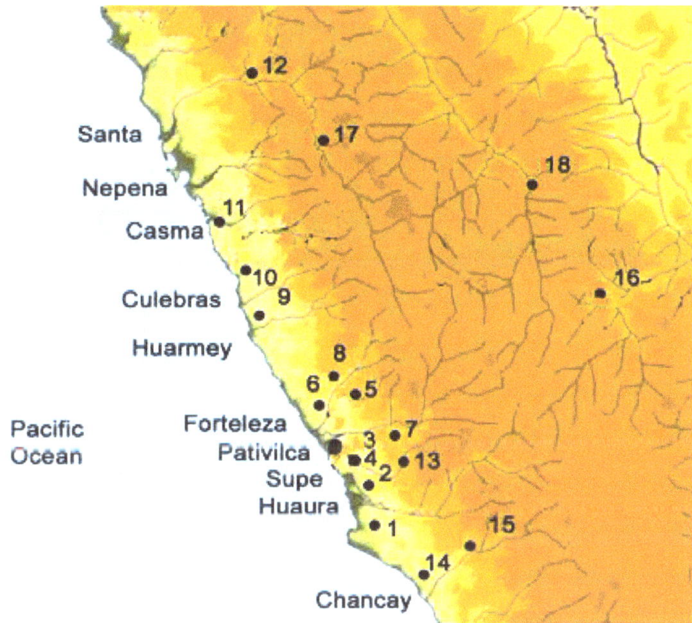

Figure 22. Location of Caral (13) in the Supe Valley on the north coast of Peru. Additional numbers indicated relate to other Preceramic, Formative, and Early Horizon sites in adjacent valleys. Color version of figure derived from [2], p. 224.

Caral is known as the "earliest city" in Peru owing to its founding date of ~2500 BCE; the site consists of many stone-faced pyramid structures, residential quarters, and several ceremonial buildings. Only recently discovered through detailed excavation efforts by Dr. Ruth Shady Solis, together with her many journal and book publications detailing site architecture [23–29], the site has received UNESCO World Heritage status. Given the discovery of a ~2500 BCE new world site with multiple elaborate pyramid structures (Figures 23–25) contemporary in time with early-dynasty Egyptian pyramid structures, Peruvian prehistory now emerges with new discoveries that indicate the presence of a previously unknown, very accomplished society that starts and underlies development of later Andean societies.

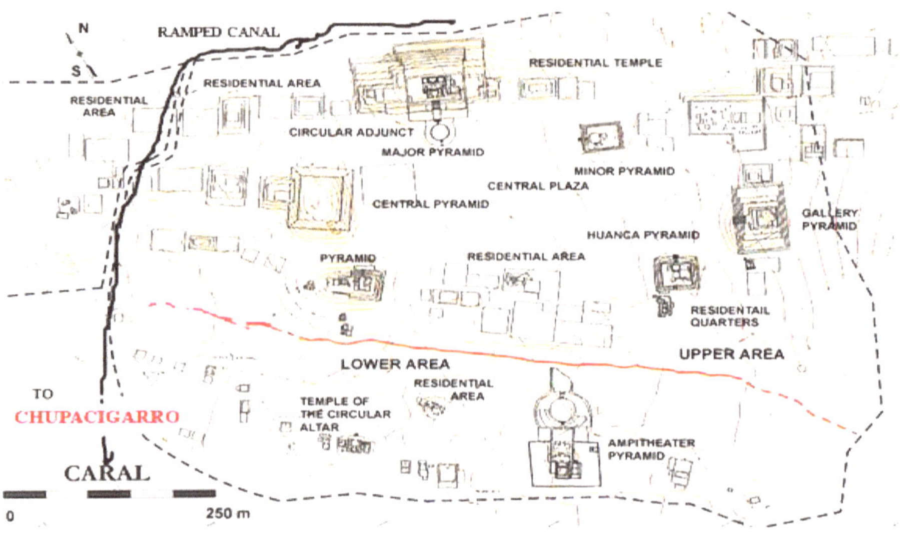

Figure 23. Details of the Caral urban center. Color version of figure derived from [2], p. 225.

Figure 24. Details of the Major Pyramid; location indicated in Figure 23. Figure derived from [2], p. 226.

The present CFD studies are comprehensive in that they examine the sources of environmental landscape changes brought about by ENSO El Niño flood and drought effects that affect Caral's history. As area field observations encounter beach ridges, landscape erosion events, and soil deposition events that alter the agricultural, marine, and living quarter landscape upon which Caral society maintained its existence for over 1000 years, these effects derive from fluid dynamics effects that CFD analysis can bring forward for the first time to match field observations.

Figure 25. The excavated Gallery Pyramid; location indicated in Figure 23. Aerial photo by author.

For ancient (and modern) societies of north coast Peru, contention with natural forces derived from ENSO flood and extended drought disasters, as well as climate change effects that alter environmental conditions that affect societal continuity, play a key role in interpreting their historical development. The present chapter concentrates on environmental changes throughout Caral's thousand-year lifespan that influence both its historical development and its demise in later years.

Contemporary with Caral proper (Figure 23) are 18 other Supe Valley sites that, together with Caral as the capital city, form an early cooperative city state sharing the same cultural and religious values made possible through intrasite trade of agricultural and marine resource products. For example, coastal sites involved in harvesting marine and marsh resources (fish, shellfish) traded their protein-rich products with inland sites involved in agricultural product farming to provide the nutrition balance for all participating Supe Valley site members. Excavated pyramid structures shown in Figures 24 and 25 indicate many levels of occupation quarters of elite members of Caral society in addition to circular arena areas (~25 m in diameter) used for religious ceremony purposes. Apparently, even for the earliest of Peruvian societies (as yet discovered), emergence of a religion and an elite class to conduct rituals comes with the development of class difference structure. Figure 23 details the presence of residential areas of lesser structural detail than those associated with pyramid structures, indicating the presence of a worker class involved in site maintenance, agricultural activities, and intersite trade activity necessary for progress of all site occupants. Further details on Caral's history and archaeology ([3], pp. 31–68; [23–29]) are given in these references. Figure 24 is the excavated Major Pyramid indicated in Figure 23, indicating multiple rooms, platforms, and ceremonial open plazas for the elite class of Caral's population; similarly, Figure 25 indicates similar details for the Gallery Pyramid, whose location is given in Figure 23. Of note are fire pits within the Major Pyramid that have circular air passageways close to the pit's bottom; this convection feature is designed to draw in air to sustain a wood fire's hot, lower-density fumes as they rise upward. As other pyramids share this similar feature, it may be surmised (or conjectured) that pit fires were used as a signaling feature to communicate information between distant pyramids. As Caral was deeply buried by aeolian sand deposits prior to Ruth Shady's excavations starting in the late 1980s, it is unclear whether the many rooms shown in Figures 24 and 25 originally had reed roofing. Given the adjoining circular ceremonial area for the Major Pyramid with an access stairway to the inner reaches of the pyramid, the site played a significant role in ceremonial activities for Caral's population. A similar, but smaller, circular ceremonial area that is associated with a temple at Chupacigarro (Figure 23) was found prior to Caral excavations.

Of interest to the site's long history is the agricultural system evolution over centuries to feed the growing population of the valley's many sites. As Peru is well known for its ENSO climate changes, particularly major El Niño flood and long-term drought effects, challenges to agricultural productivity arose over centuries of site occupation. Of note is the special water availability that the Supe Valley has compared to all other coastal valleys: from high levels of highland rainfall in the Andean Cordillera Blanca mountain range, freshwater lakes evolve. Water from these lakes proceeds down to valley lowlands through geologic faults to perpetually maintain valley ground water to with ~1 m of the valley bottomland ground surface, making multicropping agriculture possible at high productivity levels. This water transport feature, known as *amunas* in prehistory, was thought to occur from subterranean tunnels built by the ancients as informed by local valley farmers. Together with intermittent Supe River flow, the Supe Valley experiences sufficient water resources for multicropping on a year-round basis. Thus, the Supe Valley was a logical place for a complex society to initially develop given the valley's large coastal and interior valley bottomland areas and plentiful water resources that exist even to the present day. As climate challenges affect the agricultural base of a society, FLOW-3D CFD investigation of landscape changes adversely affecting agricultural production are next considered to tell the story of the collapse of Caral society after ~1000 years of successful occupation of the Supe Valley.

Figure 26 presents results from test probes to find a specific mollusk presence found only in coastal water depths of ~1 m. C14 dating of found mollusk shells then provides the shoreline position changes over long time periods (Tom Dillehay, personal communication). For example, Figure 26 indicates that a far inland coastline at Waypoint 49 accumulated rainfall washed-in soils carried by ocean currents, aeolian sand transfer, and surface soil transfer from rainfall runoff over the next ~1100 years to form a new coastline addition to Waypoint 48. Subsequently, over the next ~1600 years, the coastline eroded back to Waypoint 44 from many large-scale, mass-wasting El Niño events followed by shoreline additions up to present times, as Figure 26 indicates. This figure denotes the importance of landscape change over Caral's lifetime that influenced its agricultural land base as later discussion details. Observed at certain locations on the Peruvian north coast shoreline are inland sequences of linear beach ridges, one of which is shown in Figure 27. The origin and composition of beach ridges and landscape geometry change are next determined by a CFD coastal model (Figure 28) as these formations proceed from fluid dynamics effects.

The full-scale CFD model in Figure 28 includes the sloped coastal landscape incorporating the ~50 km distance between the Santa and Viru rivers. An El Niño flood-derived slurry mixture of soil particles, sand, pebbles, rocks, and boulders collected from adjacent hillside runoff proceeds down the two river valleys to interact with coastal and ocean currents; details of flow rates and slurry physical properties are given in [2] (pp. 48–50).

In Figure 29A,B, the scale denotes 1.94 slugs/ft^3 for water and deposition density values representative of the composition of the runoff slurry. Results shown are for a singular El Niño event. Note that northward ocean and offshore currents carry slurry mixtures northward to create lengthy subsea deposits, given the 50 km distance between the Santa and Viru valleys. With ongoing tectonic uplift and further El Niño slurry deposition events occurring over time, sequences of inland beach ridges form. Note that the heavier slurry particles (pebbles, rocks, boulders) appear to centrally deposit within lighter slurry mixtures and, with later rain and flood events washing away lighter particles, form beach ridges mainly composed of heavier stones, as Figure 27 indicates.

Figure 26. Coastal area landscape changes over time. Color version of figure derived from [2], p. 228.

Figure 27. Coastal beach ridge formation on north coast of Peru formed from a major El Niño flood event. Photo by author at Supe Valley coast in Peru.

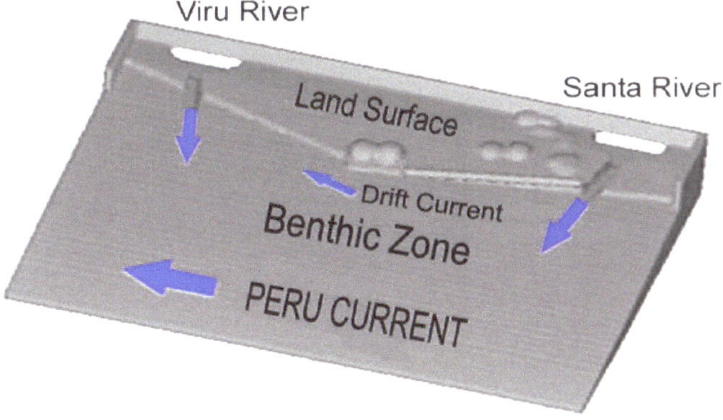

Figure 28. FLOW-3D CFD coastline model from the Santa to Viru River Valleys. FLOW-3D CFD model by author [2], p. 243.

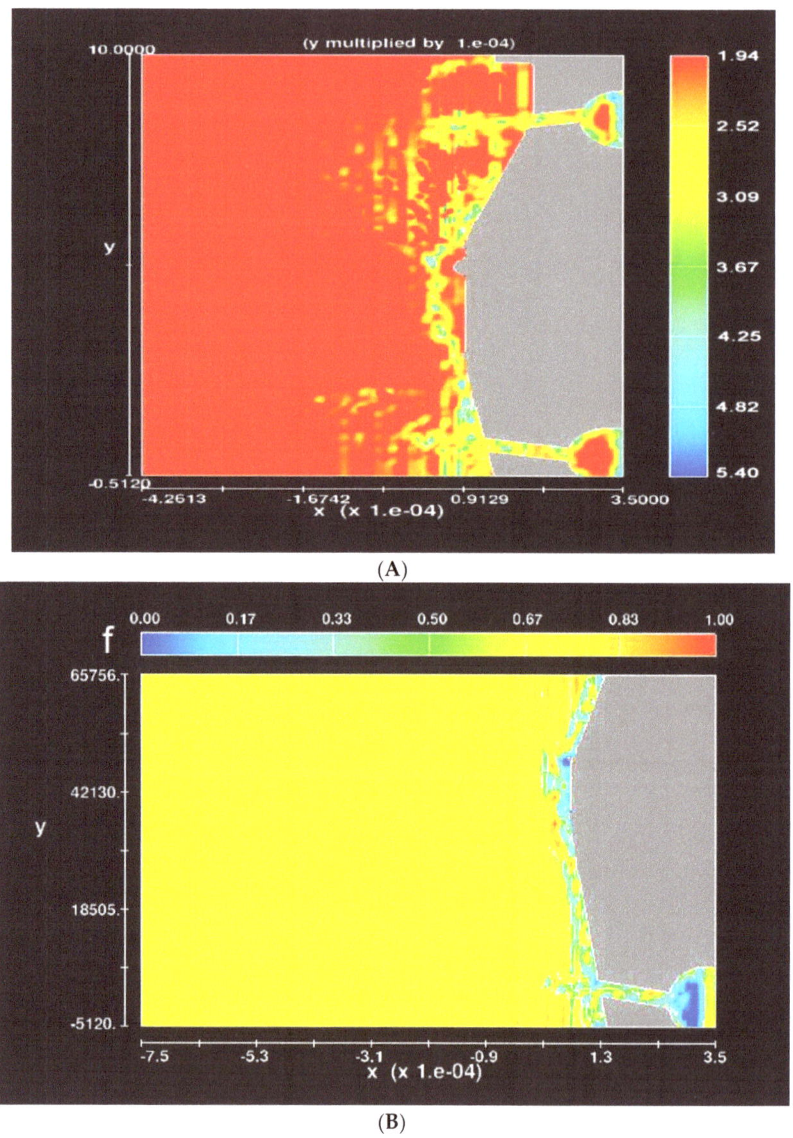

(A)

(B)

Figure 29. (**A**) Offshore slurry deposition density from an El Niño flood event from CFD calculations for a low offshore benthic slope. Scale density in slugs/ft^3 (1 slug/ft^3 = 515.38 kg/m^3) representative of slurry mixtures with high rock content. Result from [2], p. 246. (**B**) Offshore slurry deposition fluid fraction density from an El Niño flood event from CFD calculations for steep benthic slope typical of the Supe Valley offshore region. Result from [2], p. 248 (1 slug/ft^3 = 515.38 kg/m^3).

Figure 29A,B show calculated slurry deposition close to the shoreline for an offshore shallow benthic slope (Figure 29A) and for a steep benthic slope typical of the Supe Valley coastal area (Figure 29B). The steep offshore benthic slope case ultimately leads to a more concentrated linear beach ridge deposition typical of Figure 27. With coastal tectonic uplift and coastal landscape additions, earlier stranded beach ridges are found sequentially located inland from the coastline. Figure 30 shows a previously settled offshore slurry ridge (gray area) subject to a further El Niño event. From the offshore blockage ridge formed from the earlier slurry deposition event, a further flood event can create a marsh and/or an infilled area behind the ridge, as Figures 30–32 indicates. Surveys of several well-preserved coastal areas indicate a sequence of beach ridges; these ridges, C14-dated from mollusk shell slurry debris, date major El Niño flood events that influenced the landscape and often the survival history of archaeological sites in north coast areas. With respect to coastline shape changes from geophysical events, as shown in Figure 26, Figure 31 indicates bay infilling from a major El Niño event that deposits slurry in a coastal concave bay area subject to vortical water motion that traps silt-laden ocean currents. Such events, as shown in Figures 30 and 31, lead to the Figure 26 coastline shape changes.

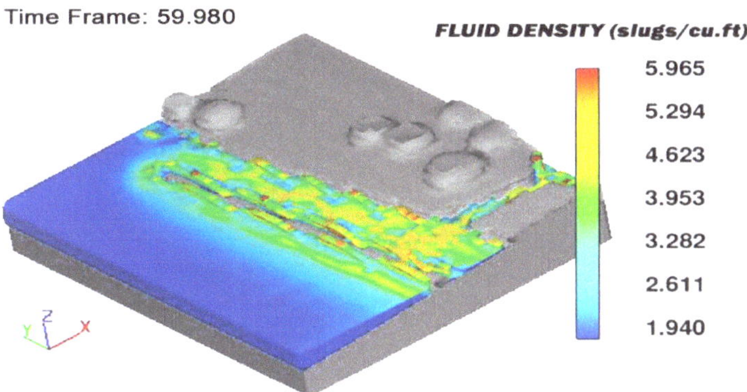

Figure 30. FLOW-3D CFD deposition results of slurry contents in Pacific Ocean coastal area between the Santa and Viru valleys. Result from [2], p. 248. (1 slug/ft^3 = 515.38 kg/m^3).

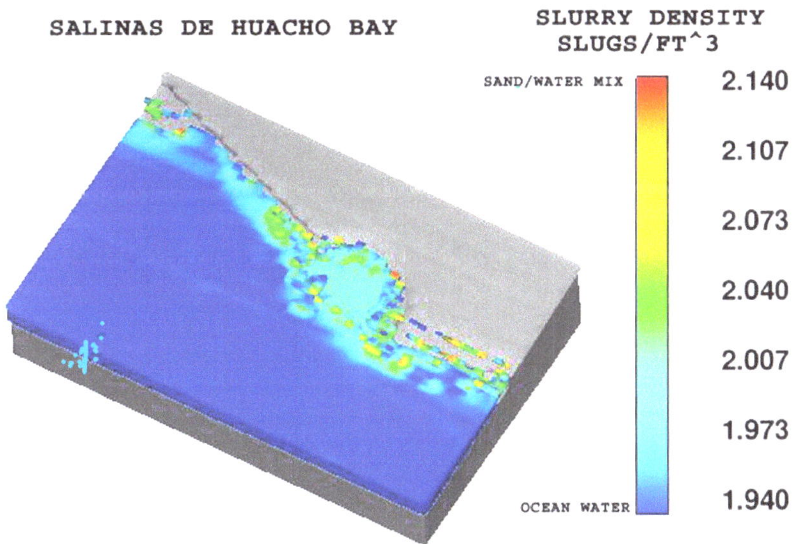

Figure 31. FLOW-3D CFD prediction of a coastline change that proceeds from a major El Niño event that traps silt laden ocean water to deposit in bay areas. Result from [2], p. 249. (1 slug/ft^3 = 515.38 kg/m^3).

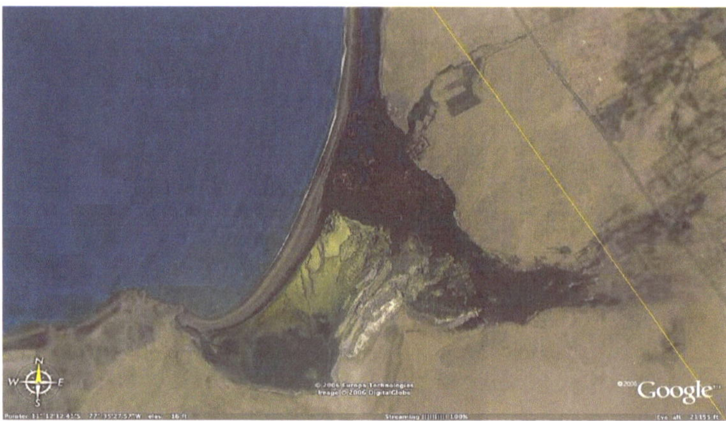

Figure 32. Extensive beach ridge formation in the Supe Valley coastline area. Existing segment of the Medio Mundo beach ridge shown. Result from Google Earth.

To this point, geophysical landscape changes originating from major El Niño flood events were demonstrated by CFD calculations to qualitatively duplicate nature's productions. The question of the extinction of Caral in the ~1600 BCE time frame is next addressed by geophysical changes that affected Caral and its satellite settlements' agricultural and marine resource base.

The presence of a major beach ridge along the northern Peruvian coastline created in the late 11th–12th century BCE (the Medio Mundo Ridge, Figure 32) originates from a major El Niño flood event and affects the continuity of Supe Valley sites. One major effect was the diminishment of the offshore fishing and shellfish gathering marine resource base due to bay infilling, as Figure 31 indicates. A further landscape change noted from this major beach ridge presence was blockage of Supe River drainage resulting in sediment deposits behind the beach ridge and marsh creation unsuitable for agriculture (Figure 32). These consequences led to a contraction of the near-coastal farming areas (Figure 33) now replaced by marsh areas unsuitable for agriculture. Given the large extent of the Medio Mundo beach ridge (~90 km), coastal infilling, as shown in Figures 30 and 31, similarly affected the Supe Valley's agricultural and marine resource base supporting Caral's settlements.

A further Supe Valley landscape transformation involved northwesterly aeolian sand transfer from beach-ridge-deposited sands in exposed beach flat areas (Figures 30, 31 and 34) inundating previously established agricultural and settlement areas. The net result over time was contraction of the coastal agricultural base to narrow inland bottomland valley areas (Figure 33) and reduction of easily netted small fish varieties and mollusk marine resources necessary to maintain the large Supe Valley population at previous levels.

With the reduction of farming areas and the marine resource base from the Medio Mundo event, society population levels could not be sustained at earlier levels. Of further note is that valley bottomlands are flat, resulting in Supe River course meander during intermittent rainfall flows. During a major El Niño event, shallow fertile farming soils are washed away, requiring many years before mountain soil runoff redeposits fertile soils suitable for agriculture. A conversation with a local farmer reveals that during the 1989 El Niño major flood event, some 50 acres of his farmland was washed away; such events occurred in Caral's historic times requiring contraction of valley agriculture to limited plateau areas, thus limiting the valley's ability to sustain large valley populations. The totality of environmental changes, as demonstrated by CFD investigation, thus precipitated contraction of Supe Valley settlements as sources of food supply declined to support the large valley population; this course of events in the ~1500 BCE time period ultimately led to the abandonment of all Supe Valley sites.

Figure 33. Present-day Supe Valley bottomland agricultural area; Caral situated on left elevated plateau. Photo by author.

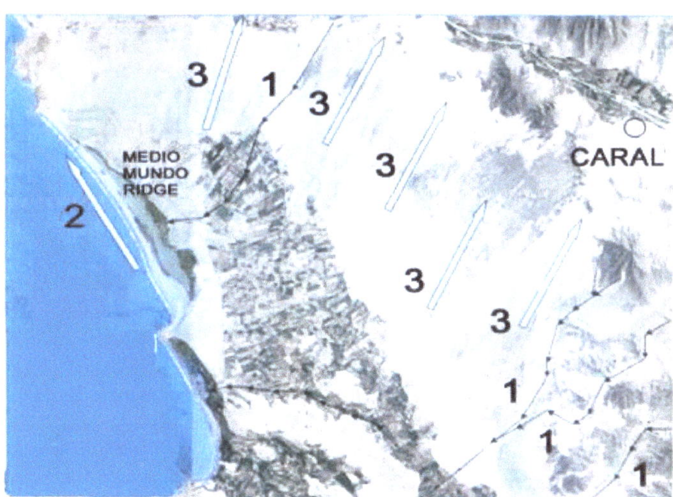

Figure 34. Major environmental changes resulting from the Medio Mundo beach ridge. Derived figure from Google Earth.

7. FLOW-3D CFD Investigation of the Pont Du Gard Aqueduct Design (Nîmes, France) and Roman Water Engineering Technology

The first century CE Pont du Gard Aqueduct bridge (Figure 35) and the water distribution castellum (Figure 36) located in southern France (Figure 37) has received wide attention by researchers over the years, focusing on its history, construction, and water engineering technology. A detailed summary of this research is given in ([2], pp. 295–318).

As little descriptive material about the water engineering necessary for the construction and operation of the aqueduct system survives from Roman times, use of CFD analysis provides new insights into the depth of Roman water engineering that underlies the totality of the water system's design.

Figure 35. The Roman Pont du Gard Aqueduct Bridge. Photo by author.

Figure 36. Interior view of the Pont du Gard water distribution castellum. Photo by author.

Figure 37. The aqueduct path from the spring source at Uzes to the water distribution castellum, Figure 36, within the city of Nîmes, France (Roman Nemausus). Figure derived from [30].

Of interest to determine Roman water engineering capabilities are several main design factors: (1) the 0.002 degree channel slope leading aqueduct water to the castellum (Figure 38); (2) the seven castellum wall exit ports and the three castellum floor ports (Figures 36 and 39); (3) details of the castellum wall's ~1.5 m low height (Figures 36 and 39); and (4) the geometric details of the castellum water entrance from the aqueduct (Figure 38) that had a vertically movable sluice plate (now lost) to control castellum water height.

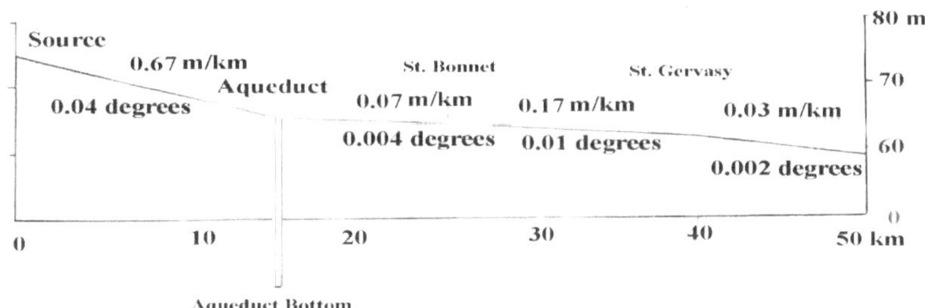

Figure 38. Aqueduct slope measurements from the Uzes spring source to the castellum. Figure derived from [30].

The goal of all water engineers, in past and present centuries, is to design a water structure that minimizes cost and construction labor while maximizing water transfer efficiency with aesthetics typical of all things Roman. Technical details of Roman water engineering used in the Pont du Gard channel and castellum not previously reported in the archaeological literature are given in ([2], pp. 295–318) and summarized in the discussion to follow.

Figure 39. The Pont du Gard castellum showing the back wall rectangular water delivery port from the aqueduct and several of the castellum front wall water exit ports to pipelines leading to city destinations. Photo by author.

From Figure 40, initial pipeline sections exiting from the castellum wall ports (Figures 36 and 39) are nearly level to facilitate junction with the castellum wall ports; downstream attached pipelines to city destinations are sloped to match the city street slope. Flow in these sloped downstream pipelines must be designed to support partial flow with air over the water surface to reduce pressurized, full flow leakage at pipeline joints (Figure 40C) such as would occur in a near-level pipeline A design. For Figure 40B, at a steeper prescribed declination angle than the landscape slope, supercritical (Fr > 1) flow exists, and hydraulic jump creation is possible as the pipeline inner wall surface is exposed to pipeline roughness over long lengths. Post the hydraulic jump, full flow exists, inducing pipeline joint leakage. Figure 40C shows the desired flow type within the sloped castellum pipelines joined to short, level castellum wall openings. These pipelines are set at a critical angle θ_c matching the landscape slope. This condition, called critical flow, induces an airspace over the partial flow, reducing pipeline joint leakage possibilities while producing the maximum flow rate that individual pipelines transport [19–21]. The key to the Roman design is to induce Fr = 1 critical flow in downstream sloped pipelines (Figure 40C) where the θ_c pipeline slope matches the landscape declination slope. This is carried out by selecting the height of the castellum from the landscape ground base height; this requirement determines the incoming aqueduct low 0.002 degree slope, as noted in Figure 38. From Figures 36 and 39, the pipeline exit ports are extremely close to the top of the castellum-containing wall. This design implies that if entry flow into pipelines is at y_c critical height ([20], p. 51; [2], p. 310), then critical, partial flow enters the sloped pipelines that are designed to continue critical flow at the θ_c declination angle (Figure 41). This Roman design explains the low wall height of the castellum (Figure 36) as the entry flow height from the aqueduct only reaches up to half of the exit pipelines' diameter and this is guaranteed by the sluice gate open height.

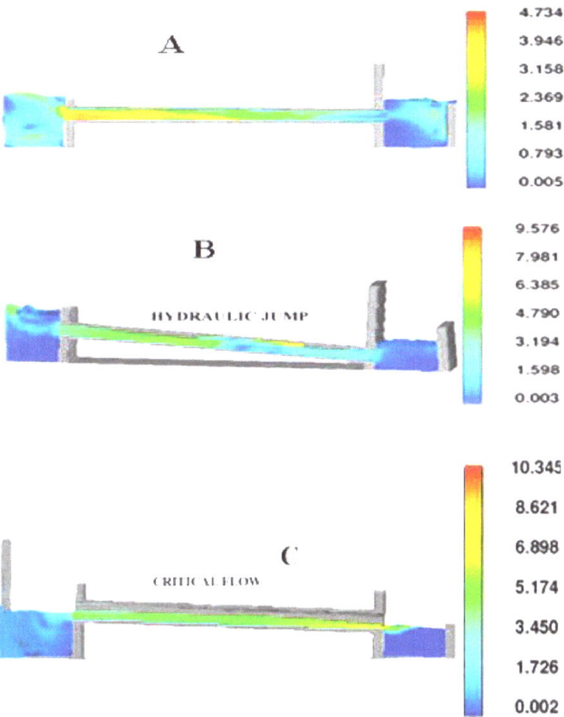

Figure 40. CFD pipeline flow characteristics for pipelines at different declination angles. Velocity is in ft/s. FLOW-3D CFD results from [2], p. 310.

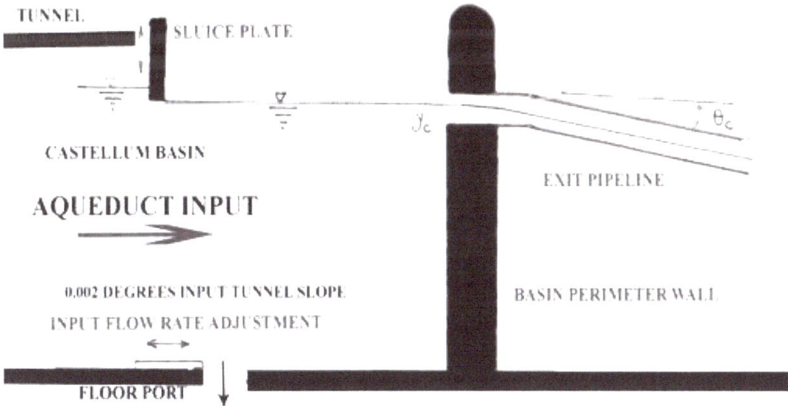

Figure 41. The Roman design of the Pont du Gard castellum (not to scale). Results from [2], p. 310.

As the input spring flow in the aqueduct may seasonally vary from the design flow rate, the movable sluice plate (Figure 41) can be set to ensure that entry flow depth into pipelines is at the y_c depth; this design flow rate condition is met by adjustable openings in the castellum floor (Figure 36) that drain away excess water to maintain the design flow rate. From Figure 38, it is noted that the aqueduct channel slope entering the castellum port (Figure 39) is practically level (0.002°). The reason for this near-level slope section (beyond raising water height, lowering entry velocity to the castellum, and determining the height of the castellum above the ground) is to guarantee that the low height of the castellum walls safely contains the input flow and that exit pipelines can be set at θ_c which is the slope of the Nîmes street carrying the pipelines to downhill city destinations.

Figure 41 summarizes the castellum design originated by Roman water engineers. Again, the vertical setting of the sluice plate is set to maintain the y_c water depth in the castellum; this is designed to ensure critical (Fr = 1) flow into the θ_c angled sloped exit pipelines for design flow rate condition. This design limits spillage from the castellum wall top surface by means of excess water removal past the design flow rate by castellum floor ports that have movable baffles to adjust the flow to the design flow rate. A further benefit to this design is to maintain the aesthetics of the castellum basin water surface—a viewer would see a calm and stable water surface worthy of the elaborate structure that once enveloped the castellum ~2000 years ago.

Given that the Roman castellum design can be understood in terms of modern hydraulic terminology, counterparts to this knowledge must exist in Roman times, albeit in terminology or documents yet to be discovered. Given Roman expertise in water science, as evidenced by many aqueducts, fountains, public bath structures, and port structures in Roman cities, water technology that clearly originated from experiments and observations over many years enabled the sophistication of the Pont du Gard castellum.

8. FLOW-3D CFD Analysis of the Moat Structure Hydraulic Engineering at 600–1100 CE Tiwanaku in Bolivia

Located in the interior of the ancient city of Tiwanaku is the prominent Gate of the Sun (Figure 42), whose iconography offers an introduction to learning the depth of spiritual beliefs that influenced the citizens of that city, the structure of their society, and the architectural patterns of ceremonial compounds that gave their priests the means to communicate with their deities. Commensurate with religious aspects of Tiwanaku society is the practical side involving urban and agricultural water supply and distribution systems to support the large city population. Here, accomplishments in the water sciences for urban and agricultural use are vital to the development and continuity of the Tiwanaku society. To this end, details of new hydraulic engineering discoveries as applied to Tiwanaku's urban and agricultural systems ([2], pp. 1–30; [31–36]) are summarized in the present chapter from CFD analysis and modern hydraulic engineering studies of their water supply and distribution systems.

The UNESCO World Heritage site of 600–1100 CE Tiwanaku located in altiplano Bolivia was the subject of early archaeological investigations [37,38] to determine its role in Andean history. Of note to these scholars was the investigation of an encompassing "moat" that surrounded the elite quarters of the city which harbored the Acapana multistoried pyramid, the Calasasyaya royalty compound, the subterranean temple noted for its central stele of a prominent deity as well as its many wall-mounted deity heads, and other administrative and royal compounds indicated in Figure 43. Later excavations within the moat area revealed an intricate drainage system consisting of dual subterranean pipelines connected to surface structures, indicating the presence of advanced water engineering not previously noted or analyzed. Later investigations [32,33,39,40] brought forward the argument that the moat had a previously unanticipated use beyond being a separation boundary between religious and administrative elite quarters and the secular, working class housing of the city population.

Figure 42. Gateway of the Sun deity iconography; site located in central urban Tiwanaku. Photo by author.

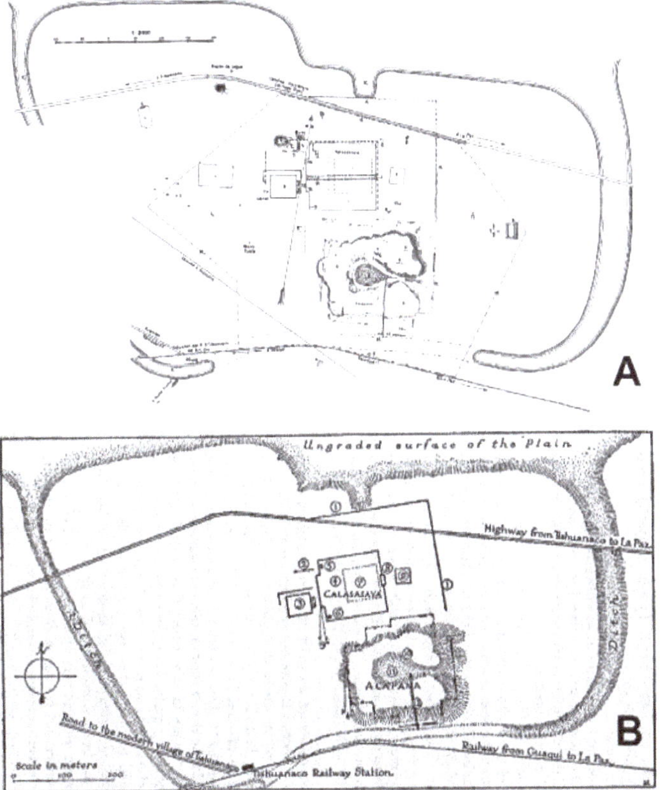

Figure 43. Early ground drawings of the moat structure by Bandelier 1911 (**A**) and Bennett 1934 (**B**) References [37,38].

Recent discovery of site aerial site photographs revealed further details related to the moat and its connection to numerous spring-supplied channels originating from nearby mountain areas. Many of these lengthy channels are connected to the moat. For the moat, estimated surface east–west length dimensions (Figure 43) range from 700 to 1000 m, and the surface north–south width dimensions are ~400 m. The moat depth is sufficient to penetrate the groundwater level in the dry season, estimated to be on the order of ~5 m [39].

The hydrodynamic engineering role played by the moat can be investigated through use of a CFD model that includes all of the discovered geometric features of the moat and their relation to the ceremonial structures (Figure 44) within the moat boundary.

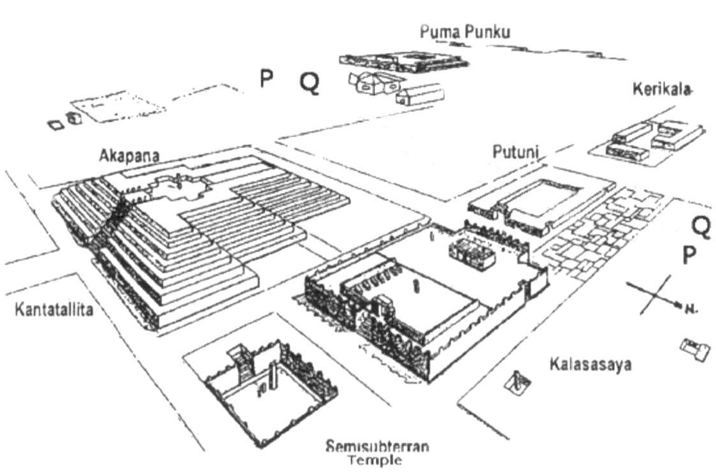

Figure 44. Details of elite and administrative structures within the moat and subterranean P, Q pipelines (from [2]).

One recently discovered feature of the moat is the Mollo Contu channel intersecting the southern portion of the moat (Figures 45 and 46) that plays a central role in interpreting the purpose of the moat. The Mollo Kontu canal carries water from a reservoir at the base of the southeastern mountain range that remains at full capacity from rainy season runoff from mountain gullies as well as water transport from nearby springs. Ground traces of this channel are apparent from the aerial photographs together with other features that include many additional channels, inner-city field systems (*cochas*), flood plains, and major (then unexcavated) architectural features. Of note are subtle moat geometry differences between earlier investigator's versions (Figure 43) and the later aerial photograph versions (Figures 45 and 46) likely due to soil transfers from yearly heavy rainfalls as well as human activity on the site. For the CFD model of this area, the preference is for use of the earlier Figure 43 versions as they represent an earlier undisturbed view of the site.

The CFD (Figure 47) model incorporates an aquifer below the ground surface with specified hydraulic conductivity, porosity, specific surface, and particle size variables included in the model aquifer description; water viscosity and density, as well as aquifer properties, are specified [22]. Shown are two subterranean stone-lined channels (P and Q, Figure 47) originating from moat arm W to arm V then continuing underground to empty in the Tiwanaku river A'–B'. The Mollo Kontu channel U empties into the W moat upper arm. The CFD model is tilted 1 to 2 degrees downward in the south to north direction as well as the east–west direction to facilitate water drainage into the Tiwanaku River that empties into Lake Titicaca. The P and Q channels are composed of ~0.75 m stone plates on all sides with a top plate to seal channel (Figure 48). Although only two subterranean channels, P and Q, are shown, indications of other subterranean channels exist to the west of those shown whose presence remains to be verified by excavation. A series of perforated stone disc plates lead vertically upward from circular openings on the subterranean channel's top plates to connect surface structure drainage to the P, Q channels to allow wastewater drainage from high status buildings. Subsidiary water input from Corocoru spring-fed channels N and O serve wetland agriculture in the K region, with excess water emptying into moat channel D at point a (Figure 47).

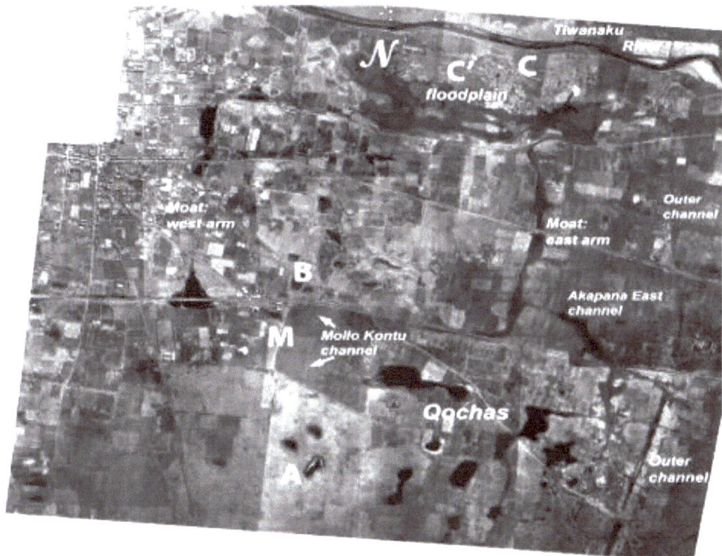

Figure 45. The recently discovered Mollo Contu canal. Figures originally derived from 1940s aerial photos from Alan Sawyer at Tiwanaku and later annotated 2010 by John Janusek.

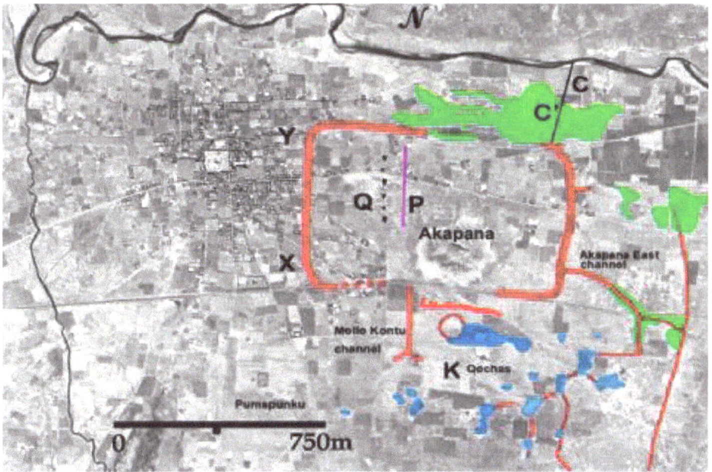

Figure 46. The recently discovered Mollo Kontu canal. Figures originally derived from 1940s aerial photos from Alan Sawyer at Tiwanaku and later annotated 2010 by John Janusek, Vanderbilt University.

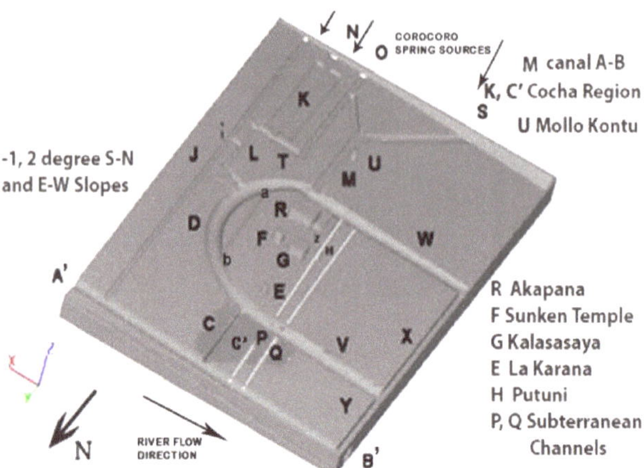

Figure 47. FLOW-3D CFD model of features in the moat location area, from [2], p. 173.

Figure 48. Excavated subterranean channel P located ~3 m below the ground surface. Photo by author.

To examine the function of the moat in wet and dry seasons, FLOW-3D CFD [15] calculations were performed to indicate the moat's purpose. Figure 49 results concern the post rainy season at a time well into the dry season. The fluid fraction (ff) detail from FLOW-3D calculations of a section of the moat east arm (D) shows the dry (blue) ground surface within the largely paved and roofed elite structure area inside the moat boundary that promotes rainfall runoff to keep the ground surface dry. The red ground area has absorbed heavy rainfall leading to a saturated ground surface now showing signs of evaporation and aquifer drainage water loss. Figure 49 shows the moat's inner surface walls conducting seepage water to the saturated bottom of the moat (dug below the water table) and the start of surface drying from the rainy season saturated (ff = 1) surface. The aquifer seepage plus surface evaporation thus helps to dry housing areas interior and exterior to the moat, inducing hygienic benefits by control of mold in housing structures. As the moat depth penetrates the saturated groundwater level, water from aquifer seepage cannot penetrate the moat bottom and thus transfers water to moat arms D and V for drainage into the Tiwanaku River. Note that ff = 1 denotes aquifer and ground surface saturation, ff = 0 denotes dry surfaces, and intermediate ff values denote moisture levels in the aquifer. Drainage water then flows downslope to marsh area C′ used for specialty crops to continue seepage drainage into the Tiwanaku River A′, B′.

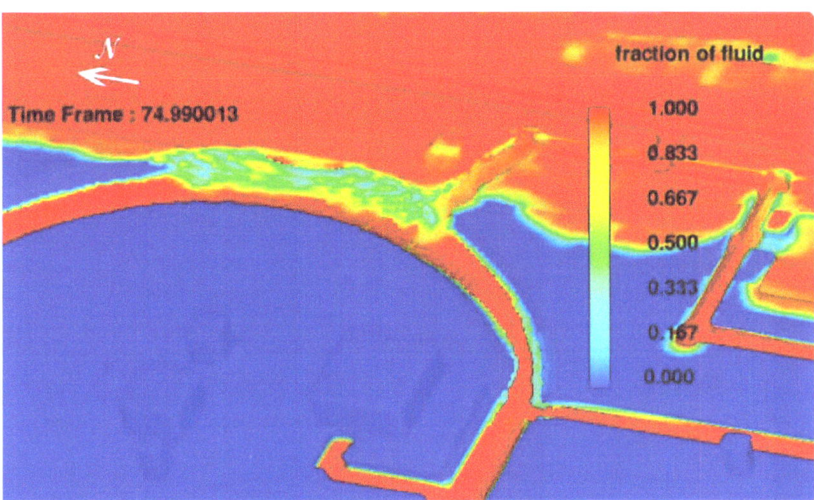

Figure 49. End of rainy season and well into the dry season fluid fraction distribution indicating aquifer drainage into the moat; from [2], p. 179.

An elevated channel C conducts excessive volumes of seepage water and rainy season surface runoff directly into the Tiwanaku River. Seepage maximums occur from saturated land areas adjacent to the moat at D. Continuous surface channel flow into the moat from the Mollo Kontu canal (M, Figure 47) and canal (a) draining excess irrigation water from (*cocha*) pasturage area K contribute further water to the moat for drainage to the Tiwanaku River. The following set of figures details the hydraulic engineering function of the moat through dry and wet seasons.

The Mollo Kontu M channel continues its flow from spring-fed mountainside reservoirs during both dry and wet seasons. The subsidiary channel L (Figure 47) drains water from pasturage area K which is supplied from channels N and O to add additional drainage water into moat arm W. Note also that straight channel J supplied from mountainside springs has a canal branch diversion channel (i) into channel L (Figure 47) to add further drainage water into the moat's D and V arms. Seepage water into the westernmost moat X channel appears to drain directly into the Tiwanaku River by a direct channel.

Figure 50 indicates fluid fraction results into the dry season with surface drying achieved by aquifer seepage and surface evaporation. Note that the Mollo Konto channel M (Figure 47) continues to provide water to subterranean P and Q channels as they originate from the northern moat wall surface W arm (Figure 51) continually supplied by the Mollo Kontu canal. Aquifer leakage continues providing water to the moat bottom to enhance drying together with surface evaporation drying. The continuous ground surface drying, as the dry season progresses, is amplified from high seepage levels continuing into later dry season times in the D region of the moat, as indicated in Figures 49 and 50. Figure 51 results are given on a depth plane of the P and Q subterranean channels and indicate the saturated (red) aquifer in regions east of moat branch D while indicating dry regions west of D. Figure 51 shows water flowing through P and C channels from the Mollo Konto source M channel. Further water input into the moat occurs from surface runoff from the elite structures within the moat (Figure 44) and occurs as areas within the moat boundary were largely paved with roofed structures, causing rainfall to runoff directly into moat arms

W–D–V–Z for drainage into A′, B′ to Lake Titicaca. Note the (red) saturated groundwater aquifer at depth outside the moat and the dry (blue) area at depth within the moat boundary due to extensive runoff from paved areas within the moat boundary.

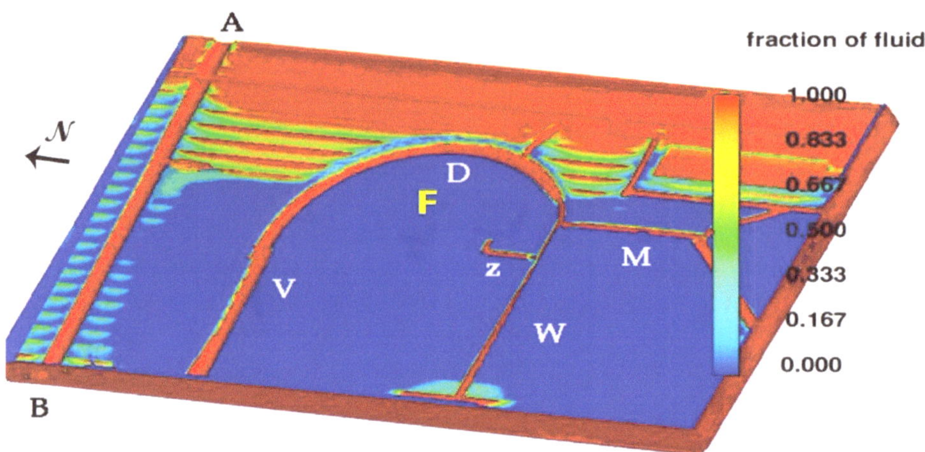

Figure 50. Dry season continuance with additional ground surface evaporation drying and aquifer drainage; from [2], p. 179. A,B represents the Tiwanaku River.

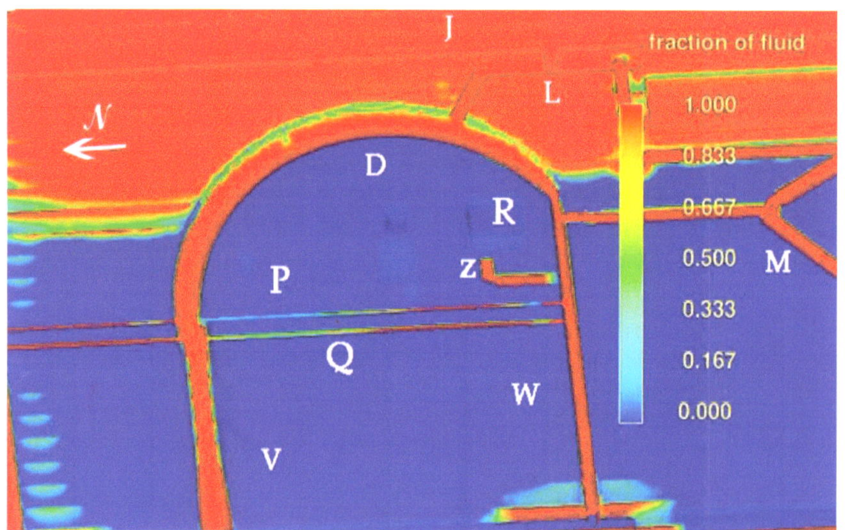

Figure 51. Dry season fluid fraction results shown at the depth of subterranean channels P and Q; from [2], p. 180.

Water input from the Mollo Konto M channel input into moat arm W provides water to subterranean channels P and Q to flush elite structure wastewater downhill from the Putini Palace (Figures 44 and 47) complex to drainage channel arm V (Figure 47) then on to the Tiwanaku River A′, B′ and Lake Titicaca. The Mollo Kontu channel (M, Figure 51) provides water through both wet and dry seasons to the structures within the elite compounds and on to other elite compounds through subsurface channels close to the ground surface (Figure 52). Z indicates a water channel connection from the Kalasasaya Platform (Figure 47, G) to the moat's W arm.

Figure 52. Excavated channel within the moat elite quarter diverting water from the P and Q source to other compound structures. Photo by author.

Dry season fluid fraction results on a subterranean plane incorporating the P and Q subterranean channels are shown in Figure 53. Moisture levels in *cocha* region K and depressed marsh area C′ indicate sustainable pasturage and agriculture due to contact with the deep water table and the discharge from the V moat arm. Water input from the Mollo Kontu M canal continually delivers water to the moat arm W that enters the subterranean channels P and Q. Aquifer drainage continues into all moat arms that, together with water transfer from P and Q subterranean channels and seepage from the C′ marsh area, deliverer drainage water to the Tiwanaku River A′, B′. Note that below the P and Q channel depth, the moat bottom penetrates the groundwater-saturated soil level (red area).

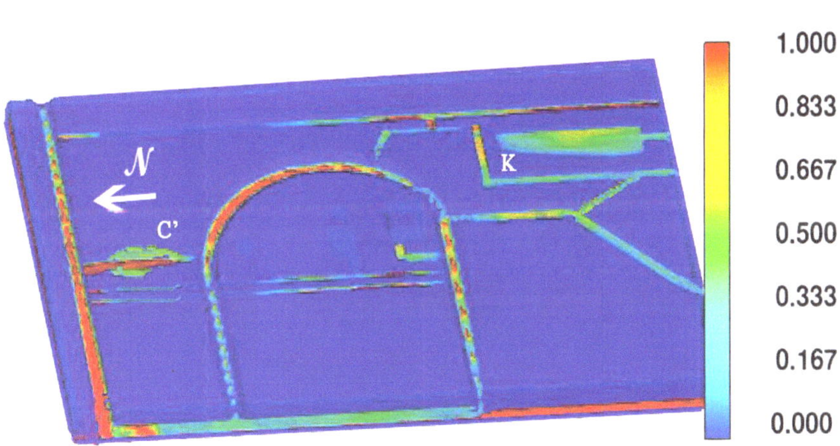

Figure 53. CFD fluid fraction results during the height of the dry season shown at the depth of subterranean channels P and Q; from [2], p. 180.

Figure 54 shows the dry season end fluid fraction results for the east arm D that indicate decreased seepage water into the moat causing accelerated ground surface drying together with evaporation drying. Mollo Kontu water continues to flow into the moat's W arm and continues to D and V moat arms on the moat's saturated bottom surface to drain into the C′ marsh area and channel C enroute to the A′, B′ Tiwanaku River. The ground surface (blue) now indicates surface dryness to depth for the region's interior and exterior to the moat boundary. Limited water in the moat bottom results from the Mollo Kontu channel input as aquifer seepage is now limited during the terminal phase of the dry season.

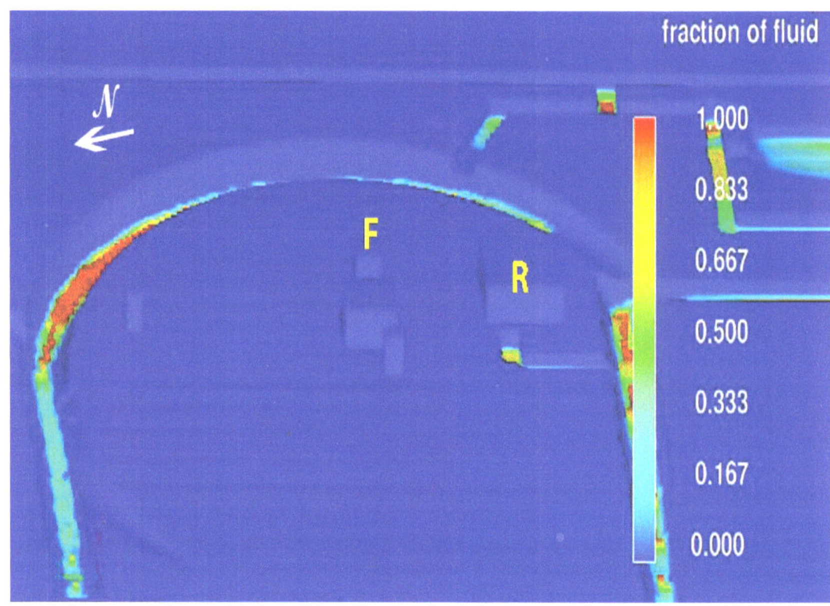

Figure 54. Late dry season fluid fraction results; from [2], p. 181.

Figure 55 details wet season conditions typically experienced by a typical moat section and the design intent of the moat as envisioned by Tiwanaku water engineers. Rainy season surface runoff enters the moat from ground surface declination angles of one to two degrees in south–north and west–east directions together with aquifer seepage and contributions from the Mollo Kontu M channel and L, Z channels (Figure 46). As the bottom of the moat is essentially saturated soil dug into the water table, no further water subtractions occur so that water collected in the south-to-north sloped moat can flow by gravity to C and C′ connections into the Tiwanaku River and on to Lake Titicaca, as Figure 55 summarizes.

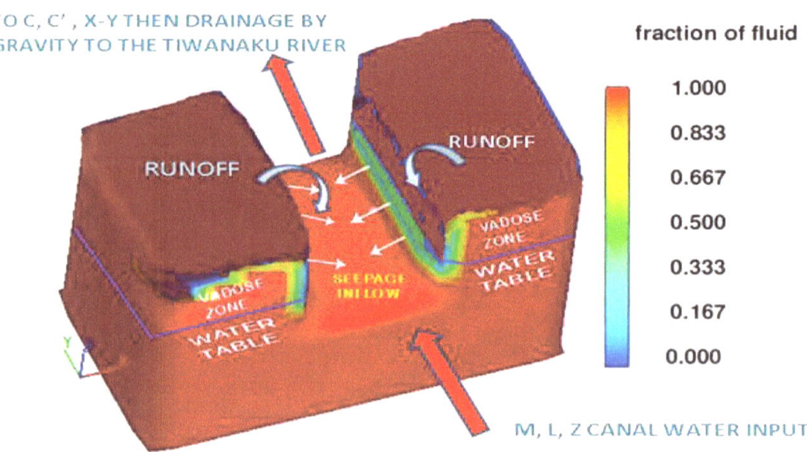

Figure 55. Summary of wet season effects on a typical moat section; from [2], p. 178.

During the dry season, Figure 56 indicates a drying ground surface aided by evaporation and ongoing aquifer seepage with continuous water flow from the Mollo Kontu channel and L, Z channels that may have partially reduced flow rates under dry season conditions. As a result, some decline in the water table may exist with a an extended near-saturated vadose intermediate region.

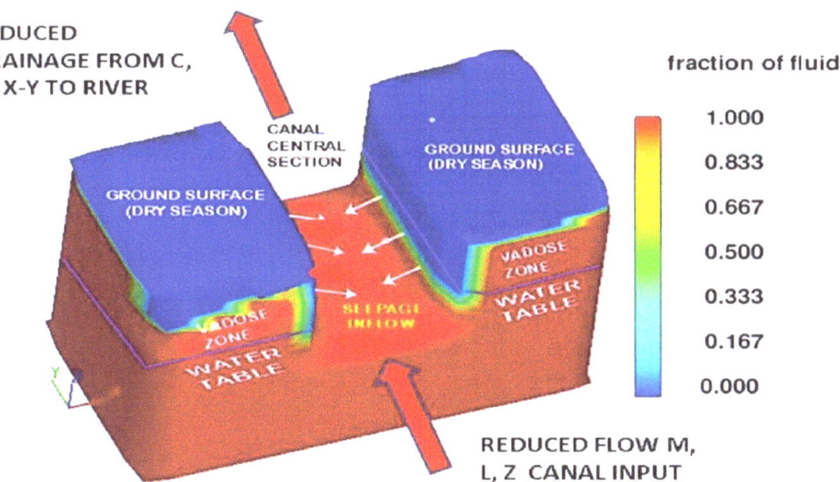

Figure 56. Summary of dry season effects on a typical moat section; from [2], p. 181.

In summary, the presence of the moat helps to accelerate seasonal ground-level dryness, limiting mold buildup in site housing structures, which is a definite health benefit. A further benefit is the near-constant high groundwater level maintained through seasonal rainfall and dry periods provided by water input from the Mollo Kontu and L, Z channels. This feature helps maintain the structural stability of the immense Akapana pyramid within the moat boundary (Figure 44) as the underlying saturated aquifer is not subject to compression volume change by weight compression of the temple. With the high groundwater height seasonally maintained, wells can be constructed in individual houses for personal use together with reservoir basins within the city to facilitate personal collection water. Further technical details on the urban and agricultural water system of Tiwanaku are found in ([3], pp. 1–30).

A contribution to the demise of Tiwanaku is attributed to long-term drought [41–44] which slowly over the course of years sank the groundwater level necessary to support raised field agriculture [41,42] supporting the economic base of urban Tiwanaku. Further authors emphasize social unrest and external and internal threats in the termination phase of Tiwanaku, but without a viable agricultural base, no society can continue its existence.

9. The Internal Reservoir at the Theater of Roman Ephesus, Turkey—Recovery of Lost Greek Water Technology

Ephesus history dates to long before the Roman occupation period. The city area was first inhabited by the local Carians, then captured by Ionian tribes in ~1050 BC. After 500 years of Ionian rule, the city was invaded by Lydian King Croesus in 560 BC but came to an end after the Persian invasion in 546 BC. Alexander the Great, defeating Persian forces at the Battle of Granicus, entered the city in 334 BC; Ephesus later passed after the death of Alexander to his subordinate commander Lysimachus in 295 BC. Under Greek rule and later Roman rule in later centuries, the city became a trade center with fleets of ships in its Mediterranean harbor serving cities under the Roman sphere of influence. With a population of ~250,000 in Roman occupation times [44], water supplies by new multiple aqueducts, together with numerous architectural changes reflecting the Roman vision of city composition occurred in the form of multiple outdoor gymnasiums, temples, commercial agoras, a coliseum, multiple elaborate baths, public lavatories, multiple public fountains, the Library of Celsus (Figure 57), a massive theater (Figure 58), an administrative center for

Roman officialdom (Hanghaus), numerous marble statues on paved streets, multiple city fountains and castellums, aqueducts, and new water system modifications, were emplaced. Newly built aqueducts led water to several city castellum water distribution structures from which multiple pipelines led to different city districts, serving public fountains, government buildings, baths, public lavatories, the theater, temples, the coliseum, gymnasiums, and private housing districts, as well as *nymphaea* with elaborate statuary incorporating aesthetic water displays. An elaborate subterranean drainage channel system collected waterborne waste material and served the Hanghaus Roman administrative center as well as almost all public use buildings. The water supplied on a continuous basis to public and private structures, when drained into the subterranean channel network, provided continuous flushing of waste material into the harbor area (Figure 59) fronting the Mediterranean sea to maintain high hygienic standards for the city's population.

Figure 57. The Roman Library of Celsus. Photo by author.

Figure 58. The Roman Theater. Aerial photo by author. H represents an earlier Greek structure adjacent to the Theater; G represents a Greek fountain house still in use in Roman times.

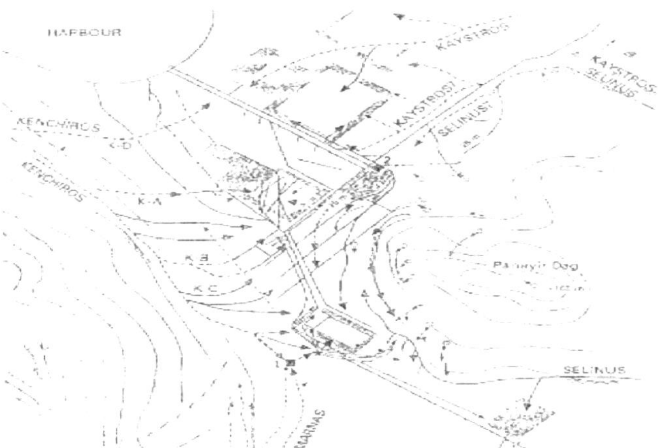

Figure 59. Multiple Roman branched aqueducts [1], p. 298.

Further details of Ephesus site history together with FLOW-3D CFD and hydraulic engineering analysis of aqueduct and water drainage systems and current research findings are found in ([1], pp. 295–337, [45]).

In the process of rebuilding the city to Roman standards, remnants of earlier Greek subterranean water channels and pipeline systems no longer of use were buried under new Roman constructions. New water increase requirements were necessary to serve elite Roman government administrative living quarters (Hanghaus), *bouleuterion* meeting quarters for legislative member assemblies, housing for the ~250,000 citizens, marketplace agoras with domestic and imported food and trade items, public baths, public lavatories, recreational gymnasium facilities, and multiple fountains with water collection basins for daily domestic use.

A series of aqueducts (Figure 59), the Kenchiros, Kaystros, Selinus, and Marnias, were the main water supply providers to the city to serve the comforts and needs of the large city population. The Kenchiros aqueduct's many branches (K–D, K–A, K–B, K–C, and others) served specific sites; remaining aqueducts had multiple branches, several of which led to castellums and settling basins to further channel water by pipelines to specific buildings and multiple fountains throughout the city.

Of specific interest is the water supply to the Theater in Roman times from the Selinus aqueduct source. Before later Roman construction, a Greek fountain house with multiple water streams from elevated lion head fixtures existed at the base of the Roman Theater. From five years' participation in excavations in the 1990s with Dr. Dora Crouch and members of the Austrian Archaeological Institute (ÖAI) to track the Roman water supply, distribution, and drainage systems, additional excavations related to the earlier Greek fountain house (located above G on the northside of the road on Figure 58) were made. An unusual structure (Figure 60) buried about ~5 m below the Roman surface was infilled with large stone blocks from Roman construction of the new Theater, and through excavation, the structure was revealed, showing a ceramic pipeline of ~12 cm outer diameter with a wall thickness of ~2 cm of earlier Greek origin positioned about ~1 m back from the back wall of the deeply buried Figure 60 structure. It is surmised that this pipeline segment originally joined with the pipeline serving the upstream Greek fountain house and was later intersected by the Figure 60 structure past which a pipeline (or channel) continued to further Greek structures. The Figure 60 structure had a stone block backwall height of ~4.45 m, a width of ~5 m, and an excavated stone lined path of ~10.3 m from excavations performed. The location of the excavation pit in Figure 60 structure is directly above H on Figure 58 and directly to the left of the lowest row of seats in the Theater at location 2. The question to be answered is: "What is the function of the Figure 60 structure?"—a question that FLOW-3D CFD analysis can answer.

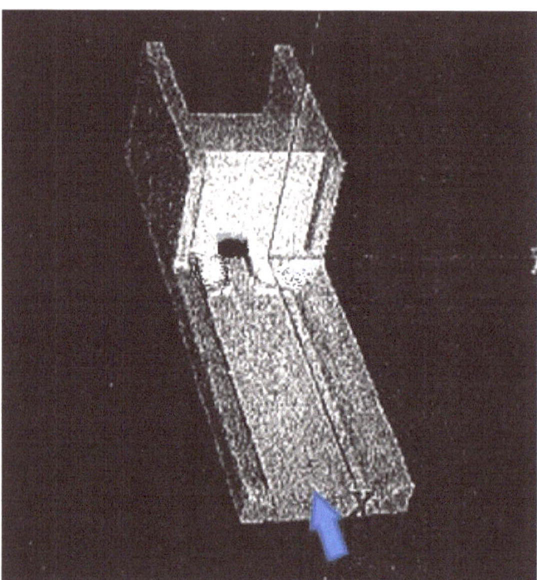

Figure 60. FLOW-3D CFD model of the deeply buried water transfer structure. Arrow denotes flow direction. Central opening is ~0.5 m by 0.5 m. Portions of back wall height were destroyed by later constructions. Note the dual ellipsoidal structures at base of wall, the leftmost one shown in Figure 61. Figure derived from [1], p. 332.

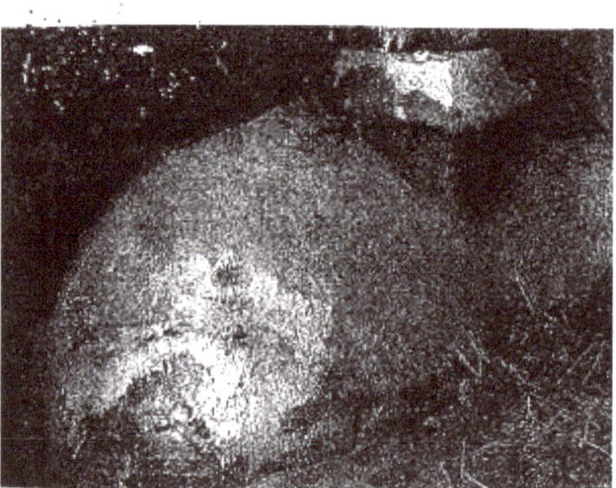

Figure 61. Photograph of the leftmost fractional ellipsoidal structure at the base of deep excavation pit. Photo by author.

The Figures 60 and 61 FLOW-3D CFD models show a rectangular opening on the backwall, ~1.3 × 1 m in size, that had ellipsoidal hardened clay sections on each side of the wall opening, as shown in Figure 61, presumably to facilitate smooth entry flow. It may be surmised that the revealed pipeline section shown past the backwall opening (Figure 62) was once part of a continuous longer pipeline that provided water to the Greek fountain house and further on by pipeline or channel to Greek structures. This pipeline was intersected by the elaborate structure represented by Figure 60. Water flow from the fountain house pipeline then flowed into the Figure 60 basin and continued through the lower wall opening to an open channel or pipeline serving Greek structures east of it, as shown in Figure 58.

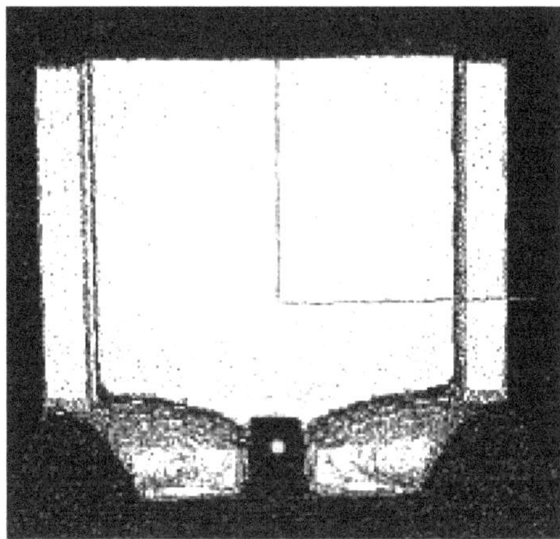

Figure 62. FLOW-3D CFD model of front view of the discontinued internal pipeline located ~ 2 m back from the wall opening.

Given that hard water through long ceramic pipelines precipitates calcium chloride sinter over long usage times that ultimately limits water transport through the progressively narrowing internal pipeline opening, it may be surmised that the Figure 60 structure was proposed by Greek water engineers as a "fix" to the anticipated water blockage problem to continue water flow to other city structures. Additionally, silt particles originating from the spring source flowing in a low declination angle pipeline can settle in low-speed water flows, further increasing flow blockage. Intuitively, the dual ellipsoidal structures adjacent to the lower back wall opening would appear to create smooth contracted streamlines from entry flow into the rectangular back wall opening to facilitate flow passage, but determining the hydraulic engineering function that Greek water engineers had in mind with the Figure 60 structure can only reliably come from CFD analysis.

Excavations revealed that this transition structure existed between the Greek fountain house and the entrance to the Figure 60 structure. To determine the intended design effect of the Figure 60 structure, CFD analysis was performed, originating the velocity vector plots given in Figure 63A,B. The upper (Figure 63A) view is at the midpoint of the entry channel and indicates circulatory flow patterns upstream of the rectangular back wall opening. Results from Figure 63A indicate that the dual vortex flow is nonsymmetrical and turbulent (the k–ε turbulence model is incorporated in the FLOW-3D CFD calculation) before entry into the rectangular opening. A further view from Figure 63B at a plane just above the bottom surface reveals a circulatory flow pattern induced by the Figure 60 structure. Figure 63A,B shows further velocity vector plots indicating chaotic flow patterns. As opposed to an intuitive estimate of the effect of the dual ellipsoidal structures to smoothly transition incoming flow through the rectangular back wall opening, an opposite effect is achieved from CFD analysis as flow appears highly chaotic before its entry through the back wall rectangular opening.

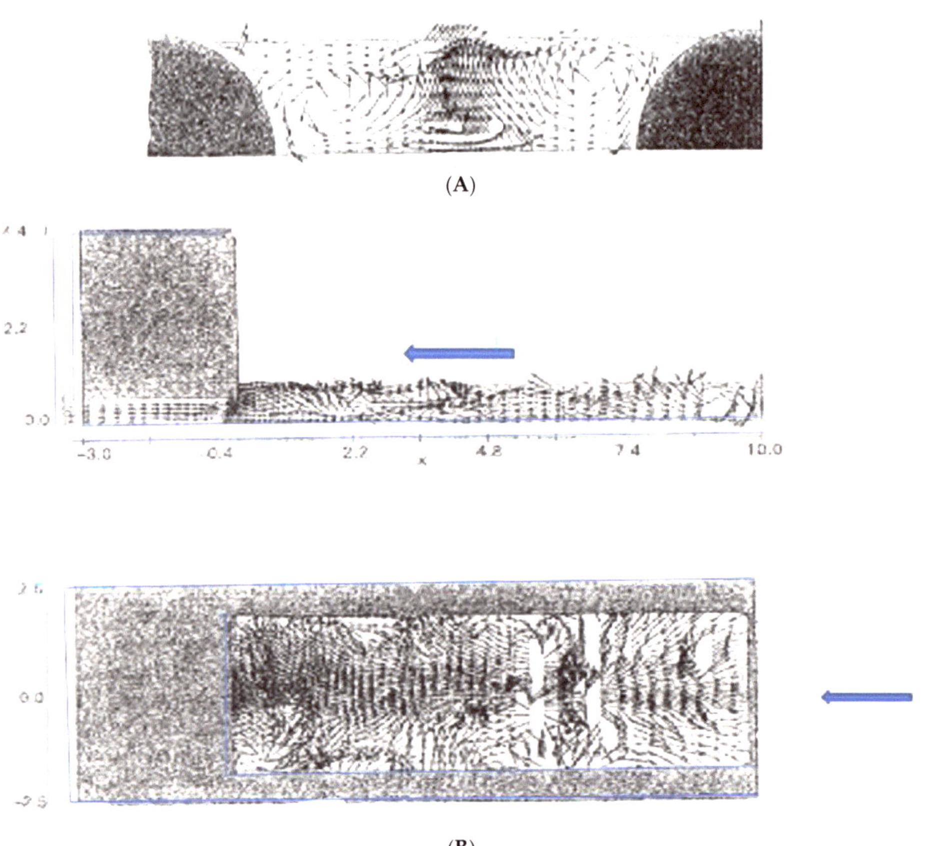

Figure 63. (**A**) Front view of flow velocity vectors prior to the back wall opening; Ref. [1], p. 336. (**B**) CFD midpoint side view (**top figure**) and top view (**bottom figure**) velocity vectors; Ref. [1], p. 335. Figures indicate a high creation level of water vortical structures and turbulence before the wall opening. Arrow represents flow direction. Scale in Av meters.

An interpretation of the intended effect of the Figure 60 shaping by Greek hydraulic engineers would be to induce chaotic flow that would entrain any settled silt and debris particles that arrived from the incoming pipeline and permit their transfer further on to a downstream settling basin where they could be removed to bring water quality to potable level for downstream occupied administrative structures.

The development of this water system is as follows: (1) the original lengthy pipeline that passes the Greek fountain house slowly transports deposited silt from fluid shear effects together with entrained silt sinter and debris particles to ultimately clog the pipeline intended to transport water to distant Greek housing structures; (2) Greek water engineers install the Figure 60 structure whose chaotic flow (Figure 63A,B) entrains arriving silt sludge, sinter, and debris particles back into circulation through a channel (or pipeline) continuing from the back wall rectangular opening to a downstream settling basin to clarify water to potable level for downstream occupied Greek structures; (3) as the Figure 60 structure is located at a low level within the earlier Greek theater, there may have been lavatory human waste also channeled into the water stream preceding the Figure 60 structure that provides a further reason for keeping particles in suspension prior to entry to a downstream settling tank; (4) with later Roman aqueduct constructions, new water supply and drainage systems (Figure 59) supplying the new Roman Theater were built over prior Greek constructions.

Excavations of the Figure 60 structure revealed large stone block infilling in Roman times. Some aspects of the earlier Greek theater design continued into Roman times with stage construction modifications to accommodate citizens that enjoyed the benefits of plays from both Greek and Roman authors.

Of note is Greek hydraulic technology involving sediment deposit removal from water structures. Figure 15 indicates a hydraulic structure at the Greek site of Priene designed to clear the major drainage system from agora waste debris and washed-in debris from city housing. Figure 60 indicates a Greek hydraulic structure at Ephesus designed to return settled pipeline transported silt and sinter debris back into highly chaotic flow (Figure 63A,B) so that it can be removed in a downstream settling tank to provide potable water to downstream structures. Similarly, hydraulic structures at Ephesus [45] supplying clearing water to public lavatories has similar turbulence-inducing structures to keep debris and human waste products in suspension before delivery by pipeline into the bay area. The sacred water basins of Demeter, located in Figure 13B at Priene's northern hills, were primarily used for fertility rituals and have a subterranean pipeline water supply downhill from the basins to ensure that sediments settle before entry to uphill ceremonial-use basins. From observed excavations within urban Priene, a pipeline supplied spring water to a first settling basin with overflow to a second settling basin, then overflow from the second basin to a third settling basin whose purified flow continued to public fountains and basins for domestic use. Sediments accumulated at basin bottoms clearly purified water for domestic use. Greek water purification technology from several example cases presented indicates the presence of advanced hydraulic engineering knowledge not previously reported in the open literature.

10. Inka Water Technology at the Royal Site of Tipon

The present chapter utilizes modern hydraulic engineering technology to uncover Inka water technology at the site of Tipon.

The site of Tipon, located in Peru, ~13 km east of Cuzco along the Huatanay River, at south latitude 13° 34′ and longitude 71° 47′, at 3700–4000 masl is known for its many unique hydraulic features coordinated in a practical and aesthetic manner to demonstrate Inka knowledge of water control principles. Details of the water supply canals and architectural features of the Tipon site are shown in Figure 64. The site has an early Middle Horizon (600–1000 AD) presence, evidenced by an encircling ~ 6.4 km long outer wall (Inka Canal Principal, Figure 64) attributed to earlier Wari control of the enclosed area.

The ~2 km^2 interior site area was under Inka control after ~1200 AD and was later converted into the royal estate of Inka Wiracocha in the early 15th century, as evidenced by the royal residence and ceremonial compounds of Sinkunakancha and Patallaqta. The site is composed of thirteen major agricultural platforms (Figure 65), with the lowermost platforms associated with nearby ceremonial centers. The agricultural platforms were mainly irrigated from water supplied from a branch of the Main Aqueduct sourced from the Rio Pukara (Figure 64); branches of the Main Aqueduct provided water to the lowermost ceremonial areas.

Each agricultural platform had a drainage channel at its base that collected post-saturation aquifer groundwater seepage as well as rainfall runoff; excess water beyond that required to irrigate the next lower agricultural platform was led to side drainage channels directed to lower site occupation and ceremonial areas. As only a fraction of each platform's water supply was used to provide the necessary moisture level for specialty crops, excess water was passed on to the next lower platform with excess water directed to an easternmost collection channel. A further portion of water from agricultural platform seepage collection channels passed through a series of interconnected surface and subsurface channels to provide water to the Principal Fountain (Figure 66) supplying the Waterfall structure that supplied water to domestic and elite residential and ceremonial areas at lower site areas. High-level springs were an additional water source to Tipon's intricate water supply and control system to guarantee that Tipon's water systems functioned on a year-round basis appropriate for a royal residence site.

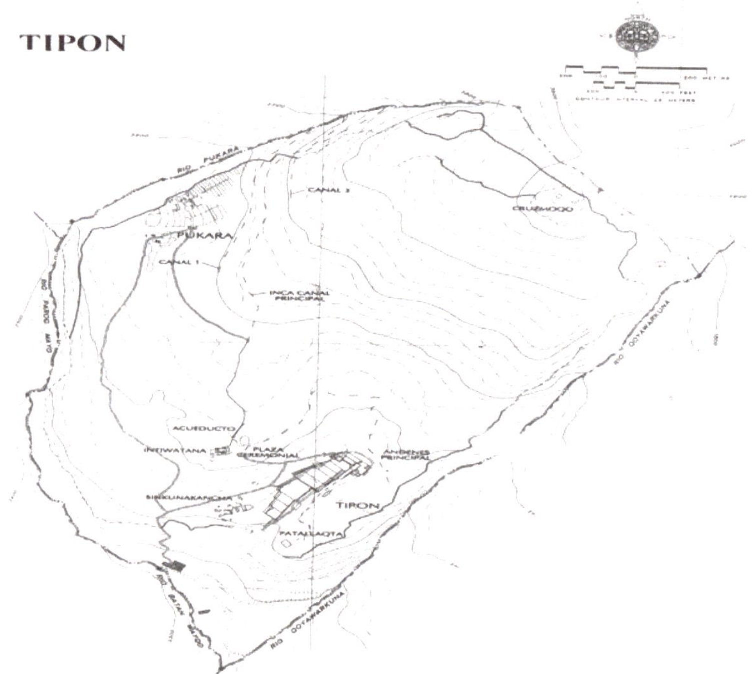

Figure 64. Site feature map of Tipon from Wright et al. 2006 [46].

Figure 65. Thirteen agricultural terraces at Tipon. Photo by author.

Of special interest is the geometry of the upstream water supply channel (Figure 66) from the Principal Fountain to the Waterfall structure shown in Figure 67. A channel contraction occurs from a 0.9 m wide, ~2.5 m long entry channel to a ~0.4 m contracted wide, ~10.5 m long channel upstream of the Principal Fountain. Both channel sections have a rectangular cross-section at the same mild low slope. The water source to the upstream wider channel section derives from eight separate water supply conduits together with a major spring, indicative of the totality of water control systems used to maintain a constant flow rate to the Waterfall during seasonal changes in water supply. Measured flow rates into the (reactivated) wide channel from two different tests [46] yielded 0.68 ft^3/s and 0.58 ft^3/s, leading to an average 0.63 ft^3/s (0.02 m^3/s) flow rate. The question arises as to the water engineering design intent of the abrupt width change of the channel section shown in Figure 66 and its effect on the aesthetics of the downstream Waterfall.

Figure 66. Channel cross-section geometry change made to induce critical flow in the contracted channel section to downstream channels supplying the Figure 67 Waterfall. Photo by author.

Figure 67. Waterfall structure supplied from the Figure 66 channel. Photo by author.

It is noted that the site of Tipon was reactivated after years of abandonment by restoring the water sources and channel systems as in Inka times. While reconstruction of certain time-degraded elements of the site have been made, the stonework associated with Figures 65–67 is original and therefore of use to determine the hydraulic engineering function intended by Inka hydraulic engineers. In the discussion to follow, use of modern hydraulic engineering practice is employed to reveal Inka hydraulic engineering practice. Although the reason behind the Inka design can be made apparent through this method of analysis, the Inka source of hydraulic engineering knowledge paralleling the modern analysis remains unknown as the Inka had no written language nor any known hydraulic test facilities that have yet been identified. The source of Inka hydraulic knowledge may therefore derive from observed nature observations of flow phenomena then codified into practice in their urban and agricultural water systems, or possibly from importation of Chimú water technicians from Inka-conquered north coast territories in the late 14th century CE.

To understand Inka hydraulic engineering in terms of modern hydraulics technology, use of the Froude number (Fr) is convenient to explain water behavior. For shallow depth D flows, the Froude number is $Fr = V/(g\,D)^{1/2}$, where V is the water velocity and g is the gravitational constant. Physically, Fr is the ratio of water velocity V to the gravitational wave velocity $(g\,D)^{1/2}$—when Fr > 1, water velocity exceeds the signaling gravitational wave velocity so that water has no advance warning of a downstream obstacle; this leads to the creation of a sudden hydraulic jump before the obstacle. In physical terms, for Fr > 1, there is no upstream awareness of the presence of an obstacle until the obstacle is encountered by the water flow, as the gravitational wave signaling velocity that informs the flow that obstacle exists $(g\,D)^{1/2}$ is much less than the V flow velocity. For Fr < 1, the gravitational wave signaling velocity travels upstream of the obstacle faster than the water velocity V to inform the incoming water flow that an obstacle lies ahead. This causes the water flow to adjust in height and velocity far upstream of an obstacle to produce a smooth flow over the obstacle. The "obstacle" for the present application is the large contraction in channel width shown in Figures 66 and 68. Fr > 1 flows are denoted as supercritical, Fr < 1 flows are denoted as subcritical, and Fr = 1 flows are denoted as critical. While the presence and characteristics of subcritical, critical, and supercritical flows can be calculated from modern hydraulic theory, a simpler method exists to determine flow types. Insertion of a thin rod into a water flow that produces a downstream surface V-wave pattern behind the rod is indicative of supercritical flow. If the water surface pattern shows upstream influence from the rod, then subcritical flow in a channel is indicated. If only a local surface disturbance around the rod is noted, then critical flow is indicated. This simple test can determine different flow regime types to promote various usages associated with the different Fr flow regimes. As a first observation of water flow patterns giving insight to the hydraulic technology that went into the Figure 66 channel design, Figure 68 gives the first clue. The parallel water surface ripple wave structure pattern normal to the flow direction is consistent with $\sin\theta = 1/Fr$, where θ is the half angle of the surface wave, so that when Fr $\approx$ 1, $\theta \approx 90°$, verifying the transverse surface ripple wave structure shown in Figure 68 [19–21]. From Figure 68, incoming water flow in the wider channel section is (Fr < 1) subcritical, and in the narrower channel section, critical flow (Fr > 1) exists. Why then, was this flow type change important to Inka water engineers?

Figure 68. Alternate view of the contracted water supply channel, showing surface ripples normal to the flow direction. Photo by author.

Figure 69 is derived from the Euler fluid mechanics continuity and momentum equations that govern fluid flow behavior and is useful to describe the flow transition from sub- to supercritical flow in the supply channel, as shown in Figures 66 and 68, where water viscosity effects have minor effects on flow patterns [47,48]. As all water motion is governed by the mass and momentum conservation equations, Figure 69 indicates flow transitions based on Froude number change due to channel geometry changes. The x-axis represents $(1/2)\,Fr_1^2$ incoming flow (IN) conditions into the wide channel section; the $(1/2)\,Fr_2^2$ conditions represent flow (OUT) conditions in the continuing narrow channel section. The W2/W1 curves represent width ratios of the incoming subcritical flow section (W1) to the downstream channel (W2) width change. W2/W1 > 1 represents channel expansion and W2/W1 < 1 represents channel contraction, such as shown in Figures 66 and 68 which illustrate different views of the same channel contraction section.

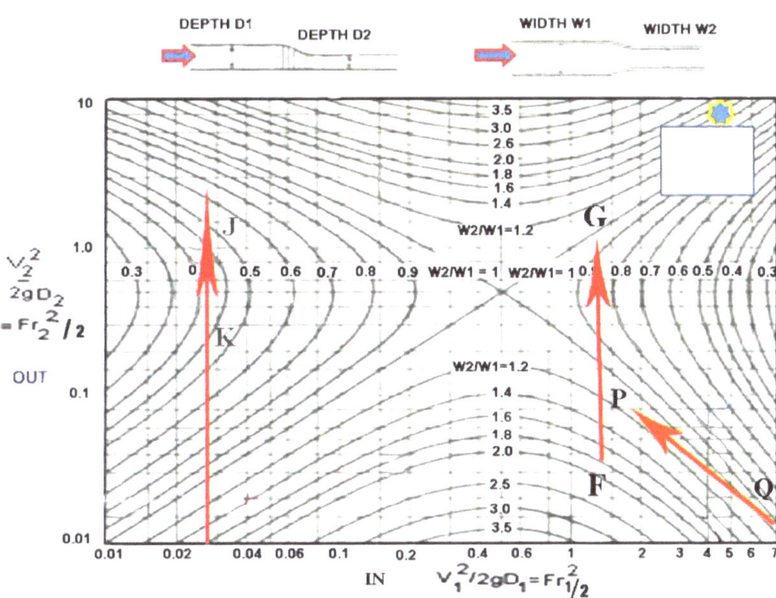

$$\frac{V_2^2}{2gD_2} = Fr_2^2/2$$

$$V_1^2/2gD_1 = Fr_1^2/2$$

Figure 69. Froude number relationship between flow entering the channel wide section (IN) and passing through the contracted channel section (OUT) from Bakhmeteff, 1932 [47].

Using the subscript notation (1) for flow conditions in the wide channel section and (2) for contracted width conditions, the Figure 69 width contraction ratio is W2/W1 = 0.44. The (1) flow entry value using the average flow rate is based on a ~0.3 ft water depth for which V_1 = 0.72 ft/s and Froude number is Fr_1 = V/(g D)$^{1/2}$ = 0.23 ($Fr_1^2/2$ = 0.03 in Figure 69). The contracted channel (2) Froude number is $Fr_2 \approx$ 1.14 ($Fr_2^2/2$ = 0.65 in Figure 69) based on ~0.3 ft depth. The channel contraction shown in Figures 66 and 68 takes water flow from subcritical flow (Fr < 1) in the wide channel to a near-critical (Fr $\approx$ 1) flow in the narrowed channel.

The flow rate per unit width in the contracted rectangular cross-section channel is q_2 = 0.68/W2 = 0.52 ft^3/s ft and the critical water depth (Henderson 1966) is y_c = (q_2^2/g)$^{1/3}$ = 0.2 ft, which is in agreement with the previous critical water depth value calculated for near-critical Fr $\approx$ 1 flow in the contracted width (2) channel.

Figure 69 illustrates the flow Froude number IN to OUT transition arrow (K to J) resulting from the wide (1) to narrow channel (2) shape change that incorporates transition from sub- to near-critical flow. This indicates that Inka engineers designed the (2) contracted channel section to support near-critical Fr $\approx$ 1 flow, but not exact critical Fr = 1 flow in the expanded part of the channel. In modern channel design practice, ~0.8 < Fr < ~1.2 is the prescribed Fr range to obtain the highest flow rates that accompany Fr $\approx$ 1 flows to avoid surface wave instabilities associated with translating large-scale vortex motion below the water surface associated with exact Fr = 1 critical flow. The Fr_2 Froude number range in the contracted (2) channel lies between ~0.8 < Fr < ~1.2, and the computed Fr_2 ~ 1.14 value is consistent with stable flow according to modern hydraulic engineering. The width reduction construction yields the narrowest supply channel (2) at the maximum flow rate per unit channel width without significant internal surface wave structures causing transient flow instabilities to be passed further downstream to the Waterfall area. In other words, the channel critical flow section does not permit any transient disturbance irregularities from the downstream Waterfall and its water supply channel from propagating upstream to influence the subcritical flow in the wider channel section to cause flow instabilities. The Figures 66 and 68 channel leads to a stilling reservoir ahead of the waterfall to limit disturbances and promote equal, stable flows into the four Waterfall channels. Any difference in equal amounts of water flowing through each of the four Figure 67 Waterfall channels would produce transient vortex structures in the stilling reservoir that constitute instabilities that the Fr = 1 contracted channel section does not allow to influence [22,48,49] the incoming Fr < 1 flow in the wider channel part of Figure 66.

The contracted Fr > 1 channel (2) therefore serves to limit any disturbances propagating upstream to disturb flow aesthetics in the Waterfall; this the design intent of Inka water engineers. Waterfall aesthetics were a prime concern at the Inka royal site, requiring advanced water engineering knowledge to accomplish. Other Figure 69 RHS arrow markings refer to flow properties of the Inka Canal Principal which involves complex analysis to determine its flow rate; results of this analysis are available ([3], pp. 195–221).

Near-critical flow is associated with the maximum flow rate per unit channel width that a channel can support [19–21]; this is closely achieved for $Fr_2 = 1.14$ conditions. Flow stability is achieved by the current Inka design; knowledge and use of this hydraulic engineering practice is evident in the Principal Fountain/Waterfall design and is vital to produce a constant, stable water delivery flow to the Waterfall area.

In summary, this contracted channel flow design is important for the following reasons: (1) the flow from the spring and channel sources channeled into the wide channel then to a downstream narrow width channel segment is associated with the maximum flow rate per unit channel width; (2) the Fr > 1 flow value in the contracted channel eliminates flow disturbances propagating upstream to the wide Fr < 1 channel; (3) the settling reservoir immediately upstream of the waterfall is intended to give symmetrical flow delivery to the waterfall: this is difficult to achieve but the Fr > 1 narrow channel prevents any disturbances propagating from this effect influencing the wider channel section input Fr < 1 flow.

Why is this size choice of channel sections important? If Inka engineers chose channel sections with wider widths throughout than those shown in Figures 67 and 68, then Fr < 1 would exist in all channel sections, making any downstream disturbances propagate in both up- and downstream directions, affecting waterfall aesthetics. Any water supply decrease (due to drought) or increase (due to excessive rainfall) disturbances would alter the original design intent, causing erratic transient Waterfall patterns that produce a nonaesthetic display. An overview of the Inka technology used for the Principal Fountain/Waterfall display reflects modern hydraulic design principles to preserve fountain aesthetics. Inka knowledge of channel width change effects on subcritical and supercritical flow regime creation to achieve stable flow at the maximum flow rate to the Waterfall is apparent in the channel design; this analysis adds a new understanding of Inka hydraulic engineering capabilities.

11. Concluding Remarks

Pythagoras (582–500 BC) stated that "Everything in the universe is governed by mathematical rules and reasons—if we understand the arithmetic and mathematical relations, then we will understand the structure of the universe and mathematics as the basic model for philosophical thought". The present nine chapters bring forward the relevance of his comments into modern times to add a further dimension to what is currently known of archaeological sites and the knowledge base of ancient water engineers not previously reported in the open archaeological and engineering literature. Of the new world sites discussed in Sections 2, 5, 7 and 9, their history, political, religious, and economic science development is well represented in the open literature—to this knowledge base, new revelations relevant to the water engineering knowledge base of pre-Columbian societies are brought forward for the first time to add a further dimension to the accomplishments of these societies. For old world sites discussed in Sections 1, 3, 6 and 8, their archaeological and cultural histories are well known through long periods of excavation and research by many investigators. New revelations at major World Heritage sites regarding hydraulic engineering discoveries at their major urban and agricultural centers developed by ancient water engineers are now brought forward by CFD analysis to add a further dimension of their engineering accomplishments to the archaeological record.

The present manuscript is designed as an introduction to the new field of paleohydrology in a readable manner for scholars familiar with hydraulic engineering practice, scholars in the field of archaeology, and a general audience interested in historical and scientific fields of study. Readers of this manuscript interested in further technical and historical detail of individual sites analyzed in this manuscript are provided references relevant to each chapter where sociopolitical, socioeconomic, and sociocultural matters are discussed in detail to give a more comprehensive accounting of site research performed by others over many years of research. References quoted under my authorship are based upon my fieldwork at each of the chapters' sites conducted over many years based upon hydraulic engineering studies. While paleohydraulics studies utilizing modern CFD and hydraulic analysis tools are new to archaeology to discover technologies utilized by ancient water engineers, work presented in this manuscript thus far presented is but a first step, as many old and new world sites remain to have their secrets revealed.

With respect to obstacles and uncertainties in the use of modern hydraulic engineering methods to uncover ancient water technologies, it came as a surprise that many World Heritage sites did not have journal publications involving modern hydraulic engineering analysis used to reconstruct ancient versions of water engineering. This is an opportunity for present-day water engineers to contribute something new to world history. As many sponsored and funded field projects involve excavation and mapping of archaeological sites and discussion of site historical relevance, those interested in paleohydrology have to consider travel to world archaeological sites with individual walkabout "tourist" field surveys and data collection of surface water systems. The option of joining sponsored projects requires background reading in depth of archaeological sites of interest then contacting archaeology departments of academic institutions to ask if they are interested in paleohydrology inputs that can add a further dimension to their research. Sites investigated in this manuscript evolve from a combination of these paths. As archaeology requires an element of uncertainty and conjecture about the significance of a finding, this, as the present paper signifies, only deepens interest in pursuing paleohydrology as way to bring forward what one's water engineering brothers of past times had accomplished and to bring their discoveries to the world's attention.

Funding: This research received no external funding.

Institutional Review Board Statement: Not applicable.

Informed Consent Statement: Not applicable.

Data Availability Statement: Additional research findings commentary are given in references.

Conflicts of Interest: The author declares no conflict of interest.

Appendix A

Appendix A.1. Petra Figure 1 Captions

1.Thommanon	12. Banteay Kdei
2. Chau Say Tevoda	13. Ta Prohm
3. Speam Thma	14. Preah Rup
4. Hospital Chapel	15. Ta Prohm Kei
5. West Gate	16. Baphuon
6. Elephant Terrace	17. Phnom Bakhang
7. Leper King Terrace	18. Bakani Chamkrong
8. South Gate	19. Preah Palitry
9. Preah Pithu	20. South Klrang
10. Tep Pranam	21. Canal Systems
11. Praasnt Kravan	

Appendix A.2. Figure 17 Angor Site Captions

1. Zurraba, M reservoirs
2. Petra Rest House
3. Park entrance
5. Dijn Monument
6. Obelisk Tomb and Triclinium
7. Siq Entrance elevated arch remnants
8. Flood bypass tunnel and dam
9. Eagle Monument
10. Siq passageway
11. Treasury (El Kazneh)
12. High Place Sacrifice Center
13. Dual Obelisks
14. Lion Fountain Monument
15. Garden Tomb
16. Roman Soldier Tomb
17. Renaissance Tomb
18. Broken Pediment Tomb
19. Roman Theater
20. Uneishu Tomb
21. Royal Tombs (62, 63, 64, 65)
22. Sextus Florentinus Tomb
23. Carmine Façade
24. House of Dorotheus
25. Colonnade Street
26. Winged Lions Temple
27. Pharaoh's Column
28. Great Temple
29. Q'asar al Bint
30. Museum
31. Quarry
32. Lion Triclinium
33. El Dier (Monastery)
34. 468 Monument
35. North City Wall
36. Turkamaniya Tomb
37. Armor Tomb
38. Outer Siq Wadi Drainage
39. Aqueduct
40. Al Wu'aira Crusader Castle
41. Byzantine Tower
42. North Nymphaeum
43. Paradeisos, Temenos Gate area
44. Wadi Mataha Major Dam
45. Bridge Abutment
46. Wadi Thughra Tombs
47. Royal Tombs
48. Jebel el Khubtha High Place—pipeline
49. El Hubtar Necropolis—buried channel
50. Block Tombs **B** dual pipeline supplied basin area
51. Royal Tombs d catchment dam
52. Obelisk Tomb s spring
53. Columbarium c cistern
54. Conway Tower D multilevel dam
55. Tomb Complex A Kubtha aqueduct
56. Convent Tomb T Water distribution tank
57. Tomb Complex -d major dam
58. Pilgrim's Spring
59. Jebel Ma'Aiserat High Place
60. Snake Monument
61. Zhanthur Mansion
62. Palace Tomb
63. Corinthian Tomb
64. Silk Tomb
65. Urn Tomb
66. South Nymphaeum
67. Rectangular Platform/Temple

References

1. Ortloff, C.R. *Water Engineering in the Ancient World: Archaeological and Climate Perspectives on Societies of Ancient South America, the Middle East and South-East Asia*; Oxford University Press: Oxford, UK, 2010.
2. Ortloff, C.R. *The Hydraulic State: Science and Society in the Ancient World*; Routledge Press: London, UK, 2021.
3. Ortloff, C.R. *Water Engineering in Ancient Societies*; MDPI Publishers: Basel, Switzerland, 2022. [CrossRef]
4. Al-Muheisen, Z. *The Water Engineering and Irrigation System of the Nabataeans*; Yarmouk University: Irbid, Jordan, 2009.
5. Bellwald, U. *The Petra Siq, Nabataean Hydrology Uncovered*; Ruben, I., Ed.; Petra National Trust: Amman, Jordan, 2003.
6. Bellwald, U. *Petra Hydrology Uncovered*; Petra National Trust Publication: Amman, Jordan, 2006.
7. Bellwald, U.; Bellwald, U. The Hydraulic Infrastructure of Petra—A Model for Water Strategies in Arid Lands. In *Cura Aquarum in Jordanien, Proceedings of the 13th International Conference on the History of Water Management and Hydraulic Engineering in the Mediterranean Region, Amman, Jordan, 31 March–9 April 2007*; Ohlig, C., Ed.; Peeters Publishers: Leuven, Switzerland; p. 2007.
8. Bellwald, U. The Hydraulic Infrastructure of Petra-A Model for Water Strategies in Arid Land. *Cura Aquarum Jordan.* **2008**, *12*, 47–94.
9. Bedal, L.-A. *The Petra Pool Complex: A Hellenistic Paradeisos in the Nabataean Capital*; Gorgias Dissertations, Near Eastern Studies 4; Gorgias Press: Piscataway, NJ, USA, 2002.
10. Bedal, L.-A. From Urban Oasis to Desert Hinterland: The Decline of Petra's Water System. The Case of the Petra Garden and Pool Complex. In *Water and Power in Past Societies, Proceedings of the 8th IEMA Conference, Buffalo, NY, USA, 11–12 April 2015*; Chapter 7; Holt, E., Ed.; SUNY Press: Albany, NY, USA, 2018; pp. 131–158.
11. Bedal, L.-A.; Gleason, K.; Scriver, J. The Petra Garden and Pool Complex. *Annu. Dep. Antiq. Jordan* **2007**, *51*, 151–176.
12. Joukowsky, M. *Petra Great Temple, 1993–1997*; Brown University Press: Providence, RI, USA, 1998; Volume 1, pp. 265–270.
13. Joukowsky, M. The Petra Great Temple Water Supply. In Proceedings of the 9th International Conference on the History and Archaeology of Jordan-Cultural Interaction throughout the Ages, Petra, Jordan, 23–27 May 2004; Department of Antiquities, Al-Hussein Bin Talal University Publication: Ma'an, Jordan, 2004.

14. Weis, L. Das Wasser der Nabataer: Zwishen Lebensotwendigkeit und Lexus: The Northwestern Project. In Proceedings of the Conference Proceedings: De Aquaductu Atque Aqua Urbium Lyciae Phamphyliae Pisidiae: The Legacy of Julius Sextus Frontinus, Antalya, Turkey, 9 November 2014; Wiplinger, G., Ed.; Peeters Publishing: Leuven, Belgium, 2014.
15. Ortloff, C.R. Water Control Devices in the Hellenistic City of Priene (Turkey), 300 BCE–100 CE. In *Water Engineering in the Ancient World*; Oxford University Press: London, UK, 2010.
16. Dillehay, T.; Kolata, A.; Ortloff, C.; Netherly, P.; Warner, J.; Eling, H.; Bonzani, R. Chimú-Inka Segmented Agricultural Fields in the Jequetepeque Valley, Peru: Implications for State-Level Resource Management. *Lat. Am. Antiq.* **2022**, *34*, 137–155. [CrossRef]
17. *FLOW-3D Flow Science User Manual, Version 9.3*; Flow Science, Inc.: Santa Fe, NM, USA, 2020.
18. Ortloff, C.R.; Crouch, D. Hydraulic Analysis of a Self-Cleaning Drainage Outlet at the Hellenistic City of Priene. *J. Archaeol. Sci.* **1998**, *25*, 1211–1220. [CrossRef]
19. Morris, H.; Wiggert, J. *Open Channel Hydraulics*; The Ronald Press: New York, NY, USA, 1972.
20. Henderson, F. *Open Channel Flow*; The Macmillan Company: New York, NY, USA, 1996.
21. Chow, Y.T. *Open-Channel Hydraulics*; McGraw-Hill Book Company: New York, NY, USA, 1959.
22. Bear, J. *Dynamics of Fluids in Porous Media*; Dover Publications: New York, NY, USA, 1972.
23. Shady, R. Caral-Supe: Y Su Entorno Natural y Social en Los Origenes de la civilization. In *Andean Civilization*; Marcus, J., Williams, P., Eds.; Cotsen Institute of Archaeology: Los Angeles, CA, USA, 2009; Volume 63, pp. 99–117.
24. Shady, R. *The Social and Cultural Values of Caral-Supe, the Oldest Civilization in Peru and America and its Role in Integral and Sustainable Development*; Proyecto Especial Arqueologico Caral-Supe; Instituto Nacional de Cultura Publaciónes: Lima, Peru, 2007; pp. 41–69.
25. Shady, R.; Leyva, C. *La Ciudad Sagrada de Caral- los Origines de la Civilización Andina y la Formación del Estado Pristina*; Proyecto Especial Arqueológico Caral-Supe; Instituto National de Cultura Publicación: Lima, Peru, 2003.
26. Shady, R. *Caral-La Ciudad del Fuego Sagrado*; Interbank Publishers: Lima, Peru, 2004.
27. Shady, R.; Creamer, W.; Ruiz, A. Dating the Late Archaic Occupation of the Norte Chico Region of Peru. *Nature* **2004**, *432*, 1020–1023.
28. Shady, R.; Haas, J.; Creamer, W. Dating Caral, a Preceramic Site in the Supe Valley on the Central Coast of Peru. *Science* **2002**, *292*, 723–726.
29. Shady, R. Sustento Socioeconómico del Estado Pristino de Supe-Peru: Las Evidencia de Caral-Supe. *Arqueol. Soc.* **2000**, *13*, 49–66.
30. Hodge, A.T. *Roman Aqueducts and Water Supply*; MPG Books Group: London, UK, 2011.
31. Kolata, A.L. Agricultural Foundations of the Tiwanaku State. *Am. Antiq.* **1986**, *51*, 748–762. [CrossRef]
32. Kolata, A.L. Tiwanaku Ceremonial Architecture and Urban Organization. In *Tiwanaku and Its Hinterland: Archaeology and Paleoecology of an Andean Civilization*; Kolata, A.L., Ed.; Smithsonian Institution Press: Washington, DC, USA, 2003; Volume 2, pp. 175–201.
33. Kolata, A.L. *Tiwanaku and Its Hinterland: Archaeology and Paleoecology of an Andean Civilization*; Kolata, A.L., Ed.; Smithsonian Institution Press: Washington, DC, USA, 1996.
34. Kolata, A.L. The Technology and Organization of Agricultural Production in the Tiwanaku State. *Lat. Am. Antiq.* **1991**, *2*, 99–125. [CrossRef]
35. Kolata, A.L.; Ortloff, C.R. Tiwanaku Raised Field Agriculture in the Titicaca Basin of Bolivia. In *Tiwanaku and Its Hinterland: Archaeology and Paleoecology of an Andean Civilization*; Kolata, A.L., Ed.; Smithsonian Institution Press: Washington, DC, USA, 1989; Volume 1, pp. 109–152.
36. Kolata, A.L.; Ortloff, C.R. Agroecological Perspectives on the Decline of the Tiwanaku State. In *Tiwanaku and Its Hinterland: Archaeology and Paleoecology of an Andean Civilization*; Kolata, A.L., Ed.; Smithsonian Institution Press: Washington, DC, USA, 1996; Volume 1, pp. 181–201.
37. Bandelier, A. The Ruins at Tiahuanaco. Proceedings of the American Antiquarian Society 21; Part 1. 1911. Available online: https://www.americanantiquarian.org/proceedings/44817263.pdf (accessed on 10 February 2023).
38. Bennett, W.C. Excavations at Tiahuanaco. *Anthropol. Pap. Am. Mus. Nat. Hist.* **1934**, *34*, 354–359.
39. Ortloff, C.R.; Janusek, J. Water Management and Hydrological Engineering at 300 BCE–1100 CE Precolumbian Tiwanaku (Bolivia). *J. Archaeol. Sci.* **2014**, *44*, 91–97. [CrossRef]
40. Ortloff, C.R. Hydraulic Engineering in Ancient Peru and Bolivia. In *Encyclopaedia of the History of Science, Technology and Medicine in Non-Western Cultures*; Springer Publications: Heidelberg, Germany, 2014.
41. Ortloff, C.R.; Kolata, A.L. Climate and Collapse: Agro-ecological Perspectives on the Decline of the Tiwanaku State. *J. Archaeol. Sci.* **1993**, *16*, 513–542. [CrossRef]
42. Thompson, L.G.; Mosley-Thompson, E.; Davis, M.E.; Lin, P.N.; Henderson, K.A.; Cole-Dai, J.; Liu, K.B. Glacial Stage and Holocene Tropical Ice Core Records from Huascaran, Peru. *Science* **1995**, *269*, 46–50. [CrossRef] [PubMed]
43. Thompson, L.; Davis, M.; Mosley-Thompson, E. Glacial Records of Global Climate: A 1500-year Tropical Ice Core Record of Climate. *Hum. Ecol.* **1994**, *22*, 83–95. [CrossRef]
44. Thompson, L.; Moseley-Thompson, E. One-half Millennium of Tropical Climate Variability as Recorded in the Stratigraphy of the Quelccaya Ice Cap, Peru. *Asp. Clim. Var. Pac. West. Am.* **1989**, *55*, 15–31.
45. Ortloff, C.R.; Crouch, D. The Urban Water Supply and Distribution System of the Ionian City of Ephesos in the Roman Imperial Period. *J. Archaeol. Sci.* **2002**, *28*, 843–860. [CrossRef]

46. Wright, K.; McEwan, G.; Wright, R. *Tipon: Water Engineering Masterpiece of the Inca Empire*; American Society of Civil Engineers Press: Reston, VA, USA, 2006; ISBN 0-7844-0851-3.
47. Bakhmeteff, B. *Hydraulics of Open Channels*; McGraw Hill Book Company: New York, NY, USA, 1932.
48. Lugt, J. *Vortex Flow in Nature and Technology*; John Wiley & Sons Publishers: New York, NY, USA, 1983.
49. Kosok, P. *Life, Land and Water in Ancient Peru*; Long Island University Press: New York, NY, USA, 1965.

Article

Paleohydraulics and Complexity Theory: Perspectives on Self Organization of Ancient Societies

Charles R. Ortloff [1,2]

[1]   CFD Consultants International, Ltd., 18310 Southview Avenue, Los Gatos, CA 95033, USA; ortloff5@aol.com
[2]   Research Associate in Anthropology, University of Chicago, Chicago, IL 60637, USA

Abstract: Complexity theory provides a path toward understanding the development of ancient Andean societal progress from early settlements to later high population states. The use of modern hydraulic engineering methods to develop an understanding of the technical achievements of ancient societies (paleohydraulics), when combined with complexity theory, provides a path toward understanding the role of hydraulic engineering achievements to guide population increase and societal group cooperation on the path from early kin settlements to later statehood. An example case illustrating the paleohydraulics-complexity theory connection is presented for advancement of the pre-Columbian Bolivian Tiwanaku (600–1100 CE) society through their seasonal control of groundwater levels in urban city areas. This feature provided well water availability for city housing, public fountains, city hygienic and health benefits from the control of habitation dampness levels, water on a year-round basis for intra-city specialty crops, and the structural foundational stability of monumental religious structures. Commensurate with this application, Tiwanaku raised-field systems utilized groundwater control technologies to support multi-cropping agriculture to support growing population demands. Paleohydraulics theory together with complexity theory is applied to other major South American ancient societies (Caral, Tiwanaku, Chimú, Wari, Inka) to illustrate the influence of advanced hydraulic engineering technologies on advances from early origins to statehood.

Keywords: complexity theory; paleohydraulics; archaeology; societal development; water engineering; South American sites

Citation: Ortloff, C.R. Paleohydraulics and Complexity Theory: Perspectives on Self Organization of Ancient Societies. *Water* **2023**, *15*, 2071. https://doi.org/10.3390/w15112071

Academic Editor: Giuseppe Pezzinga

Received: 19 January 2023
Revised: 16 May 2023
Accepted: 18 May 2023
Published: 30 May 2023

1. Introduction

The 2023 Special Issue publication *CFD Investigation of the Water Supply and Distribution of Ancient Old and New World Archaeological Sites to Recover Ancient Water Engineering Discoveries* presented CFD results related to the hydraulic engineering discoveries of several ancient New and Old World societies; the present companion paper adds discussion of the relevance of several of these water engineering discoveries to the social evolution of ancient pre-Columbian societies of South America as derived from complexity theory perspectives.

The investigation of hydraulic engineering systems developed by ancient water engineers was examined with the use of CFD computer models and modern hydraulic engineering theory [1–8] to provide insight into the technologies used by their water engineers. In several of these references, modern water technologies used to examine UNESCO World Heritage sites in previous centuries brought forward new revelations with regard to the water engineering used by their hydraulic engineers for the first time. Remaining to be examined are the implications of the use of advanced water technologies on the socio-political, socio-economic, and socio-cultural aspects of ancient societies to develop higher levels of societal attainment. The present paper provides a path to address these issues using an application of complexity theory to investigate socio-political and socio-economic developments related to population growth and the societal advancement of pre-Columbian Andean societies. The logical structure of complexity theory's equation systems provides the means to relate water engineering advancements to the social structure development of ancient societies. To illustrate complexity theory's use, the focus is on ancient societies of Pre-Columbian South America, as these societies' have a significant dependence on water engineering advances in order to survive the threatening ENSO El Niño and La Niña floods and droughts in coastal Peruvian ecologies, in addition to survival threats to societies in high altitude Andean mountainous environments. To supplement complexity theory's analysis predictions, the historical record of several Andean societies is detailed in order to examine the effects of climate change challenges.

Publications focused on this topic [1–16] include the societal structure and water supply and distribution systems of the Peruvian north coast Preceramic Caral (2500–1500 BCE) and the Late Intermediate Period Chimú society (800–1400 CE), the Middle Horizon Period sites of Andean Tiwanaku and Wari (600–1100 CE), and the Late Horizon Inka Andean sites (1400–1532 CE).

One source of ancient hydraulic engineering knowledge appears through lessons provided from nature observations translated into practical use for urban and agricultural water supply and distribution systems. One such lesson is that a stream originating from rainwater runoff over densely packed soil will carve a least-resistance cross-section channel shape to transport water at the highest flow rate possible for a given land slope. This observation provided an irrigation canal cross-section design used by ancient Chimú water engineers for several of their irrigation canals sourced from valley rivers such as those found in the field systems adjacent to the Chimú capital Chan Chan [2,5–7,17,18] on the north coast of Peru. Approximate half-hexagon cross-section, stone-lined canals are found in the Chimú Pampa Esperanza field systems north of the capital city of Chan Chan [3,5]; this canal design is the optimum straight-sided wall construction approximation to promote the low wall friction drag necessary to create high canal flow rates according to modern hydraulic engineering theory [19–21]. Further advanced Chimú hydraulic engineering is demonstrated from the ~900 CE Chicama-Moche Intervalley Canal construction, where the design water flow rate is made a near-critical maximum value by local canal cross-section shape changes as the lengthwise canal slope varies; this technology is close to modern hydraulic engineering practice [6,7,19–22]. The Moche Valley Chimú capital city of Chan Chan experienced extended drought in the ~1000 CE time period, as evidenced by Pampa Esperanza contracting canal cross section profiles [1] (pp. 25–26). This climate change effect effectively limited nearby canal supplied irrigation agriculture sourced by the Moche River and progressively lowered the groundwater level. To restore the water supply for urban and agricultural use, the 75 km long Chicama-Moche Intervalley Canal was constructed earlier to bring water from the high flow rate northern Chicama River to the adjacent southern Moche Valley agricultural field systems to increase agricultural production [1] (pp. 53–56) for Chan Chan's increasing population requirements. With field system water seepage into the aquifer from this supplemental water supply to augment the groundwater level, wells within several of the city's ten royal compounds had water availability resulting from the continued depth excavation to supply the royalty occupying the compounds. Supplemental to this drought restoration plan was the excavation of lengthy, wide, and deep subsurface troughs down to the water table located at high elevations between Chan Chan and the Pacific coastline to provide for additional crop growth from the moist trough bottom surface. As the ~1000 CE drought intensified over time with water table shrinkage to even lower depths, the original inland troughs were abandoned, and a new trough of similar proportions was constructed closer to the Pacific shoreline to intersect the declining water table to continue the agricultural production. These troughs were abandoned as the drought continued, with final troughs constructed even closer to the shoreline to intersect traces of the water table. This trough sequence is still observable from field observations after 1200 years of abandonment time. These comprehensive drought-defensive water system constructions indicate the Chan Chan's royal administration's knowledge of the canal supplied irrigation technology for surface agricultural field systems together with the water table hydraulic engineering applications necessary to sustain Chan Chan's population through a long-term drought anomaly. The ancient water engineering involved is typical of a modern design approach and indicates a deep knowledge of the diverse levels of water technologies necessary for societal continuity during climate change challenges to survival. With the ENSO El Niño and La Niña water supply variations (mainly drought and flooding cycles), the Chimú coastal society had to develop advanced water supply technologies to sustain their population by the creation and development of advanced hydraulic engineering practices.

Further use of advanced hydraulic technology is exhibited at the Late Horizon Inka royal site of Tipón located close to the Inka capital of Cuzco, where a water supply open channel undergoes a sudden major width reduction to create a critical flow in the narrower downstream channel supplying water to a display waterfall. This design feature creates critical flow [19,21] in the contracted downstream channel section that limits disturbance effects (channel bends, width changes and vortex flows in downstream basins) from propagating upstream to destroying flow symmetry in four waterfall spillage channels [2] (pp. 69–95), and preserves flow aesthetics, as would be expected for a compound occupied by Inka royalty. As many Inka urban, royal and administrative palace sites have elaborate aqueducts, reservoirs and channel water distribution systems [21–25], it is expected that future excavation activity will reveal further details of the engineering knowledge base equivalent to that shown at the Tipón site.

A further demonstration of advanced Andean hydraulic engineering knowledge relates to the raised-field agricultural system and urban water supply system of the Bolivian (600–1100 CE) Tiwanaku [1–3]. Here, the raised fields consist of vast areas of raised agricultural berms separated by swales excavated down to the groundwater level. The water level in the swales was maintained at a constant level in the dry season by a system of canals from a nearby river to supply water to berm plant root systems supporting agriculture. In the rainy season, drainage canals returned the excess water back into the Tiwanaku River at a downstream location to maintain swale water height. The net result of this water management was double-cropping through seasonal changes that was necessary to support Tiwanaku's 20,000–30,000 urban population. At the Tiwanaku city center, the water supply from the mountain-based springs was guided by channels into a deep 'moat' encircling the elite residence city area with Akapana, Putini, and Kalkasasaya Temple religious and administrative housing monuments [23]. In the rainy season, runoff water channeled into the moat infiltrated the groundwater aquifer to maintain the water table height constant by drainage channels carrying excessive moat runoff and seepage water to the nearby Tiwanaku River. In the dry season, perpetual springs distant from the urban center supplied channeled water into the moat to keep the water table depth constant [2] (pp. 1–30). With the water table maintained close to the urban ground surface on a year-round basis, specialty llama and alpaca pasturage areas within the city were maintained together with the control of dampness levels within city housing structures for hygienic and living quarter mold prevention health purposes. A further advantage of the maintained constant depth water table was the structural stability of the massive Akapana Pyramid and Kalasasaya structures within the moat water that were maintained at a constant level through seasonal changes in rainfall availability. Here, the water saturation of the monument's compacted aquifer underlying these structures prevented [2] structural deformation. Groundwater height control provided the continuous tappable water source for wells observed from excavations in urban housing structures [26] (p. 165). Together, the water control systems used for urban and agricultural sustenance provided high living standards for the 20,000–30,000 city population during seasonal changes in rainfall [2,8,9,14,26,27]. Combined with pastoral alpaca and llama protein food sources together with Lake Titicaca marine food sources and food imports from satellite towns [28], Tiwanaku city residents had a secure food supply base that was largely immune to transient climate change effects. A further indication of the water technology is evident in agricultural field systems south of urban Tiwanaku. Two adjacent canals exist a few meters apart from each other: one negative low slope canal takes water from the Tiwanaku

River to supply raised-field systems during the dry season, while the other canal with an exceptionally low positive slope extracts excess water from the raised-field system during the rainy season to maintain swale water height. To construct side-by-side adjacent canals with opposing slopes (one positive; the other negative) indicates that Tiwanaku water engineers were capable of slope measurements on the order of ~10^{-4} degrees, which is comparable to Roman accuracies [2] (p. 89). Of further note, raised-field agriculture was sustained through freezing altiplano (~3500 m altitude) temperatures by the capture of radiative solar heating of the raised-field swale water by several degrees centigrade over the sub-zero freezing temperature. This water temperature increase was aided by darkened swale decomposed bottom plant material enhancing the capture of heat within swale water channels between planting berms [8]. The IR radiative water heating of swale water prevented ice formation, as the latent heat withdrawal necessary for freezing was not available, resulting in the survivability of hardy crops in the altiplano winter season. These and other hydraulic engineering discoveries are a sampling of the water control technologies underlying the agricultural and urban water supply base of major Andean cities that played a key role in their societal development. Historically, innovative water supply and distribution technologies emerge as solutions to existing ENSO climate change challenges and characterize innovative societies that reimagine modified present water technologies to make them better for future generations.

Technical hydraulic engineering accomplishments discovered through use of CFD and modern hydraulic analysis at many other Old and New World sites are presently known [1–4]. It is clear that hydraulic engineering advances supporting population growth by sustainable agriculture influence the advancement of living standards of many different societies and provide the means of advancement from rural townships to cities, where different creative elements of society in close communication can mutually provide the benefits of a high standard of communal living.

The evolution of a society based upon technical achievements in water management requires an administrative structure to manage and oversee the use of water technologies that are vital for societal continuity and growth. The degree of societal evolution and its accompanying administrative structure involves the management of land, water, labor and technology resources specific to different Andean ecological environments where different societal systems and water availability and supply systems exist. For several of the early Andean Aymara kingdoms (such as the Pukara, Chiripa, Taraco, and Alto Ramirez) from the ancestral . . . to the Tiwanaku, moiety administrative systems composed of two complementary units with differences in ritual, political and economic status existed to influence the later Tiwanaku societal structure composed of a major capital city and surrounding settlements of Lukurmata, Pachiri and Khonko Wankané [10,26–28]. For the Inka governance of its provinces from Ecuador to mid-Chile, ~40,000 multi-level administrators responsible for the control of diverse province populations from Ecuador to mid-Chile conformed to central authority dictates originating in Cuzco [24], the highland capital of the Inka empire. For the north coastal Peruvian Chimú, nine separate royal adobe wall compounds related to each successive ruler's governance period existed, with ruler successions originating from dual corporate organization [15]. Such elaborate administrative architectural structures are features of the top-down resource control necessary for societal advancement. The discussion of the importance of technical advances overseen by elite administrative water management as related to social structure development are summarized by many authors in the literature of ancient South American societies [1–3,10] (pp. 364–366); [23] (pp. 23–25); [29–37], but questions remain as to the specifics of how ancient Andean societies put into practice their water technological knowledge to influence their cultural patterns, their socio-economic and socio-political security base, and their food supply base, and how water technology advancements influenced their societal development from small communities to city status. For the Peruvian north coast Chimú state where complex hydraulic technology was demonstrated in river-supplied, intra-and intervalley irrigation canal designs, the earlier lower population Moche (300–800 CE) society occupying similar coastal areas exhibited lesser levels of hydraulic knowledge for limited agricultural areas used by their lower population level. Clearly, the question of population size and the food supply base to support it, as made possible by the available water engineering technology, was a vital consideration in societal advancement. The Late Intermediate Period Chimú developed certainty about food resource production for their large population that required a higher level of technical inventiveness to provide irrigation water over vastly increased agricultural areas. For Andean highland societies, Lane demonstrated the use of dams to trap and conserve rainfall water supplies during seasonal changes in water availability, with cooperation between local communities in charge of the construction and maintenance of local water resources. Diverse types of water engineering structures and management systems thus appear in different Andean societies due to differences in local ecologies, differences in socio-political, socio-economic and socio-cultural societal structures, as well as from different responses to ENSO climate variations that affected societal continuity. Given the unpredictability of nature's climate and weather variations, and their effect on agriculture and societal sustainability, nature as a benevolent guide may seem doubtful as a reliable path that yields benefits to human society given periods in Andean history characterized by extensive droughts and catastrophic El Niño flood events. What is common to many advanced Andean societies, however, is the will of individuals to create a management system to govern the use of labor for cooperative rather than individual group land management of agricultural systems. This step requires the creation of a leadership class to guide the collective action of the population toward societal change with economic benefits for all. This class existence is demonstrated by major administrative structures found in most Andean societies that had achieved statehood status. An obstacle to societal cooperation vital to coordinating activities toward a positive, universally accepted goal originates from challenges to universal acceptance by segments of the populace to leadership dictates. This path holds for societies in their formation stages, as well as for societies already established by a family succession of rulers. For both cases, acceptance by the populace is key to their success and continuance, as many water control projects require the enlistment of vast labor resources together with the logistics to manage support facilities: all requiring the cooperation of the masses for a successful outcome. Given that different Andean societies evolved different management and administrative structures, the path toward cooperation for major infrastructure projects likely involved agreements between groups with different interests, religious and social structures, and different historical backgrounds and viewpoints as to how infrastructure improvement projects should be carried out; how cooperation toward the creation of a society sharing common goals is achieved is a subject within the realm of complexity theory analysis to investigate. To examine the creation of a managerial class that governs its populace in a cooperative manner given diverse cultural pattern differences in different Andean stratified societies, complexity theory provides a methodology to examine the interaction factors behind societal structural changes and development.

2. Societal Development According to Modern Complexity Theory

To understand the mechanism behind societal cooperative efforts, complexity theory [36] provides insight into how technical advances can be the catalytic impetus behind strategic decisions that arise from alternative strategies that lead to societal advancement. Provided that the attractiveness of a particular strategic decision advanced by elite management or individual groups can be translated into economic and labor saving benefits for individuals, cooperation within the populace to willingly perform tasks involving cooperative labor investment from all society members is assured. This process involves discussions between individuals and groups where advocates of different points of view, not necessarily those of elite management, vie for acceptance for their point of view. When the majority of a population accepts a proposed concept that coincides with official policy and customs (or vice versa), then the means for a universal cooperative effort from all society members is assured. The steps towards this end involve dictates that are accepted by the populace (or accepted by passive non-response) and are not necessarily ordered by elite decree but accepted by most of the population given the economic assurance and lifestyle benefits to individual society members. While many discussions on societal evolution give catalytic socio-political, socio-economic, and socio-cultural events as drivers to change [37], complexity theory examines leadership factions within societal groups espousing different solutions to problems that promote beneficial change. This is done in present-day democratic societies through elections, with prominent leaders espousing different political philosophies, but in many ancient societies, some form of collective agreement promoted by influential leaders influenced societal change based upon the promotion of new visions of achievable economic progress that are acceptable to all.

In historical reality, nonaccepting β groups opposing a leadership α strategy group continue to revise their strategy to overtake α leadership by either peaceful or confrontational means. Proposed solutions have time limits, as new α and β proposals are rapidly advanced in response in an effort to garner acceptance from the general populace. For this result to happen, acceptance by the populace must incorporate a benefit change $A_\alpha(t)$ and a strategy change $\alpha(t)$ evident to the general populace promoting group acceptance change. Complexity theory formulation provides a governing equation, Equation (1), that, upon solution, provides the time evolution path toward understanding how gradual changes occur in a society commensurate with technical advancements that have labor saving and economic benefits shared by all rather than just by an elite group. Solutions to Equation (1) can represent paths for time-dependent societal change or solutions for evolved changes from one strategy to another strategy. From a more comprehensive viewpoint, the decision processes involved in the creation of a cooperative union rely on economic advantages apparent to all; this is stated by Durant [38] (pp. 51–57) as " ... there is a contest among individuals, groups, classes and states for food, materials and economic power that are rooted in economic realities- in such cases, motives of leaders may be economic, but the result is largely determined by the passions of the masses." From this statement, there is a political reality familiar to those who study past and present history describable by complexity theory that investigates the management path toward creating decisions accepted by the masses that is key to societal change (for better or for worse).

For water engineering technologies of pre-Columbian societies applied to agriculture and urban use, several of which are described in the Introduction, different experiments and insights put into practice with the evaluation of results over time provide an effective means to their incorporation to update earlier systems. The role of a leadership class to guide or enable decisions [39,40] is noted as a principal factor in the advance of societal development. Given incomplete and preliminary details about the structure of Andean societies' management and decision-making capability that resides within elite (or non-elite) groups to guide technology-based decisions, the origins of societal progress from this source is subjective, given that different societies' hydraulic science knowledge depth can only be assayed from the analysis of several existing archaeological remains of their water systems. However, some degree of self-organization leading to progress, whether originating from a top-down elite management group or from bottom-up localized groups, must prevail in order to develop the water control systems for urban and agricultural use, as noted from the archaeological record. The choices of different adopted agricultural strategies are based upon the benefits evident to all participants in terms of decreased labor leading to increased agricultural productivity. As an agricultural surplus is essential to the growth and continuity of a society, particularly for formative Andean societies emerging from allyu and multi-generation community roots [12], it is of interest to examine societal complexity development from a complexity theory perspective based on the selection of an adopted agricultural strategy among various choices put forward by different elements of society. The selected strategy forms the bonding structure of shared interests to move forward on rules and practices accepted by all agreeing society members. This goal, once achieved, overcomes contrary population elements conditioned by class inequality, notions of superiority among different kin groups, the newness of the urbanism experience, the uncertainty of economic realities, male/female visions of what is important, differences in religious and behavioral standards, previous sociopolitical trajectories, and the usual vagaries of human nature that at times go against practical realities. In many Andean historical cases, a non-egalitarian society managing agricultural resources was the preferred outcome, and projects were conceived and executed by a common consensus without elite class management. For the present discussion, a consensus path toward the incorporation of water engineering advances that benefit all society members that promotes the development of an elite management structure is examined by complexity theory. This proceeds through individuals leading societal subgroups endeavoring to convince members of competing groups to join them to form a collective that shares similar goals. For established hierarchies, continuance in power depends upon their ability to retain favor with their population by succeeding in overcoming interior challenges from both interior and exterior sources as well as those from nature in the form of major climate change challenges.

To explore self-organizing societies and those already established by royal decree using complexity theory, for an agricultural strategy α, then A_α represents the benefits of choice, and α and X_α represents the number of people having adopted strategy α at a given time. The α strategy contains two factors: α_1, the convincing power of α group leaders based upon the promotion of economic advantage to all society members, and α_2, the technology advance in water engineering underlying the conversion of groups to α perspectives based upon labor-saving technical, health and economic benefits to dissenting X_β group participants. Sample α_2 technologies related to hydraulic engineering applied to advances in urban and agricultural systems are given in the Introduction for ancient New World societies, for example. The fractional number of individuals X_β considering an alternate strategy β with benefits A_β out of the total population (N) is then $X_\beta/(X_\alpha + X_\beta)$. In practical terms, X_α may represent an elite class population imposing its will on non-conforming X_β subgroups. Routledge [41] (p. 27) expresses this division as "how sovereign power is carried out against all odds and against the material interests of at least some of those involved", a route by which X_α can dominate X_β opposition. While this situation characterizes many historical examples [42], a trend toward ultimate acceptance of an A_α strategy based upon demonstrated successful economic and societal stability advantages for all promoted by an X_α elite minority can convince the X_β minority to accept elite management control. In terms of complexity theory, those individuals wishing to leave strategy β in favor of strategy α will be proportional to the number of X_β people multiplied by the relative attractiveness of strategy α, which is $A_\alpha/(A_\alpha + A_\beta)$. As $X_\alpha + X_\beta = N$ (the total population), then for the recruitment rate (a) per individual solicitor and t = time, the differential equation governing the rate of change of the X_α population favoring α strategy follows that of Nicolis and Prigogine [36] (p. 239),

$$dX_\alpha/dt = a\, X_\alpha \left\{ [A_\alpha\, X_\beta/(A_\alpha + A_\beta)] - [A_\beta\, X_\alpha/(A_\alpha + A_\beta)] \right\} \tag{1}$$

or, after $X_\alpha + X_\beta = N$ substitution,

$$dX_\alpha/dt = a\, X_\alpha\, [N\, A_\alpha/(A_\alpha + A_\beta) - X_\alpha] \tag{2}$$

This equation gives the rate of change of the number of people selecting option α. Note that the economic benefits derived from A_α water technology improvements that increase agricultural production and minimize labor input increase the supportive X_α population, as well as help convert the other X_β working classes not directly involved in agricultural production. For cases where the attractiveness of a particular strategy is evident over time due to positive economic results enjoyed by all classes of society, i.e., when A_α is large compared to A_β, then a greater number of the population participates in its use, as is evident by the larger values of dX_α/dt. Here, $0 < A_\alpha < 1$, $0 < A_\beta < 1$.

Nondimensionalizing Equation (3) with $X_\alpha/N = X_\alpha'$ and $X_\beta/N = X_\beta'$, then with $t/t_0 = t'$, then solving Equation (2) after dropping the ' notation,

$$X_\alpha(t) \sim e^{\Omega a t}/(e^{\Omega a t} - 1) \tag{3}$$

where $\Omega = [1 - (A_\beta/A_\alpha + A_\beta)]$. This long-term $t \gg 0$ solution gives the effect of a favorable agricultural and/or water engineering advance A_α strategy to increase the population X_α that endorses the α strategy over a long period of time. From Equation (3), $t \gg 0$, $X_\alpha(t) \to 1$ (by L'Hospital's Theorem), indicating total dominance of strategy α by the majority X_α population. Here, $X_\alpha = 1$ denotes the total acceptance of strategy α by the population.

For an example case from the Tiwanaku archaeological record, the population increase supporting the A_α strategy was directed by the elite management as benefitting the urban population over increasing time results, as it produces economic benefits for all. The food supply security at Tiwanaku proper would be instrumental to induce other population centers in the Moquegua Valley, coastal Chile Azapa, and southeastern Cochabamba, as well as sites flanking the western edge of Lake Titicaca [10,26,43], to adopt the benefits of Tiwanaku's Bolivian capital's α strategy, which would induce trade and cooperation between the Tiwanaku capital and the satellite towns [32]. This inducement to share the beneficial agricultural and cultural benefits of association with Tiwanaku proper would then expand the influence of all things Tiwanaku and lay the foundation for the creation of the Tiwanaku empire (as noted from the archaeological record). From Equation (3), the population segment X_α endorsing A_α benefits dominates any minority X_β opposition, as cooperation (compared to coercion) takes time to accomplish, given the success of the recruitment rate (a) from the increasing X_α population segment to overcome the opposition of X_β individuals. Satellite towns and villages share the leadership and benefits of the elite Bolivian central Tiwanaku center to form an expansive state noted as $\Omega < 1$ societies. The recruitment rate (a) from the central authority that peacefully projects the benefits of incorporation into the empire implies the existence of a Tiwanaku leadership that convincingly influences large segments of their city and satellite populations; this is opposed to command conformance by edict, such as would be expected from a dominant hierarchical management structure that is not responsive to opposition to their rule.

If a decision is made by an autocratic minority to enforce strategy α with benefits A_α, then $\Omega \to 1$ as $A_\beta \to 0$ and the conversion to a total X_α society is complete, as Equation (3) predicts. This then represents a step forward from rural village life to a coordinated population of different mutually supportive work classes coming together in a city structure, all supporting a common vision of economic benefits derived from common interests. An interpretation of Equation (3) is that when a successful, agricultural system exists together with a progressive technical advance α_2 and common interest α_1 cultural strategies, then the top-down acceptance ($X_\alpha \to 1$, $X_\beta \to 0$) of a mutually beneficial situation for all participating labor groups occurs and a managerial group acceptable to all society members is put in charge of implementing the successful α strategy with the population growth assured by adequate food supplies. Individual allyu kin groups possessing the ability to generate a successful agricultural strategy A_α using local control only has also been demonstrated for Tiwanaku in its early formative stage [44]. This was followed by elite management governance in later stages [26], where communal labor was evidenced in major construction projects, communal festive bonding feasts, advanced agricultural practice, and military expansion policies—all in a city environment. City formation is then a natural occurrence, as different agricultural workers and labor specialty groups working in cooperative union are most efficiently productive when in a close contact city communication environment. While elements of society may not initially conform to changes from established norms that are promoted by the leadership, the provided economic benefits of a more secure existence prevails, and then community consensus ultimately prevails when society members agree that they benefit from new policies [33] and close cooperation. Further research [32,40,44,45] emphasized collective ritual behavior in state societies as a common bonding A_α strategy element of self-organization. In this respect, paccha and huaca rituals and festive ceremonies [2] (pp. 6–12), [10,23,46] are additional key bonding elements observed in Andean societies that are best served in a city environment. Provided that leadership guiding the acceptance of an A_α strategy with mutual economic and social benefits is apparent to a majority body of the population, then, as Equation (3) indicates, elements of a collective, cooperative society appear in city form as a preliminary step toward statehood and empire when neighboring satellite societies adopt similar rules of existence to those of the main governing center city. Note that while knowledgeable agricultural management leads to sustainable population increases, geometric population increases with linear food supply over time leads to Malthusian extinction; thus continuous agricultural surplus and the water engineering technology to support it is necessary to sustain population growth. Thus emphasis on hydraulic technology as it is related to agricultural production is a critical factor in Andean societal development. A further option arises in times of crisis, where communities yield to absolutist forms of leadership authority [18,40,41] (p. 17). In cases of extended drought or flooding emergencies inducing the collapse of agricultural systems, populations cede total control to authority figures that provide the basis for collective action to organize labor to begin the restoration of damaged water supply and distribution systems. In such cases, a new β strategy with an A_β solution is the only option with an X_β group under immediate control that may displace a former governance X_α officialdom. From a solution of Equation (2), $X_\alpha \sim e^{-a\,A_\beta\,t}$, indicating the decline of the former X_α leadership class and its outdated α strategy under an emergency condition affecting agricultural production as A_β declines. In certain cases, the existing X_α leadership may utilize features of earlier agricultural increase projects to maintain their leadership role in drought crisis times to restore a semblance of agricultural production and restoration to continue their leadership role. The Chicama-Moche Intervalley Canal originated by the Chimú leadership is a prime example [6,7] of this leadership strategy, as severe drought affecting the agricultural fields of the capital city of Chan Chan mandates mandatory enlistment of the population to serve as laborers to construct this new water supply system to reinvigorate the capital's agricultural base in record time to eliminate the population collapse in the capital city.

Outside of hierarchical governance societies lie societies that resist the establishment of hierarchical leadership and proceed with organized labor that resist the emergence of local leaders [47]. For this case, there is only one A_α strategy, and X_α remains fixed and not subject to change in time. From Equation (3), for a society that defers to egalitarian rule and prefers community consensus government, for $t \gg 0$, only X_β exists, confirming the coherent society rule by consensus rather than by an $X\alpha$ elite class.

Equation (1) is also applicable to pre-state societies characterized by a clan hierarchy without hereditary rank where community, rather than individuals or elite groups, provide focus for community rituals and the initial materialization of a leadership class. This social structure is found in Late Preceramic societies of the Peruvian north coast and central highlands, where complex architecture and agricultural systems begin to appear that are derived from community cooperation [48] (pp. 45–47), [49]. In Late Preceramic and Formative sites, (3000–1500 BCE), the site of Caral in particular, the trade of marine resources available from coastal fishing communities for agricultural products available from inland sites becomes the start of a more complex societal system relying upon irrigation technologies and enlarged agricultural field systems under a central management, as trade comes with traders versed in commercial dealings [8,16,50]. From the central administrative center at Caral, X_α leadership controlled eighteen other Supe Valley sites sharing the same α and A_α values by trading resources specific to each site to compose a version of a cooperative pre-state society. With the evolution of early societies, complex architectural structures originated, indicating the creation of religious and administrative centers as formal steps necessary for later statehood. The initial reliance by individual Tiwanaku allyu clan groups exploiting lakeside surface field plots later translates into the cooperative labor reliance of many groups that effectively manage raised-field agriculture over a vast scale (~19,000 hectares). This efficiency improvement is based upon α_2 groundwater control refinements related to spring-sourced canals and drainage channels to control seasonal raised-field swale height in order to preserve agricultural output as well as changes in berm width patterns to extract more agricultural product per unit raised-field land area [3]. The implementation and installation of technical improvements relies on the recruitment of large labor resources guided by a centralized, top-down control that additionally managed planting and harvesting cycles [10,26]. Thus, a single agricultural α strategy guided by a central authority guided Tiwanaku's large city population effectively as an outgrowth of reliance on competing individual kin groups, with individual strategies only accountable to their individual welfare. For the Inka, established elite management X_α controlled all aspects of land resettlement, labor assignments, and land allocation for populations under their control [22,23]. Population conformance to Inka hierarchical control was achieved through proficiency in the technical aspects of different ecology agricultural systems together with multi-level personnel control enforced by state controlled kuracas governing the agricultural resources of the expansive Inka empire [23].

As the archaeological record shows for later stages of major Andean societies, vast arrays of ceremonial structures and corporate administrative compounds exist within large urban compounds, signaling the arrival of top-down decision making authority consistent with the $\Omega \to 1$ strategy. Archaeological theory, as applied to early ancient Peruvian and Bolivian societies, proves difficult with regard to the extraction of populace attitude details of societal change over time, but alternatively relies on the building of large ceremonial and administrative complexes requiring cooperative group infrastructure to complete, as the proof of societal advances through emergence of an elite rulership class takes place. Was there an initial form of authoritative coercion involved in the decision processes ($\Omega < 1$) involving the accession by societal leaders to organize *corveé* labor resources according to Equation (2), or did a $\Omega \to 1$ governing hierarchical structure originate from democratic consensus, or from the assumption of power by a hierarchical elite to rapidly stimulate the efficient progress towards a more efficient use of land, water, labor, and technology resources? As administrative, religious, ceremonial, and elite class compounds separate from secular classes remain in the archaeological record as a source of interpretation of societal structure that created them, many questions remain unanswered with regard to the transition process of early Andean societal structures. The results of complexity theory indicate at least one path toward consensus and cooperation through the recognition and acceptance of economic benefits that can be experienced by all society members, a logic which in any society has the strength of consensus building.

The examination of Inka society (1400–1532 CE), for which many accounts are available from Spanish chroniclers related to their social, economic, and political organization, indicate a well-ordered and well-organized society with many rules, rites, and rituals accepted as orthodoxy by a population controlled by a top-down Inka hierarchy. Other earlier Andean societies, such as the Chimú, Wari and Tiwanaku, followed this path with ultimate acceptance of the stability, prosperity and predictability that top-down management provided. As populations increased in these societies over time, the archaeological record shows that top-down management was vital for organizing labor intensive projects in the form of water resource management for agriculture and urban use, as well as religion-based monumental structures. While societal structure variations exist in the path toward empire, many of the major Andean societies that are dependent upon agricultural success appear to follow a variation of Wittfogel's vision of a hydraulic society [51–53].

3. From Theory to Reality

In discussions to follow that probe the source and progression of the Andean path to statehood and empire, the case is made for the conversion from bottom-up to top-down management strategies based upon economic and labor-saving improvements of a population that are mutually shared between different labor groups in a society. While only two strategies (α_1 and α_2) are considered for the present discussion, the further extension to more competing strategies and strategy variants involving different supporting segments of the population are summarized in Nicolls and Prigogine [36], (p. 240, Equation 6.15) by the extension of Equation (1) to multiple clan and kin groups X_i supporting different A_i strategies with different success levels of recruitment strategies. Although the idea of systematic planning with anticipated results leads to the development of planners within a top-down societal management structure, a further progress path derives from a trial agricultural system that produces marginal benefits over time and is then replaced from lessons learned by a new system whose A_i benefits are only apparent over an extensive trial run period under stable climate conditions. This progress path requires a stable environment over time to demonstrate benefits over a previously used agricultural system. Given nature's climate and weather vagaries, this learning path is dependent upon lessons from prior experimental trials which may not have time to demonstrate a replacement agricultural system. Thus, emergency trial innovative solutions may proceed from new climatical environmental and ecological conditions such as long term drought and periodic El Niño flooding affecting the agricultural and social structure environment. The creation of the Chicama-Moche Intervalley Canal that was designed to bring water from the Chicama River to the later post-drought period reinvigoration of the desiccated Moche Valley Chimú agricultural lands is a prime example of an emergency solution requiring a massive labor input controlled by the top-down Chan Chan royal administration to preserve the seat of government power and the surrounding agricultural resources supporting the capital city. Under stable climate conditions, there is time to derive agricultural surplus progress from sequential agricultural experiments to induce top-down societal management structure's planning and execution of a reliable system that is based upon the knowledge gained from past experiments. Under drastic climate change, this state is replaced by mandatory survival projects, lest civilization collapse. The Andean archaeological record is replete with examples of societies that manage through technological innovation to survive climate change events; the record is also replete with less innovative civilizations that were overwhelmed by the magnitude of a climate change event that disappear from the archaeological record. The hydraulic technology available to the Tiwanaku, the Chimú and other major Andean societies served as the basic element to progress to a more complex social structure capable of more complex projects serving their population with improved economic and security benefits.

Excavations of Tiwanaku raised-field agricultural areas west of Lake Titicaca at two meter depth revealed traces of an earlier raised-field design that had shorter water channel lengths between narrower planting berms than later, more productive raised-field designs. This α_2 advantageous berm wavelength and swale width change made over centuries derived from trial raised-field configuration differences over time that improved agricultural productivity; this an example of trial-and-error advances designed to gain knowledge in a stable climate period. The emergence of an organized agricultural system over long time periods generates a population increase (as Equation (3) predicts), with a top-down management center emphasizing religious practices designed to influence the deities' oversight of the progress of a society. Significant technical advances in hydraulic and hydrological sciences are observed in Tiwanaku, Wari, Chimú and Inka archaeological history that are indicative of an active branch of government focused on higher levels of technical improvement. As Andean history is characterized by several infrequent, but long-lasting, unstable climate periods [2,9,14,49,54–56], many defensive strategies evolved to protect agricultural continuity; these strategies resulted from the memory of past destructive climate events and the innovative defensive agricultural design solutions that evolved and were incorporated into later agricultural systems of major Andean societies.

4. Tiwanaku Societal Collapse: ENSO Drought According to Complexity Theory Models

From Tiwanaku social organization [8,10,12,14,26], an initial allyu kin societal structure prevailed by using early versions of raised-field agriculture. The consolidation of allyu groups led to the reorganization of raised-field systems on a larger, more integrated scale that promoted increased agricultural productivity with lower labor input. This suggests that a cooperative agricultural strategy advantage A_α predominated over previous strategies so that the X_α majority conversion prevailed. As demands of a more secure food supply and provision for increasing population became evident ($\Omega < 1$), a higher level of management expertise prevailed ($\Omega \to 1$) to expand raised-field productivity. As Tiwanaku hierarchical rule managed more complex water projects involving urban groundwater control [4] (pp. 97–103) and used labor organizations to carry forward complex water control projects, consensus from allyu kin groups was achieved, resulting in a higher level of economic security for Tiwanaku society. From the 10th to post 11th CE centuries, Tiwanaku city abandonment gradually occurred as the population decreased due to the long-term drought-induced contraction of agricultural capacity [9,51–56]. The decrease in population size is represented by $N \sim K \exp\{-kt^2\}$, where K is the initial population size, k is the rate of progression of the drought, and (a) is the society participation level. For $\Lambda = a\,K\,A_\alpha/(A_\alpha + A_\beta)$, Equation (2) becomes:

$$dX_\alpha/dt = \Lambda\,X_\alpha\,\exp\{-kt^2\} - a\,X_\alpha{}^2 \tag{4}$$

where $\Lambda = a\,K\,A_\alpha/(A_\alpha + A_\beta)$. The solution to Equation (4) for the population decline of the ruling elite $X_\alpha(t)$ is:

$$X_\alpha(t) = \exp\{\Omega\,\mathrm{erf}(t)\}/\{(a/k^{1/2}) \int_0^t \exp\{\Omega\,\mathrm{erf}(t)\}\,dt + [X_\alpha(0)]^{-1}\} \tag{5}$$

where $\Omega = (\Lambda/2)\,(\pi/k)^{1/2}$ and where $X_\alpha(0) > 0$ and $X_\alpha(0)$ is the population at the start of the precipitous drought at time $t = 0$. The erf(t) term is the Error Function given by:

$$\mathrm{erf}(t) = (4/\pi)^{1/2} \int_0^t \exp\left(-s^2\right)\,ds \tag{6}$$

At $t = 0$, the initial condition $X_\alpha(0)$ is recovered on both sides of Equation (5). The $X_\alpha(t)$ population consists of the ruling class of society; the X_β population is the nonsecular working trade class segment, with both classes sharing a similar fate, as climate conditions affecting agricultural production led to population decline. The solution shown in Figure 1 from Equation (5) indicates the decline of the $X_\alpha(t)$ ruling over time for k and K values representing 10th–11th CE century extended drought conditions. The governing leadership class $X_\alpha(t)$ segment diminishes in a damaging way as the disappearance of leadership elites disassembles guidance figures directing the fate of the population. Rituals and celebratory ceremonies were no longer performed, and trust in leadership's authority was compromised.

Under these conditions, social fragmentation and political unrest were followed by the abandonment of trust in deities and their connection to the elite class [57]. From on-site observation, an excavated deity statuary was defaced and a deity representation carved into a stone block was found as part of a building wall with the deity face reversed into the interior part of the wall during the extended drought period. The occurrence of human sacrifices in the main areas of Tiwanaku in this period was noted [10,26] as offerings to deities designed to restore stability. Figure 1 provides the numerical evaluation of Equation (5) for $K = 1$, $k = 1$, $a = 1$ and $\Omega = \pi^{1/2}/2$. Drought timing and intensity is regulated by the increased values of K and k; the $a = 1$ notation denotes the full involvement of all members of the population that were subject to drought conditions.

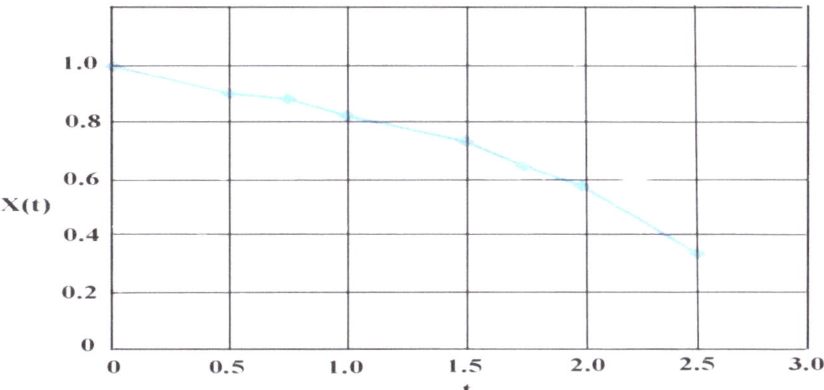

Figure 1. Typical decline in ruling class $X_\alpha(t)$ with time t (in years) from Equation (5) using typical K and k values that prescribe the drought intensity and duration.

According to the archaeological record of the Tiwanaku society, the population decline was accompanied by the dispersal of small kin groups abandoning the city to areas with localized water resources to permit limited forms of cocha agriculture. The lure of city life with the advantage of unity and prosperity derived from sharing common goals together with managed land areas that guaranteed adequate food supply supplemented by trade from satellite societies occupying different ecological realms provided the initial draw for rural populations to initially band together to form a city. As the later period drought intensified, the advantages of city life receded as the stable food supply diminished and key management personnel were no longer available to make leadership decisions to manage labor and resource allocation effectively. Social fragmentation and political discontent followed with the abandonment of formerly beneficial deities and rituals that were no longer effective in maintaining societal harmony [57]. The abandonment of near-lake-edge raised-fields and the transfer of farming to more distant areas from the Titicaca Lake edge proceeded as the distant water table remained sufficiently high to support the vestiges of raised-field agriculture and to limit open-field farming [11]. The decrease in Lake Titicaca's height from extended drought [13] affected near-lake raised-fields directly, as the groundwater profile declined with the lake surface shrinkage and decline. Rainfall infiltration over previous centuries into plains near mountainous areas distant from Lake Titicaca transferred groundwater vital for cocha farming to portions of the urban area. The presence of cocha farming pits excavated down to the declined water table supporting lower population groups was noted, as the drought intensified consistent with the dispersal of the urban population into smaller groups [10] (p. 266). In the final stages of drought, the Tiwanaku urban center became abandoned by ~1100 CE, and the later return of rainfall norms formed the basis of the next phase of population development, which was characterized by multiple, dispersed defensive sites that guarded precious water resources.

5. Visions of the Last Days of the City of Tiwanaku

The final stages of urban Tiwanaku during the extended drought are described [58,59], and further details of the social structure disassembly indicated from complexity theory usage derived from the solution of Equation (5) is shown in Figure 1. The establishment of extended drought conditions that led to the gradual demise of urban Tiwanaku and its raised-field agricultural systems [3,9,13,14] is substantiated by geophysical data originating from Thompson's ice core data [54–56].

Figure 2 summarizes the drought initiation and duration by integration of the nine-year moving averages of the Cordilla Negra Quelccaya mountain ice cap thickness data [56], where the sequential deposited ice layer thinness from drilling core data denotes rainfall decline. The extensive drought from ~1000 to 1300 CE slowly lowered water table levels in raised-field swales, promoting the loss of crops to freezing events together with water unavailability to plant root systems [8,14]. This extended drought likewise affected the demise of the contemporary highland Peruvian Wari society, as well as influencing the contraction of canal irrigation systems of the north coastal Peru Chimú society [3] (p. 253) indicating its influence over large parts of both Peru and Bolivia over several centuries. Given the different geophysical ecologies and the different levels of susceptibility to drought of major Andean societies in the 1000–1300 CE time period, and the different defensive responses to preserve their agricultural base, some societies, including the Tiwanaku and Wari, vanished from the archaeological record, while others, particularly the Chimú, managed to sustain their existence by the resourceful use of alternate water sources provided by the use of the Chicama-Moche Intervalley Canal [6,7].

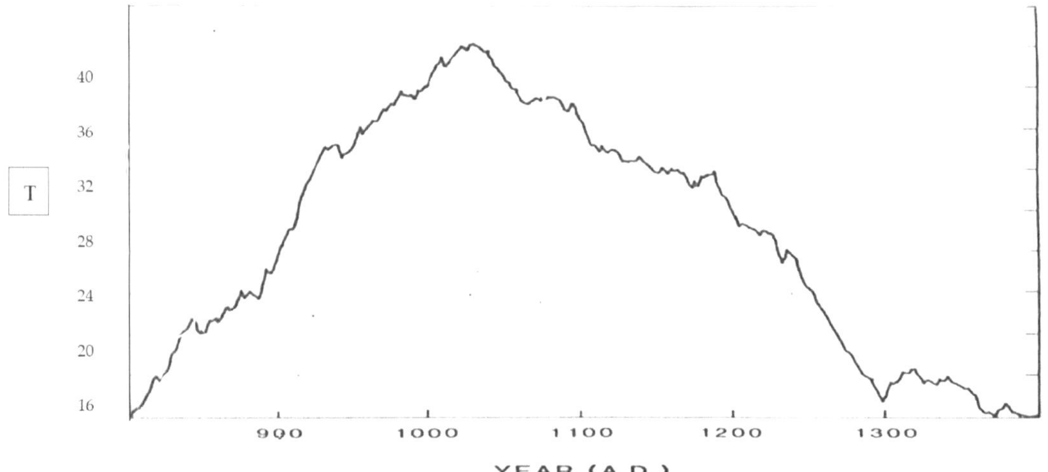

Figure 2. The moving average of Huascaran Mountain's ice core yearly deposit layer thickness begins a decline at ~1000 CE, indicating the initiation of severe drought conditions. The ordinate scale T is in centimeters.

Information related to the last days of Tiwanaku city life is available from multiple stable isotope methods involving the analysis of skeletal remains dating from the ~1100 AD time-period, which corresponds to the city abandonment dates [58,59]. The dietary change from previous norms was experienced by the city population as the drought intensified; these changes included the absence of fish from the diet, and there are no reported instances of child remains that contained fish or marine food sources which might be indicative of a change in the Lake Titicaca marine resource base to supply a protein source to the city population.

As Tiwanaku's working class population decreased with the loss of skilled fishermen as well as the X_α management direction (Figure 1), the population decline and outward migration from urban Tiwanaku continued with the loss of specialized industries' key members, together with population numbers sufficient to continue with the sowing and harvesting of food resources.

The decline in fish and marine resources available to citizens may represent the allocation of high nutrient marine food types to the most productive members of society that remained, with the hope that their skills may be available to restore elements of the previous way of post-drought life; alternatively, fish species resource depletion may result from higher salinity, as the Lake Titicaca lake level declines, limiting fish catch amounts as well as nutrient values. The results from [60,61] indicate the presence of maize as a food source in the ~1000 CE time period, indicating the importation from different Tiwanaku satellite areas where maize could be raised and transported in dried kernel form. Recent studies [58–61] detailed the societal collapse of contemporary Middle Horizon Wari sites due to food shortages in the same time period leading to the collapse of that society, the capital of which was centered in the Ayacucho region of Peru. Figure 3 indicates that severe drought conditions present in Peruvian and Bolivian highland societies as well as in Peruvian north coastal areas [3] (pp. 24–26) at ~1000 CE, together with an earlier ~600–700 CE ~30-year drought period on the north coast of Peru, altered the survival fate of several Andean societies, confirming the extent and duration of the drought on societal survival and extinction [62]. In some cases, societies survived by altering their farming methodologies (or by the conquest of other societies with significant land and water resources), while other societies disappeared from the archaeological record [2,3,59,60].

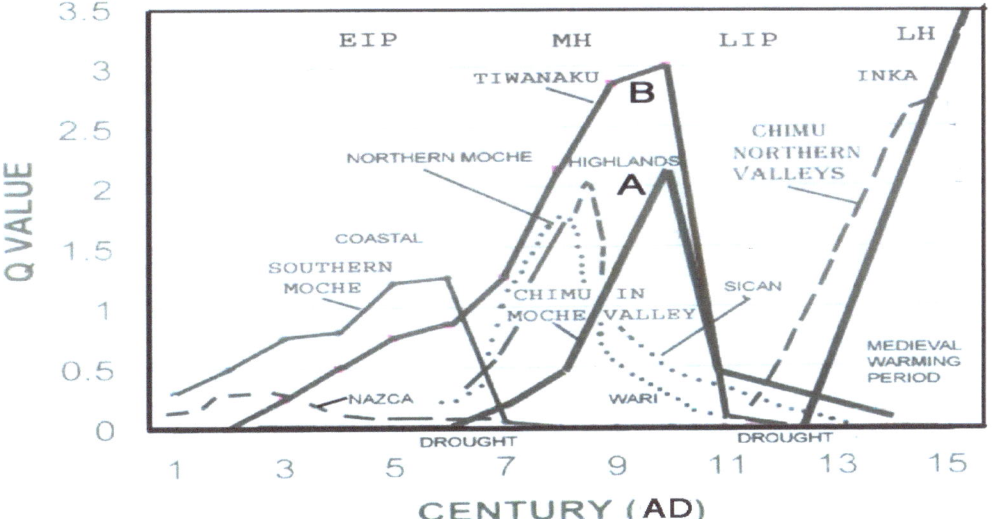

Figure 3. Drought led to societal decline for both the Tiwanaku and Wari polities as well as influencing the Peruvian north coast Moche, Sican and Chimú Societies (Ortloff and Moseley 2009). The Q Value ordinate denotes the survivability value of a civilization as defined in [18].

Of interest in Figure 3 is the post drought recovery in the 13th–14th century CE that led to Inka expansion and control over vast areas of coastal and highland western South America, with no state level polities from other societies left to contest their dominance. For the Late Horizon

For the Inka, the conquest of Peru by Pizarro's conquistadors in 1532 basically eliminated the earlier Inka ~100 year dominance of Pacific coast and inland sites of South America from (present-day) mid-Columbia to mid-Chile regions, and was subsequently replaced by Spanish administrative control of these areas. Based upon the research of the effects of the on the demise (or continuation) of many Peruvian and Bolivian empires [18] from Early Horizon to late Horizon times, several late Intermediate Period droughts lasting several hundred years were characterized by continuous small percentage declines in rainfall levels that had different effects on different sites' survival depending upon the ecology of their agricultural system types. For the Bolivian altiplano highland site of Tiwanaku, drought manifested itself as a slow decline in the ground water level, affecting raised field dependent agriculture [3,9,14]; for Peruvian Chimú coastal sites dependent upon rainfall runoff from the Cordillera Blanca mountain areas to rivers supplying canal based irrigation systems, drought had a more severe effect over a shorter period of time [1,7] and limited agricultural production, as Figure 3 indicates.

6. Complexity Theory and Societal Evolution

Returning to the complexity theory analysis of societal change, Equation (1) is based on only two competing groups: X_α and X_β. Given human nature, new challenges and sociopolitical changes in a society alter previous norms and produce many additional competing group factions: $X_\alpha, X_\beta, X_\gamma, X_\delta, \ldots$ that vie for dominance by a single group. Equation (1) can be modified to accommodate additional X_i factions, promoting different strategies whose success in winning converts relies on the persuasive powers of group leaders to achieve converts from other groups. Given human nature, the proposed collective group economic benefits are sometimes not perceived by some as individual member benefits, which leads to subgroups choosing not to retain former group identities and practices. With respect to Tiwanaku society, Goldstein's interpretation [28] of the evolved societal structure as determined from the formative processes that led to state development is given as " . . . the dynamic interplay among myriad counterposed factions in both state and periphery representing different factions interacting to produce behavioral procedures that define their society". This statement is consistent with complexity theory models that allow for group competition through the specification of their X_i, A_i and α_i terms that may challenge the ruling and other societal classes that resist challenges to their continuance. Andean history demonstrates cases where climate change affecting agriculture was viewed as a deity's rejection of the leadership, with subsequent modifications of the authority of the leadership class [9,57].

7. Early Versions on the Optimum Structure of a Society: Perspectives from Complexity Theory

Perspectives on societal structure development have a history that was understood by scholars from previous centuries. The discussions of the optimum structure of a society from philosophers of the Greek golden age, particularly discussions between Plato and Aristotle [63] (pp. 331–332), relate the views of land-apportioned individuals enlisted to, and committed to, voluntary communal labor to achieve community goals; this was expressed as "friend's property is common property" according to Plato's *Ideal State* from his *Laws*. Aristotle's modification (from his *Politics*) of Plato's propositions involves private ownership of property with the proviso that "friend's goods are common goods" means that the results of individual labor (primarily for agricultural products) are available to one's neighbors upon request. Imbedded in Aristotle's view is the consideration that all workers will not (or cannot) contribute equal amounts of voluntary labor but will share an equal amount of the agricultural production. Another view from Epicurus' *Rules for Living* would have a "state run by wise men" as the ideal state. Further research [63,64] traced the conditions of state success and the accompanying organization of power necessary for this to be achieved in Greek city states: their research on this topic proceeds from Greece's Golden Age under Pericles, where the involvement and approval of the citizenry was vital for government practices. For Roman and Hellenistic societies, the importance of water for urban and agricultural use is key to societal advancement [65–72]. A later view from the turmoil of the 17th century [65,66] saw the state and its rule of law as the antidote to "war of all against all", and the "fear of violent death at the hands of another" as the way to peaceful coexistence between enemies. The crucial role the state plays in the domestication of humanity is foremost in Hobbes' *Leviathan* or *The Matter, Form and Power of a Common-Wealth Ecclesiastical and Civil Society*. Further thoughts on the role of the city as a civilizing entity was expressed in later centuries, referring to the Roman presence in the British Isles [67] as " . . . to live well instead of merely living, was the membership in an actual physical city". From thoughts and writings about how societies could be ideally composed and the leadership necessary to achieve this end, history has demonstrated much experimentation along these lines, with various degrees of success for equality and individual freedoms being integral to state formation. Scholars over the centuries have developed many theoretical paths to explain societal development. These include the theory of evolution (society proceeds in a definite direction to a higher levels); unilinear evolution theories (all societies pass through similar successive stages to the same end); and multilinear evolution (changes occur in several ways but do not progress in the same direction). Later scholars, such as Childe [68], laid down ten traits (the presence of cities, labor specialization, agricultural surplus, monumental architecture, a ruling elite, science, writing, standardized artwork, long distance trade, and solidarity based on residence rather than kinship) that define state formation, although all of these elements need not be present for the transition from farmer collectives to chiefdom-level societies. Childe's emphasis on a processional viewpoint was that archaeology is a branch of anthropology and that the conditions of societal development result from the creation of material items that lead to surpluses from agricultural production, tradable export products, complex irrigation networks, and implements the supporting production that was once concentrated in leadership elites, and that this formed the basis for societal advances. Childe's views accepted that archaeologists delivered 'cultures' that were based upon a material culture were influenced by Marxist views that material things were more important than ideas for societal advancement, and that social advancements derive from material conditions with modes of production that produced equal economic benefits for all. Childe's view was that Marxist diffusionism functioned in a society where humans were not inventive or inclined to cultural evolution, and thus were susceptible to cultural change through economic benefits equally available to all society members. The Marxist perspective involves class struggle to achieve its goals; this contrasts with Childe's views of a material culture evolving naturally without class struggle. With respect to complexity theory trends as described for pre-Columbian South American societies, Childe's Marxist materialist-influenced perspectives appear relevant, as much in the way of the advancement of their societies depended upon invention and the use of advanced irrigation technologies (a_2) to deliver the agricultural surplus necessary for population growth, together with binding religious and social rituals (a_1).

Liverani [69] (p. 6) concluded that state formation was not inevitable from early kinship groups, but progress relied on an agricultural surplus that was "not for consumption within the family but for the construction of infrastructures and for the support of specialists and administrators, the very authors of the social revolution itself". Progress supporting this transition is based on the communal use of agricultural surplus and the creation of an administrative class supporting and dedicated to the creation of a state's infrastructure; this view parallels complexity theory's equation systems predictions that rely on the creation of a vibrant agricultural base and leadership formation for Andean societies that is likely applicable to other early world societies. Progress toward statehood from complexity theory perspectives involves strategy cooperation on all matters between diverse groups to achieve social harmony and progress. Complexity theory thus provides the path whereby advances in technology and social cooperation rituals lead to the creation of a governing leadership elite class that concentrates the derived surplus from trade and agricultural resources to direct the advance from early diverse kin group settlements to city status. This path formally proceeds from complexity theory, requiring the cooperation of society members to be governed by commonly accepted rules. From Equation (3), $X_\alpha(t) \sim e^{\Omega a\, t}/(e^{\Omega a\, t} - 1)$, society evolves when different kin groups coalesce into united cooperative X_α members whose growth in time ($X_\alpha \to 1$) over increasing time t, Equation (3)) depends upon $a = a_1 + a_2$, where a_1 is composed of shared social rituals and strategies (religious practices, social binding feasts, expansionist policies, shared behavioral beliefs, agreed upon leadership rulers) and a_2 is composed of hydraulic engineering advances that provide food supply security and surpluses for societal security, as well as for advanced urban living standards.

From Greek and Roman authors and for pre-Columbian societies, societal advances emphasize both a_1 and particularly a_2 considerations given the importance of water for urban and agricultural use. From Wittfogel's viewpoint [51–53], hydraulic engineering advances in dynastic Chinese eras provided the basis for societal stability, as they provided defenses against destructive Spring season flood events [2] (pp. 340–349) as well as providing innovative irrigation networks to support large populations. Wittfogel's emphasis on water technologies created by dynastic Chinese emperors to forge their empires derived from their absolute rule that controlled vast labor resources from the obedient working classes This is but one means to characterize progress toward statehood and empire creation in Chinese dynastic times, but is not applicable to other societies that have different paths to statehood.

The examination of the archaeological record of the societal structure of the Bolivian Tiwanaku society from its elementary kin group structure to the later state level appears to follow elements of Childe's and Liverani's precepts. This follows from the agricultural surplus provided by hydrological science advances using raised-field agriculture that led to the creation of an elite governing class, as predicted by complexity theory. Other authors [30,31,70,71] emphasize scientific and engineering material observations and advances at world city sites which were previously kept as seemingly unrelated and independent bits of information, only to be later shown to relate to each other through an encompassing theory that tied together the previously scattered information bits into "a close and effective relationship." This path of key technical and engineering developments is vital to the evolution of modern western science [30] (pp. 224–254), given many examples from 18th century Enlightenment Period origins [71], and is a key provision described by complexity theory's Equations (1) and (5) that incorporate α_2 considerations. In a similar manner, the sophisticated hydraulic technologies exhibited by Andean societies [1–5] followed a similar evolutionary path as hydraulic engineering technology proceeded through cumulative incremental verification steps before their use to advance stable societal and agricultural progress. This path is consistent with complexity theory's 'A and α' conditions that are specified in Equation (1). From complexity theory, the process of social formation that permits different views to compete for dominance $(A_\alpha, A_\beta, \dots)$ by different competing elements of the population $(X_\alpha, X_\beta, \dots)$ defines the culture change that emerges. This model characterized early Tiwanaku allyu kin groups that were deciding on more efficient ways to maximize agricultural production by collective labor enlistment that raised the economic status of all participating members of different kin groups. This procedure produced natural leaders that formalized their evolution to elite status, and whose importance increased as continued success resulted from the economics of the labor projects that they conceived and managed. In summary, complexity theory provides a logical explanation path, through its governing equations and solutions, to explain and understand the formation over time of advanced world societies from elementary levels characterized by the evolution of a leadership class utilizing a wide spectrum of engineering advances to promote high living standard levels to their populations.

8. Conclusions

The use of complexity theory provides a formal means to examine the basis of societal progress in Andean (and other) societies, and is new to archaeological and anthropological analysis. Equation (1) relates the expansion of major ancient Andean societies to hydraulic and hydrological science that maximized agricultural production and improved urban living standards; this path is but one among many that contribute to societal development. With the benefits of the increase in agricultural productivity, societal sustainability and security led to the development of a managerial class designed to oversee and develop society. From different elements of society with different capabilities, ambitions and viewpoints, effective elite management could enlist labor from all societal classes to cooperate in industrial scale agricultural projects that provided economic benefits for all, as predicted by complexity theory formalism.

Over ~4000 years, from the early Preceramic to Late Horizon times, Andean societies' use of advanced water supplies and distribution technologies provided one foundational element for the self-organization into hierarchical management polities, as noted from the archaeological record. Complexity theory thus provides a formal path to describe how different elements of societies working together in cooperative union can produce a

Water 2023, 15, 2071

path to higher levels of societal development, in addition to the creation of an overseeing managerial class to direct progress toward this end. Complexity theory thus has a universal application with regard to understanding the progress and evolution of societies given the different challenges and constraints that come with historical evolution.

Funding: This research received no external funding.

Data Availability Statement: All data contained in this manuscript is available and can be shared with the public.

Conflicts of Interest: The author declares no conflict of interest.

References

1. Ortloff, C.R. CFD Investigations of Water Supply and Distribution Systems of Ancient Old and New World Archaeological Sites to Recover Ancient Water Engineering Technologies. *Water* **2023**, *15*, 1363. [CrossRef]
2. Ortloff, C.R. *The Hydraulic State: Science and Society in the Ancient World*; Routledge Press: London, UK, 2020.
3. Ortloff, C.R. *Water Engineering in the Ancient World: Archaeological and Climate Perspectives on Societies of Ancient South America, the Middle East and South-East Asia*; Oxford University Press: Oxford, UK, 2010.
4. Ortloff, C.R. Water Engineering in Ancient Societies. *Water* **2022**, *12*, 3562. [CrossRef]
5. Ortloff, C.R. Chimú hydraulics technology and statecraft on the North Coast of Peru, AD 1000-1470. In *Economic Aspects of Water Management in the Prehispanic New World*; Scarborough, V., Isaac, B., Eds.; Research in Economic Anthropology, Supplement 7; JAI Press: Greenwich, CT, USA, 1993; pp. 327–367.
6. Ortloff, C.R.; Moseley, M.E.; Feldman, R. Hydraulic Engineering Aspects of the Chimú Chicama-Moche Intervalley Canal. *Lat. Am. Antiq.* **1982**, *47*, 572–595. [CrossRef]
7. Ortloff, C.R.; Feldman, R.; Moseley, M.E. Hydraulic Engineering and Historical Aspects of the Pre-Columbian Intravalley Canal System of the Moche Valley, Peru. *J. Field Archaeol.* **1985**, *12*, 77–98.
8. Kolata, A.L.; Ortloff, C.R. Thermal Analysis of the Tiwanaku Raised-Field Systems in the Lake Titicaca Basin of Bolivia. *J. Archaeol. Sci.* **1989**, *16*, 233–263. [CrossRef]
9. Ortloff, C.R.; Kolata, A.L. Climate and Collapse: Agro-ecological Perspectives on the Decline of the Tiwanaku State. *J. Archaeol. Sci.* **1993**, *16*, 513–542. [CrossRef]
10. Kolata, A.L. *Tiwanaku and Its Hinterland: Archaeology and Paleoecology of an Andean Civilization*; Kolata, A.L., Ed.; Smithsonian Institution Press: Washington, DC, USA, 1996; Volume 1.
11. Graffam, G. Beyond State Collapse: Rural History, Raised Fields and Pastoralism in the South Andes. *Am. Anthropol.* **1992**, *94*, 882–904. [CrossRef]
12. Albarracin-Jordan, J. Tiwanaku Settlement Systems: The Integration of Nested Hierarchies in the Lower Tiwanaku Valley. *Lat. Am. Antiq.* **1996**, *3*, 183–210. [CrossRef]
13. Binford, M.; Brenner, M.; Leyden, B. Paleoecology and tiwanaku agroecosystems. In *Tiwanaku and Its Hinterland: Archaeology and Paleoecology of an Andean Civilization*, Kolata, A.L., Ed., Smithsonian Institution Press. Washington, DC, USA, 1996, Volume 1, pp. 89–108.
14. Kolata, A.L.; Ortloff, C.R. Agroecological perspectives on the decline of the Tiwanaku state. In *Tiwanaku and Its Hinterland: Archaeology and Paleoecology of an Andean Civilization*; Kolata, A.L., Ed.; Smithsonian Institution Press: Washington, DC, USA, 1996; Volume 1, pp. 181–201.
15. Netherly, P. The Management of Late Andean Irrigation Systems on the North Coast of Peru. *Am. Antiq.* **1994**, *49*, 227–254. [CrossRef]
16. Shady, R. The social and cultural values of caral-supe, the oldest civilization in Peru and America and its role in integral and sustainable development. In *Proyecto Especial Arqueologico Caral-Supe*; Instituto Nacional de Cultura Publicación: Lima, Peru, 2007; pp. 1–69.
17. Moseley, M. *Chan Chan: Andean Desert City*; University of New Mexico Press: Albuquerque, NM, USA, 1982.
18. Ortloff, C.R.; Moseley, M.E. Climate, Agricultural Strategies and Sustainability in PreColumbian Andean Societies. *Andean Past* **2009**, *9*, 1–27.
19. Morris, H.; Wiggert, J. *Open Channel Hydraulics*; The Ronald Press: New York, NY, USA, 1972.
20. Flow Science. *FLOW-3D CFD User Manual V. 9.3*; Flow Science, Inc.: Santa Fe, NM, USA, 2020.
21. Henderson, F. *Open Channel Flow*; The Macmillan Company: New York, NY, USA, 1996.
22. Baudin, L. *A Socialist Empire: The Inkas of Peru*; Van Nostrand, Inc.: Princeton, NJ, USA, 1961.
23. D'Altroy, T. *The Inkas*; The Blackwell Publishing Company: Oxford, UK, 2003.
24. Hyslop, J. *Inka Settlement Planning*; University of Texas Press: Austin, TX, USA, 1990.
25. Lumbreras, L. *The Peoples and Cultures of Ancient Peru*; Smithsonian Institution Press: Washington, DC, USA, 1974.
26. Kolata, A.L. *The Tiwanaku: Portrait of an Andean Civilization*; Blackwell Publishers: Oxford, UK, 1993.
27. Erickson, C. Lake titicaca basin: A precolumbian built landscape. In *Imperfect Balance: Landscape Transformations in the Precolumbian Americas*; Lentz, D., Ed.; Columbia University Press: New York, NY, USA, 2000; pp. 311–356.

28. Goldstein, P. *Andean Diaspora: The Tiwanaku Colonies and the Origins of South American Empire*; University of Florida Press: Gainesville, FL, USA, 2005.

29. Marcus, J.; Stanish, C. (Eds.) *Agricultural Strategies*; Cotsen Institute of Archaeology, University of California: Los Angeles, CA, USA, 2006; pp. 1–13.

30. Koestler, A. *The Act of Creation, the Evolution of Ideas*; The Macmillan Company: New York, NY, USA, 1964; pp. 224–254.

31. Bartlett, F. *Thinking*; G. Allen and Unwin: London, UK, 1958.

32. Carballo, D.M.; Roscoe, P.; Feinman, G.M. Cooperation and Collective Action in the Cultural Evolution of Complex Societies. *J. Archaeol. Method Theory* **2014**, *21*, 98–133. [CrossRef]

33. Giddens, A. *The Constitution of a Society*; University of California Press: Berkeley, CA, USA, 1984.

34. Lumbreras, L. Childe and the urban revolution: The Central Andean experience. In *Studies in the Neolithic and Urban Revolutions*; Manzanilla, L., Ed.; B.A.R. International Series 349; Archaeopress: Oxford, UK, 1987; pp. 327–344.

35. Lane, K. Engineering resilience to water stress in the late prehispanic North-Central Andean highlands (~600–1200 BP). *Water* **2021**, *13*, 3544. [CrossRef]

36. Nicolis, G.; Prigogine, I. *Exploring Complexity*; W. H. Freeman and Company: New York, NY, USA, 1989; pp. 238–242.

37. Sabloff, A.; Sabloff, P.L. (Eds.) *The Emergence of Premodern States: New Perspectives on the Development of Complex Societies*; SFI Press: Santa Fe, NM, USA, 2018.

38. Durant, W.; Durant, A. *The Lessons of History*; Simon and Schuster: New York, NY, USA, 1968.

39. Plowman, D.; Solansky, S.; Beck, T.; Kulkani, M. The role of leadership in emergent, self-organization. *Leaders. Quart.* **2007**, *18*, 341–356. [CrossRef]

40. Spencer, C. Human Agency, Biased Transmission and the Cultural Evolution of Chiefly Authority. *J. Anthropol. Archaeol.* **1993**, *12*, 41–74. [CrossRef]

41. Routledge, B. *Archaeology and State Theory*; Bloomsbury Academic Press: London, UK, 2013.

42. Toynbee, A. *A Study of History*; Weathervane Books: New York, NY, USA, 1972; pp. 348–350.

43. Stanish, C. Archaeological survey in the Juli-Pomata region of the Titicaca Basin, Peru. In *Report to the National Science Foundation*; NSF: Washington, DC, USA, 1991.

44. Feinman, G.; Neitzel, J. Too many types: An overview of sedentary prestate societies in the Americas. In *Advances in Archaeological Method and Theory*; Schiffer, M., Ed.; Academic Press: New York, NY, USA, 1984; Volume 7, pp. 39–102.

45. Stanish, C. *The Evolution of Human Cooperation: Ritual and Social Complexity in Stateless Societies*; Cambridge University Press: Cambridge, UK, 2017.

46. Swenson, E.; Jennings, J. (Eds.) Place, landscape and power in the Ancient Andes and Andean archaeology. In *Powerful Places in the Ancient Andes*; University of New Mexico Press: Albuquerque, NM, USA, 2018.

47. Langlie, B. Building Ecological Resistance: Late Intermediate Period Farming in the South-Central Highland Andes (CE 1100-1450). *J. Anthr. Archaeol.* **2018**, *52*, 167–179. [CrossRef]

48. Stanish, C. The Origin of State Societies in South America. *Annu. Rev. Anthropol.* **2001**, *30*, 41–64. [CrossRef]

49. Moseley, M. *The Inka and Their Ancestors*; Thames & Hudson: New York, NY, USA, 2001.

50. Shady, R. *Caral-La Ciudad del Fuego Sagrado*; Interbank: Lima, Peru, 2004.

51. Wittfogel, K. *The Hydraulic Civilizations*; University of Chicago Press: Chicago, IL, USA, 1956.

52. Wittfogel, K. *Oriental Despotism: A Comparative Study of Total Power, Water Management in the Ancient World*; Philsophical Transactions of the Royal Society: London, UK, 1957.

53. Wittfogel, K. *Oriental Despotism*; Yale University Press: New Haven, CT, USA, 1957.

54. Thompson, L.G.; Mosley-Thompson, E.; Davis, M. Late Glacial Stage and Holocene Tropical Ice Core Records from Huascaran, Peru. *Science* **1995**, *269*, 46–50. [CrossRef] [PubMed]

55. Thompson, L.G.; Davis, M.E.; Mosley-Thompson, E. Glacial records of Global climate: A 1500-year tropical ice core record of climate. *Hum. Ecol.* **1994**, *22*, 83–95. [CrossRef]

56. Thompson, L.; Moseley-Thompson, E. One-half millennium of tropical climate variability as recorded in the stratigraphy of the Quelccaya Ice Cap, Peru. In *Aspects of Climate Variability in the Pacific and Western Americas*; Peterson, D., Ed.; Geophysical Union Monograph; American Geophysical Union: Washington, DC, USA, 1989; Volume 55.

57. Janusek, J. *Identity and Power in the Ancient Andes*; Routledge Press: London, UK, 2004.

58. Tung, T. *Violence, Ritual and the Wari Empire*; University of Florida Press: Gainesville, FL, USA, 2012.

59. Tung, T.; Vang, N.; Culleton, B. *Dietary Inequality and Indiscriminant Violence: A Social Bioarchaeological Study of Community Health during Times of Climate Change and Wari State Decline*; Institute of Andean Studies: Berkley, CA, USA, 2017.

60. Kurin, D. The bioarchaeology of social collapse and regeneration in Ancient Peru. In *Bioarchaeology and Social Theory*; Martin, D., Ed.; Springer: Berlin/Heidelberg, Germany, 2016; ISBN 978-3319284026.

61. Reilly, S.; Roddick, A. Plants in Transit Communities: Circulating Tubers and Maize in the Lake Titicaca Basin, Bolivia. *Lat. Am. Antiq.* **2022**, *33*, 791–808. [CrossRef]

62. Shimada, I.; Schaaf, C.; Thompson, L.; Moseley-Thompson, E. Cultural Impacts of Severe Droughts in the Prehistoric Andes: Application of a 1500 Year Ice Core Precipitation Record. *World Archaeol.* **1991**, *22*, 247–270. [CrossRef]

63. Romilly, J. *The Rise and Fall of States according to Greek Authors*; The University of Michigan Press: Ann Arbor, MI, USA, 1991.

64. Freeman, K. *Greek City-States*; Norton & Company, Inc.: New York, NY, USA, 1950.

65. Hobbes, T. *Man and Citizen (De Homme and De Cive)*; Gert, B., Ed.; Hackett Publishing Company: Indianapolis, IL, USA, 1991.
66. Hobbes, T.; Tuck, R. *Leviathan*; Cambridge University Press: London, UK, 1996.
67. Collingwood, R.; Meyers, J. *Roman Britain and the English Settlements*; Clarendon Press: Oxford, UK, 1936.
68. Childe, G.V. The Urban Revolution, London. *Town Plan. Rev.* **1950**, *21*, 13–17. [CrossRef]
69. Liverani, M. *Uruk: The First City*; Bahrani, Z., van de Mierop, M., Eds.; Equinox: London, UK, 2006.
70. Sarton, G. *Introduction to the History of Science*; Williams & Wilkins Company: Washington, DC, USA, 1931.
71. Sloan, K. (Ed.) *Enlightenment: Discovering the World of the Eighteenth Century*; British Museum Publications: London, UK, 2003.
72. Angelakis, A.N.; Margeta, J.; Salgot, M.; Zheng, X.-Y. Evolution of Hydro-Technologies Focusing on the Hellenistic World and Relevant Associations. *Water* **2023**, *15*, 1721. [CrossRef]

Article

The Aqueducts of Lugdunum

Paul M. Kessener

Faculty of Art, Radboud University, 6525 XZ Nijmegen, The Netherlands; lenl@euronet.nl

Abstract: Not long after the *Colonia Copia Felix Munatia Lugdunum*, in present day Lyon, France, was founded in 43 BCE by Lucius Munantius Plancus on the 300 m high Fourvière hill overlooking the Saone and Rhône rivers and the plains to the north and east, it became the capital of the Gallia provinces, growing to be with some 50,000 inhabitants the largest town in Gaul. In the early days, the *colonia* on the west valleys surrounded Fourvière hill and depended on local springs, wells, and rain cisterns for its water provision, which soon became insufficient for the growing city. A first aqueduct was constructed in 20 BCE, bringing waters from a spring some 10 km north of the town. In the decades to follow, another three aqueducts were added. All of the aqueducts were equipped with one or more pressure lines and installations of (inverted) siphons, totaling nine, to cross valleys that were thought too deep or too wide for a bridge. Today, the *Métropole de Lyon* counts over a million inhabitants; it is after Paris and Marseille the third largest town in France. Since 1998, Lyon has been listed on the UNESCO World Heritage List, among others, because of the historic architecture in its urban settlements over 2000 years of age. This manuscript recounts the history and present remains of the four aqueducts and their nine extraordinary siphons, and is dedicated to Dr. Jean Burdy, who, with his team over many years of research of earlier literature and of investigations and discoveries of the physical remains of valley sites, produced a great number of publications saving the Lyon aqueducts from oblivion and leading to restorations in recent times.

Keywords: Lyon; Lugdunum; Roman aqueducts and water provision; siphons; lead pressure conduits; hydraulic towers

Citation: Kessener, P.M. The Aqueducts of Lugdunum. *Water* **2024**, *16*, 2117. https://doi.org/10.3390/w16152117

Academic Editor: Athanasios Loukas

Received: 30 May 2024
Revised: 10 July 2024
Accepted: 11 July 2024
Published: 26 July 2024

1. Introduction

While the promontory of the 300 m Fourvière hill (the elevation or altitude of geographical points is in the following expressed in meters (m) above average sea level. In French: la cote est exprimée en altitude normal) dominates with its steep 130 m ascent the Saone and Rhône rivers and the plains to the north and east, the lands to the west present mountainous areas with abundant springs and rivers that became important for the water provision of Roman Lugdunum (Figures 1–3). Over a period of several decades, four aqueducts were built, one after the other, taking their waters from springs in the mountainous lands to the north, west, and south, and having lengths ranging from 25 to 86 km, each aqueduct being equipped with one or more siphons, with a total of nine for all aqueducts.

However, lands in this region are from all sides separated from the Fourvière hill and its associated heights by an over 100 m deep valley stretching from Vaise to Oullins (Figures 1–4). This obstruction for the passage of aqueducts could only be overcome by a closed conduit under pressure: a siphon [1]. All four aqueducts of Lugdunum crossed the valley in this way. The aqueducts then had to pass towards the Fourvière hill (A in Figure 4), for which again a depression, less deep, had to be crossed, shown as the Trion depression B in Figure 4.

The 30 m deep Trion depression was again crossed by a siphon for the Gier and the Brevenne aqueducts, but not for the aqueduct of Mont d'Or, as this aqueduct came in at level 265 m, corresponding to the bottom of the Trion depression (Figure 5).

Figure 1. France, Lyon.

Figure 2. Roman theatre on the Fourvière hill built around 15 BCE (https://en.wikipedia.org/wiki/Ancient_Theatre_of_Fourvi%C3%A8re (accessed on 1 March 2024).

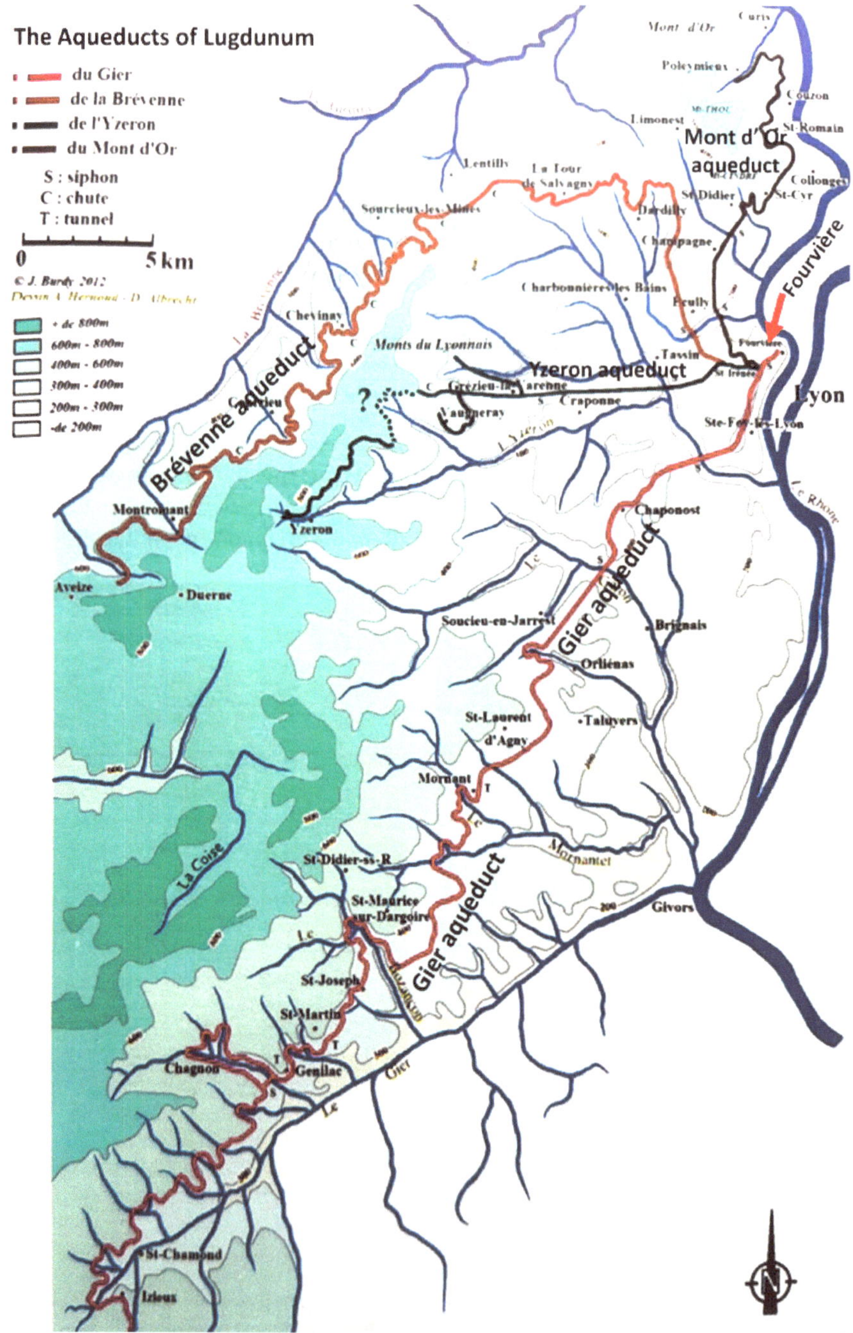

Figure 3. The four aqueducts of Lyon, the 26 km Mont d'Or aqueduct to the north, the Yzeron aqueduct 27 km (question mark '?': unconfirmed section) and the Brevenne aqueduct 70 km to the west, and the 86 km Gier aqueduct to the south [2], p. 19.

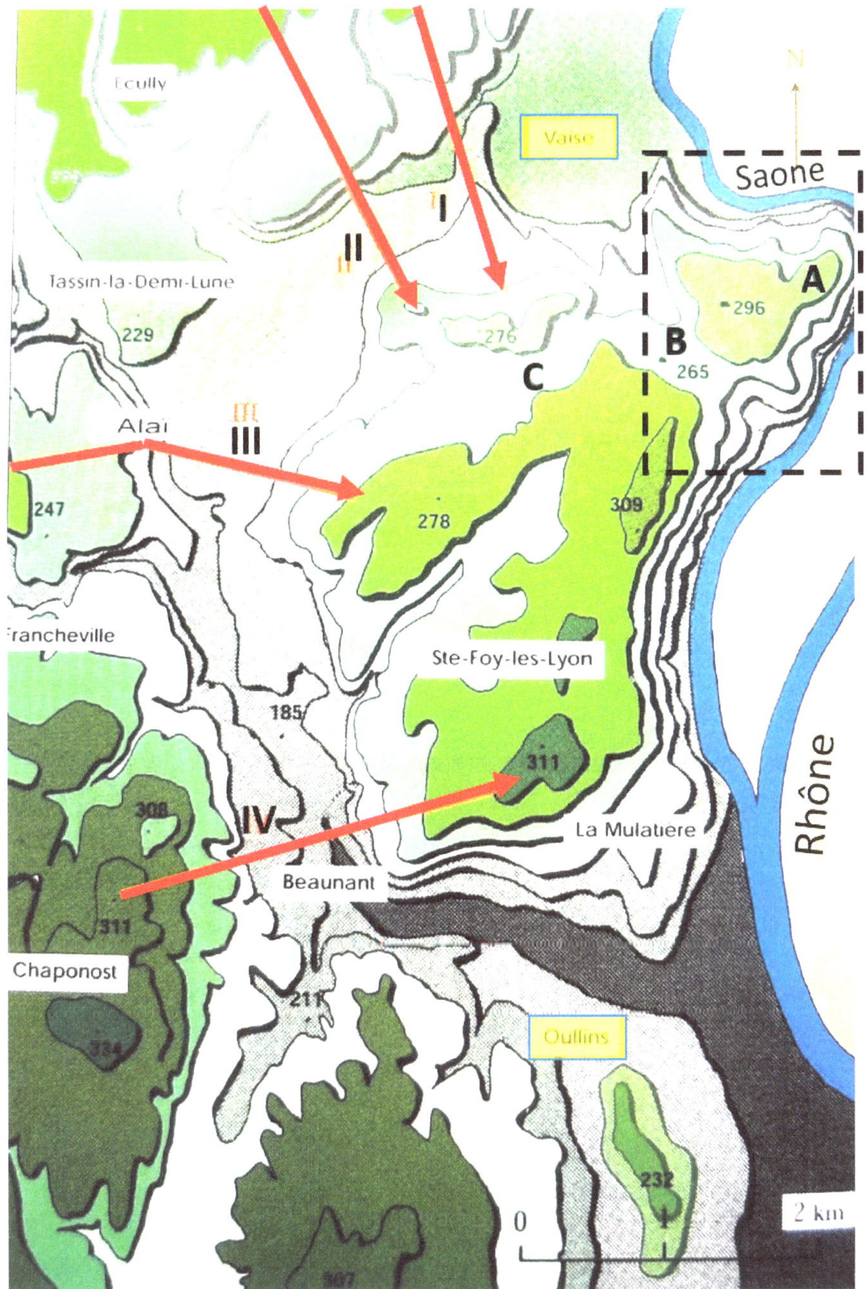

Figure 4. A Fourvière promontory, B Trion depression, C area 'Point du jour' is indicated. The deep valley from Vaise to Oullins surrounding Fourvière and associated heights necessitated a siphon for each aqueduct (red lines): I for the aqueduct of Mont d'Or, II for the aqueduct of Yzeron, III for the Brevenne aqueduct, and IV for the Gier aqueduct [2], p. 43.

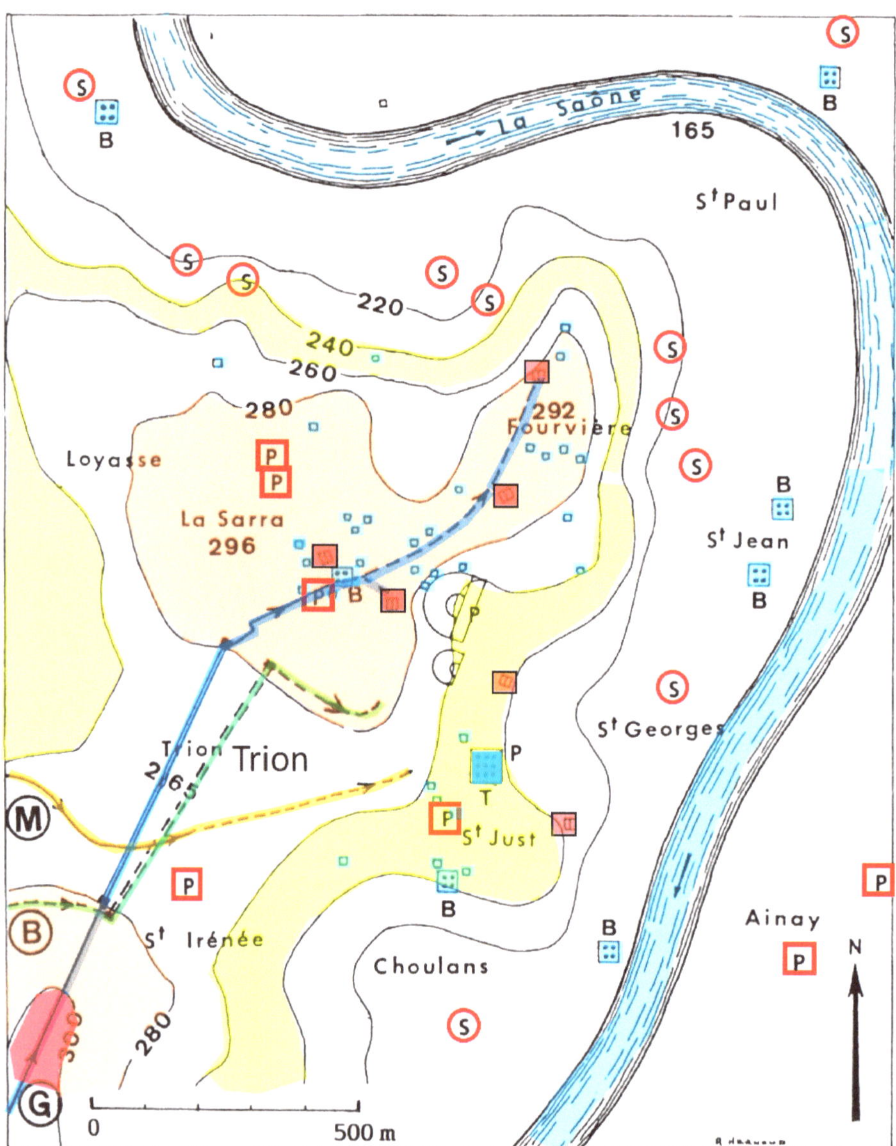

Figure 5. Inset of Figure 4. Fourvière hill with three arriving aqueducts. G encircled: Gier aqueduct, dotted course speculative. The latter two aqueducts crossed the Trion depression with a siphon. S: spring; P: well; B: bath building; T: large thermes; small blue squares: rain water cistern; black squares with redish fill and small symbol: reservoir [3], p. 18, adapted.

2. History

The water provision of Lugdunum has attracted the attention of writers in the past who produced books and images of parts of the aqueducts. First mention of Lyon's Roman aqueducts date from the 16th century, referring to channels passing close to, or through, villages. Then, in 1759, Guillaume Marie Delorme gave a lecture at the *Académie de Sciences, Belles-Lettres et Arts de Lyon* on his '*Recherches sur les Aqueducs de Lyon construites par les Romains*', discussing the remains of the Gier aqueduct, and of parts of the Brevenne and the Yzeron aqueducts. A year later, the lecture was published in a 60-page book which was well received by water scholars of that era (Figure 6). Continuing his researches in later years, Delorme produced a great number of engravings measuring 55 × 82 cm, a total of 126, all of which depict the investigated remains of the Gier aqueduct, as well as two engravings of the Brevenne aqueduct and one of the Yzeron aqueduct. After the death of Delorme in 1782, and of his pupil and friend Catherin-François Boulard in 1794, the engravings were in 1803 offered to the city of Lyon, which, however, refused to buy them.

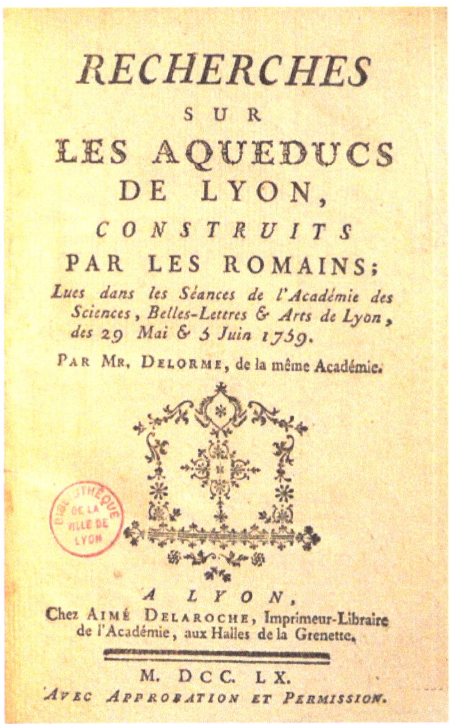

Figure 6. Front cover of the 1760 publication by Guillaume-Marie Delorme about his investigations of the aqueducts of Lyon constructed by the Romans.

Only mentioned in 1818 in a letter, the engravings were considered lost. After two centuries, they miraculously reappeared in October 2003 in Paris, now to be acquired by the city of Lyon with the subvention, among others, by the Water Company *la Générale des Eaux*. Today, the engravings are kept at the City Archives of Lyon (Figure 7). In 2015, a publication about these high-quality engravings was issued by the organization *L'Arraire* [4], p.17.

In the 19th century, several authors such as Maudet de Penhouet, Alexandre Flachéron (who proposed a functional restoration of the Brévenne aqueduct), Paul Gasparin, François Gabut, Achille Raverat, and Louis Bresson contributed to the knowledge of the Lyon aqueducts, with some publications, however, being lost while parts of presented data proved incorrect in later days [5]. Then, in 1909, the thesis by Camille Germaine de Montauzan (1862–1942) appeared describing the Roman aqueducts of Lyon, their water provision and hydraulics engineering, their organization and administration, and the construction techniques together with legal matters—this a great work and reference book of 436 pages [6].

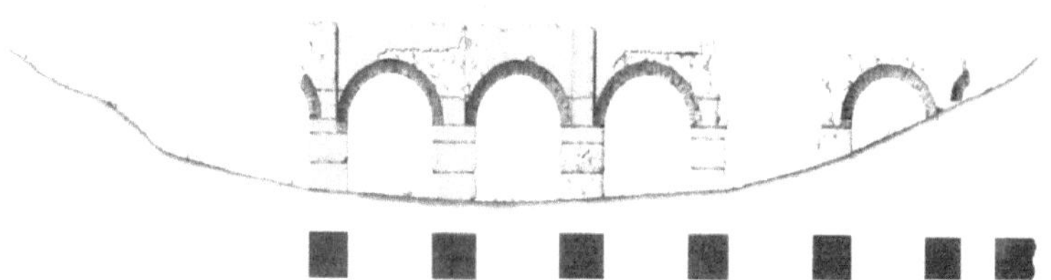

Figure 7. Engraving by Delorme, of remains of a bridge of the Gier aqueduct at village of Mornant. Today, the three arches in the centre have collapsed [4], p. 55.

It was only in the late 1960s that aqueduct enthusiasts as Jean Burdy, Louis Jeancolas, Henri Bougnol and others started research again, now supported in the 1980s by AFAN (Association pour les Fouilles Archéologiques Nationales) and INRAP (Institut National de Recherches Archéologiques Préventives). Jean Burdy published over the years 1987–1996 four monographs, one for each aqueduct, presenting the results of the investigations and leading to his doctorate in 1996 at the Lyon University (Burdy [3,5,7–9]. On numerous occasions, Jean Burdy reported of his findings, among others, for the *Union des Sociétés Historques de Rhône*, *the Société Natiionale de Antiquaires de France*, the *Académie des Sciences, Belles-Lettres et Arts de Lyon*, and at the international conferences of the German Frontinus Gesellschaft and the Deutsche Wasser Historische Gesellschaft DWhG. In November 2023, a conference on the historic water provision of Lyon named *'Premieres Assises de Patrimoine Hydraulique'*, 22–25 November 2023 was organized. It was attended by Jean Burdy, then at the age of 93.

3. The Aqueducts

To the north of Fourvière, the Mont d' Or aqueduct took waters from the spring of the rivulet of Thou, and to the west, the springs of the Yzeron river and the Brevenne river fed, respectively, the Aqueduc de l' Yzeron and the Brevenne aqueduct, and to the south near the springs of the Gier river, the 75 km eponymous aqueduct found its origin (Figure 3).

4. The Aqueduct of Mont d'Or

Near the Poleymieux village in the up to 600 m high mountain range of the Mont d'Or, at about 372 m elevation and 10 km north of Lyon, the 26 km aqueduct took its origin (no. 0 on the map of Figure 8). It was built ca. 20 BCE under the reign of Augustus (Burdy 1987, 16). The old water catchment system had long since been replaced by an underground chamber connected to a gallery that fed a washing basin; today, a modern installation performs the task of water collecting (Figure 9). First remains of the channel appear 650 m downstream at the north east, no. 1 on the Figure 9 map, while the intact channel was found at no. 3, measuring 50 cm wide and 65 cm high, covered by stone slabs (Figure 10). The channel then, after some turns, changed direction to the south where at Couzon au Mont d'Or the channel appeared again, without its cover, (no. 8, Figure 11) at a level of 311 m. Channel remains were spotted on several locations further downstream as indicated on the map, for instance, at St. Romain au Mont d'Or, no. 12. (Figures 12 and 13), where an additional spring was tapped, and at St. Didier au Mont d'Or (no. 27).

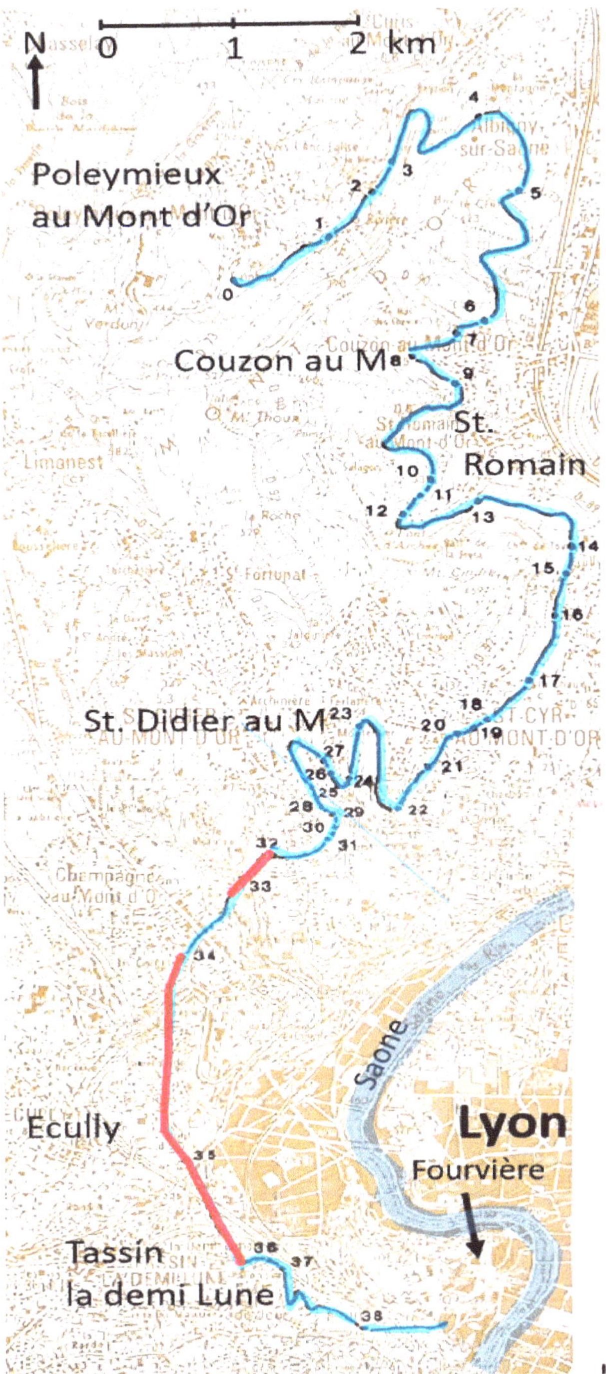

Figure 8. Aqueduct de Mont d'Or. The numbers 1–38 on the map represent finds and remains of the aqueduct channel and related objects. Red: siphon (after [7], pp. 40–41 adapted).

Figure 9. Poleymeiux au Mont d'Or, no. 0 on the map, deserted washing basin where traditionally the channel started. The white building in the back houses the modern water catchment installation [7], p. 46.

Figure 10. Channel of the aqueduct of Mont d'Or, at no. 3 on the map [2], p. 20.

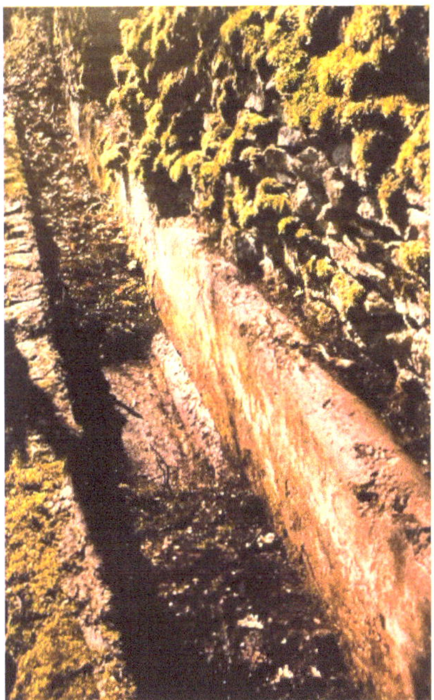

Figure 11. Channel near Couzon au Mont d'Or at no. 8 on the map [2], p. 20.

Figure 12. Channel near St. Romain au Mont d'Or, no. 12 on the map [7], p. 64.

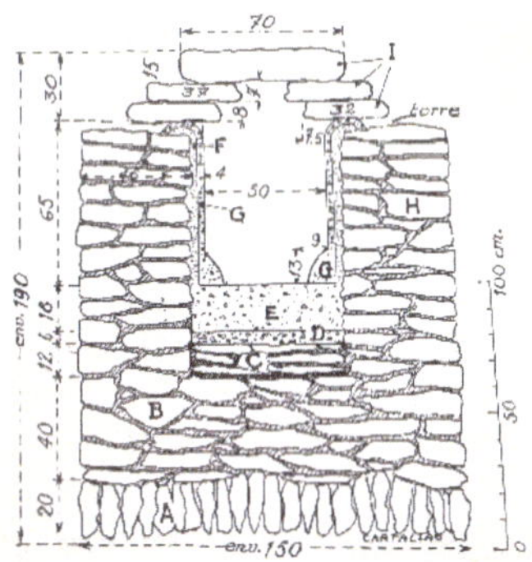

Figure 13. Reconstruction of the channel of the Mont d'Or aqueduct [7], p. 23.

Downstream of point 31 on the map, the deep valley of the Limonest stream was crossed by a 30 m deep and 420 m long siphon, a header tank at 285 m (no. 32 on the map) and a receiving tank at a supposed 277 m height (Figure 14) (for siphons/pressure lines in ancient aqueducts, see [1]). The remains of the about 12 arches of the 100 m long and 17 m high siphon bridge (point no. 33) have all disappeared (Figure 15).

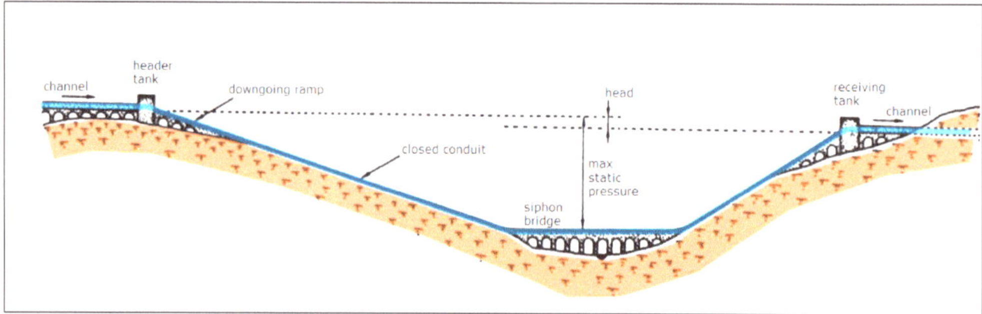

Figure 14. Elements of an (inverted) siphon, a pressure line according to the principle of communicating vessels. Down in the valley reigns highest water pressure that the pipeline must be able to withstand [10].

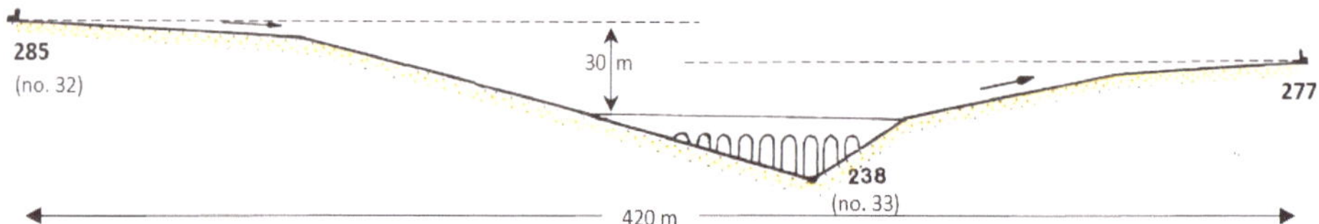

Figure 15. Siphon of the 'ruisseau de Limonest', 420 m long and 30 m deep, with a siphon bridge 100 m long and about 17 m high, header tank at point no. 32 of the map [7], p. 26.

At no. 34 on the map, the header tank of a second and much larger siphon '*de ruisseau des Planches*' was located, 3500 m long and 70 m deep, with a 200 m long and 25 m high siphon bridge, the remains of which were torn down in 1967 (Figure 16) [7], p. 90.

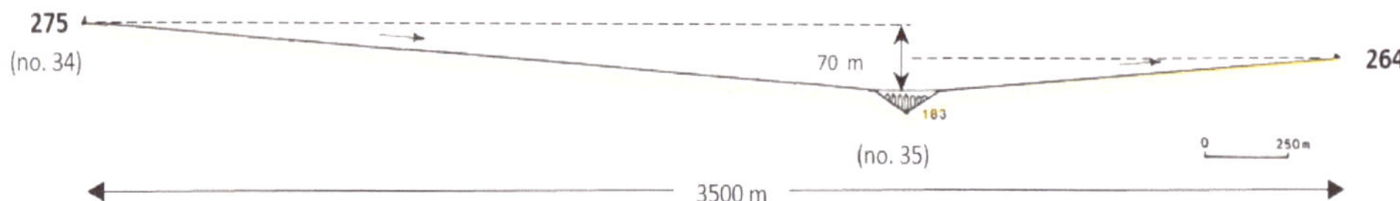

Figure 16. Siphon of the 'ruisseau des Planches', 3500 m long and 70 m deep, with a 200 m long bridge 20 m high. Header tank at 275 m (point 34 at the map), receiving tank at about 264 m [7], p. 27.

The conduits of the siphons are thought to be made of lead and several pipes laid out parallel, maybe six or eight, for the great siphon as Jean Burdy envisages, but no traces of the pipes have remained [7], p. 27; [2], p. 21. Downstream of the siphon and barely visible are some remains of the channel that are to be found at points no. 36 and 37 on the sides of an excavation for a train track end in the 19th century that destroyed a 120 m long aqueduct section. Point 38 relates to a finding of the channel in perfect condition in the Rue de la Favorite, probably discovered during activities for building construction in the first half of last century (Figure 17). From there, the channel took a slightly east/north-eastern course via the *Point du Jour* and the *Trion* area's at a level of about 260 m directed toward the Roman theatre and the thermes (see Figures 4 and 5).

Figure 17. Channel found in the Rue de la Favorite (http://archeolyon.araire.org/AqueducsLyon/MontdOr/MalbumK.html, accessed on 1 March 2024).

The capacity of the Mont d'Or aqueduct, that is, how much water was transported per day to Fourvière, depends on the dimensions of the channel and its slope, the roughness of the channel walls, and the level of the water running in the channel. The channel's slope, as depicted in Figure 18, appears rather constant between the steep section at Poleymieux and the two siphons. The wall roughness of a newly built channel is less than that of an incrusted one, which translates to a different roughness coefficient in calculating formulas. The smaller this coefficient, the faster the water will flow.

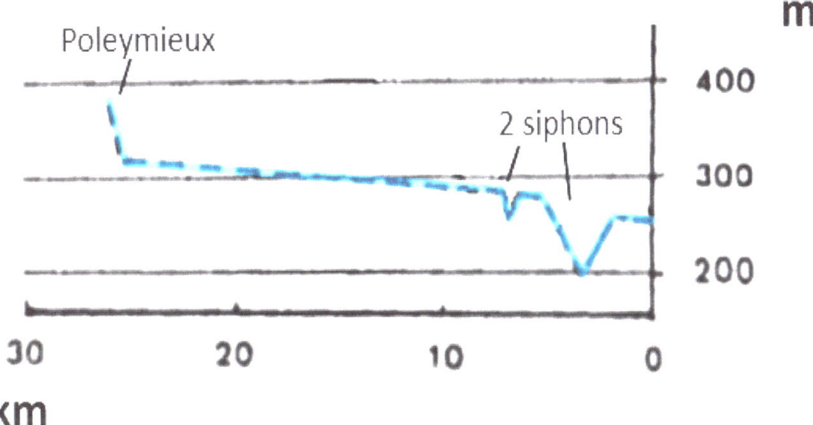

Figure 18. Profile of the Mont d'Or aqueduct, except for the initial section at Poleymieux, leave out Figure 7 is rather constant at about 1.4–1.7 m/km [9], p.14, adapted.

Jean Burdy has calculated the capacity in cubic meters per day for the channel 50 cm wide, and for a presumed water level of 30, 40, and 50 cm for a slope of 1.4 m/km with the Bazin formula (see for Bazin and similar formula's for open channel flow for instance https://theconstructor.org/fluidmechanics/chezys-and-mannings-formula-in-open-channel-flow/558411/, accessed 1 March 2024). Taking the seasonal water delivery of the springs feeding the aqueduct into account, Burdy estimated an effective capacity of 2000–6000 m^3/day [7], p. 36–37.

5. The Gier Aqueduct

The aqueduct of Gier, taking its name from the river that provided its waters, was probably constructed under Hadrian (117–138 CE) [2], p. 28. It is the longest and latest aqueduct bringing water to Lugdunum. Its channel, running from south to north, is 86 km long, including an 11 km section circumventing a complicated valley at Chagnon in the south that was bypassed by means of a siphon (Figure 19). The Gier channel took its waters from a tributary of the Gier river near Izieux village about 42 km south of Lyon. A low dam is thought to have collected the water, of which no trace has remained due to modern construction for the water provision of nearby St. Chamond. The acidic water entering the Gier channel was free of calcium carbonates, as the supplying area is not karstic but granite. As a result, the Gier channel exceptionally remained without any calcareous incrustations (calcareous incrustations are common in aqueducts and may provide information about the period during which that water flowed, as well as of many other aspects [11,12].

Without counting the four siphons, the 86 km Gier channel including the 11 km detour at Chagnon, and ran for 90% in an earth covered trench some 2 m wide and 3–4 m deep that followed the terrain level (Figure 20). There were 11 tunnels incorporated in the aqueduct, 10 of which have been identified, and 1 being hypothetical near St. Chamond. The tunnel lengths range from 80 m (the so-called *Cave du Curé* near Chagnon, (Figure 21)) to 825 m (near Mornant); the total length for all tunnels is 3.4 km (Figure 22).

Starting at a level of 405 m, the Gier aqueduct ended on the Fourvière hill at 299.5 m, the only of the four aqueducts that could serve the entire town. The average slope is about 1 m/km, ranging from 0.6 to 1.85, not taking the siphons and the Chagnon detour into account [9], p, 284. The slope of the Nîmes aqueduct is an extremely low 7 cm/km for 10 km downstream of the Pont du Gard [13], p. 189.

For maintenance, repairs, and cleaning, the Gier channel was accessible from above by means of vertical inspection shafts (french: 'regards', german: 'Einstiegschacht') of rectangular shape ($2 \times 2\frac{1}{2}$ feet to 3×3 feet), the entrance covered by thick stone slabs (Figure 23). A total of 91 of these inspections shafts have been found, also connecting to tunnels, having a regular interdistance of 77 m which would count, if systematically installed, to up to 1000 *regards* for the entire aqueduct [2], p. 36.

The Gier aqueduct incorporated over 30 bridges and some 10 channel sections on arches and walls, of which important remains survive. At the '*Plat de l'Air*' near Chaponost, the channel was carried on an impressive row of 92 well preserved arches that run along the local traffic road (Figure 24). Directed towards the north-west it turns eastwards, ending at the header tank of the siphon that crossed the valley of the Yzeron river.

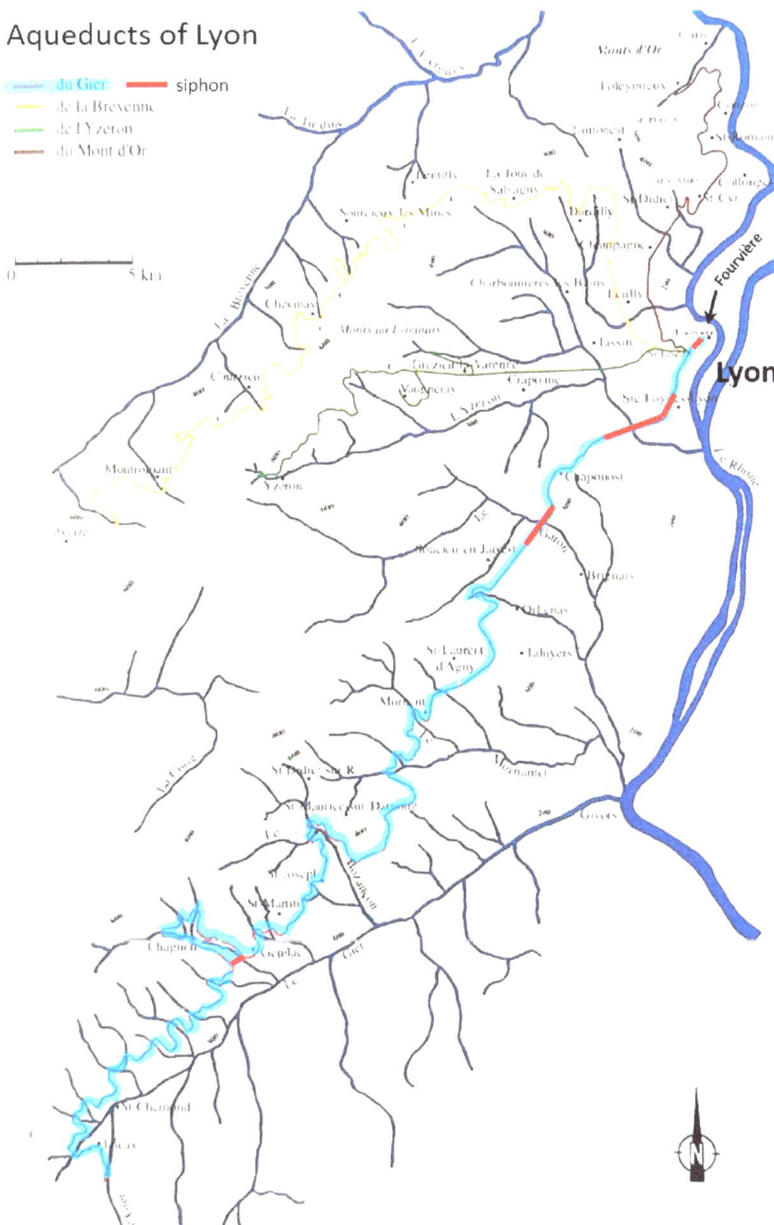

Figure 19. The aqueducts of Lyon, with the Gier aqueduct highlighted in blue, its four siphons in red. The Gier aqueduct took its waters from a contributary of the Gier river near Izieux village in the south where a low barrage is thought to have been installed, of which no trace remains because of the installation of water provision works for St. Chamond village 3 km to the north [14], p. 64, adapted.

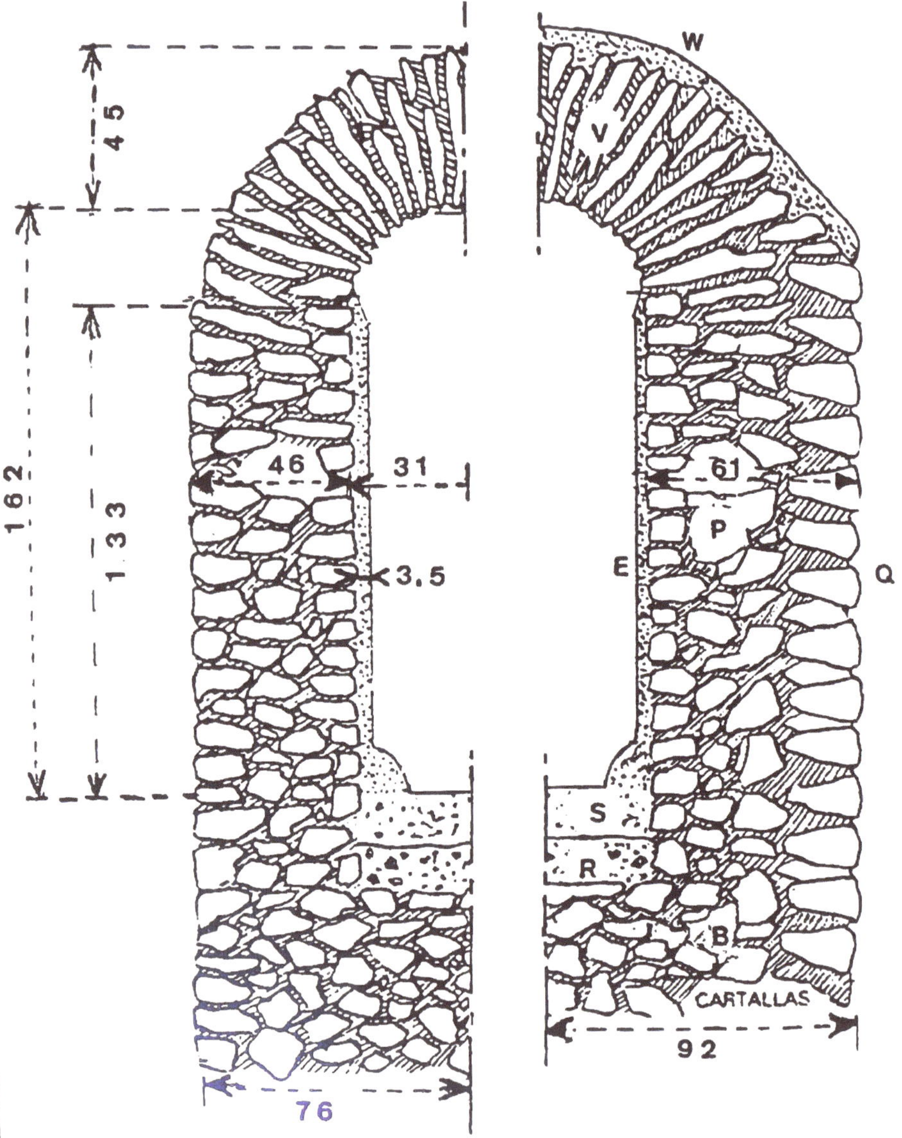

Figure 20. Gier aqueduct, channel construction. Left: trenched channel, right: channel on or above ground level. Channel width 55 cm, height to top of vault on the inside 162 cm. The walls of the channel exposed to atmosphere are with 61 cm, 15 cm thicker than of the trenched channel, while its barrel vault is covered on the outside with a water tight cement, as are the floor and inside walls for both types of channel [9], p. 190.

Figure 21. Cave du Curé, a tunnel excavated in hard rock, in which the aqueduct channel was constructed. Words go that the curé is to have used the tunnel to store his wine bottles (http://archeolyon.araire.org/AqueducsLyon/Gier/GalbumI1.html, accessed on 2 March 2024).

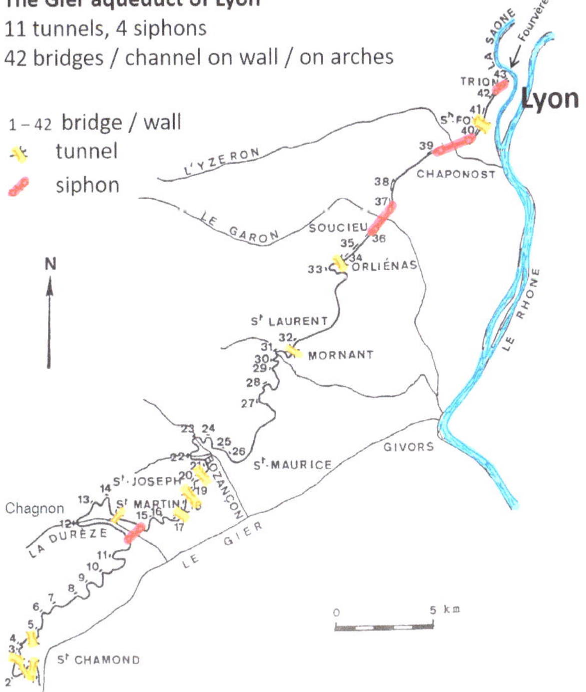

Figure 22. Gier aqueduct. Tunnels, bridges, and channels on wall or on arches [9], p. 226, adapted.

Figure 23. Inspection shaft near Mornant, covering slab broken [2], p. 36.

Figure 24. Gier aqueduct, Plat de l'Air–Chaponost. Channel on arches 15 m high over a distance of 600 m. The piers are 1.85 m square, interspace 4 m [2], p. 25; [9], p. 154. The header tank of the Yzeron siphon where the row of arches ends is visible in the left top of the photo. The arches have undergone restoration in 2022–2024.

6. The Siphons

The aqueduct counted four siphons to cross extended and deep valleys. These siphons were an extraordinary and singular construction, with up to 11 leaden pipes running parallel down the valley and up again on the other side. Each siphon will be shortly discussed, downstream from source to city.

7. The Durèze Siphon

Of the siphon crossing the Durèze river near Genilac, the header tank has been well preserved, but no trace remains of the receiving tank (Figure 25). On the valley side, the header tank shows openings for the lead conduits going down, from which the size of the conduits may be estimated at about 23–25 cm outer diameter. A gap from destruction in this side is suggestive for another three openings, a total of ten, of which according to researchers like Germain de Montauzan, one or two openings were obstructed (Figure 26).

Figure 25. Map with start and ending of the Chagnon detour of 11 km (1, 6, 10) and the 700 m siphon crossing the valley of the Durèze river (2 header tank, 5 location of receiving tank), 3 and 4 are remains of the 19 m high siphon bridge. The southern side of the valley drops precipitously down for its lowest half (about 50 m over a distance of 100 m (north of 'les Mûres'), (http://archeolyon.araire. org/AqueducsLyon/Gier/GcarteH.html, accessed on 4 March 2024).

The back side of the 8.3 × 4.15 m construction shows the opening for the channel that provided the 6.4 × 2.25 m reservoir from which the lead conduits took water. The down-going rampart is destroyed over its greater part.

The question is, of course, why is there a siphon at this point bypassing the 11 km channel that circumvents the complicated valley at Chagnon? Were both constructed simultaneously? This 11 km channel has the moderate slope of 0.5 m/km or 5 cm per 100 m, not an uncommon gradient for aqueducts which would translate into a somewhat reduced water velocity combined with a high water level in the 133 cm high channel, but not necessarily resulting in an insufficient capacity (the channel of the Nîmes aqueduct had a slope of 7 cm/km for 10 km downstream of the Pont du Gard (Hodge 1992, p. 190)). Because of the acidic water of the Gier springs, no calcareous incrustations were left on the channel walls, so we do not know the historic water level in the channel (Burdy calculated from formulas the theoretical capacity of the Gier Aqueduct for water levels of 30 cm to 70 cm in the 133 cm high channel, resulting in an average capacity of 12,000 m^3 per day (for which the water height in the channel at Chagnon would have been some 70 cm (Burdy 2017, p. 48; Burdy 1993, pp. 287–288)). The capacity of the siphon must be deemed sufficient as well (see Burdy 1996, 293–294. Burdy states that further investigations might resolve the issue whether the 11 km circumvention or the Durèze siphon was installed first).

Figure 26. Valley side of the 8.3 m wide header tank of the Durèze siphon, with three and four openings for leaden conduits still present, while three more openings were located in the destructed part in the centre. Through the centre part, the back wall is visible with the opening where the arriving aqueduct channel ended. In the left side wall, a partly destructed opening (not visible) is interpreted as an overflow facility [9], p. 76.

On 27 April 1887, an interesting find was done during new road works close to Chagnon, when a stone slab 158 × 62 × 20 cm was found carrying an inscription, since then called '*la Pierre de Chagnon*' (Figure 27). The inscription reads 'By the order of Caesar Traianus Hadrianus Augustus no person is allowed to work, sow, or plant in the space of the terrain that is destined for the protection of the aqueduct'. Remarkably, on 10 September 1996, a similar inscription stone was found, along the channel at the village of St. Joseph, broken in four pieces and incomplete, but shown to be inscribed with the same text (Figure 28).

The inscription reminds of the text of Frontinus, in 97 CE Curator Aquarum of Rome, who refers to a decree of the Senate stating '... that around the springs, the arches, and the walls on both sides a space must be left free of 15 feet, and that around underground channels and roofed channels inside the town and in adjacent built-up area outside the town on both sides a space of five feet must left free ...' subject to a penalty of 10,000 sesterties. No such recommendation about is mentioned on the stones of Chagnon and St. Joseph.

Figure 27. The 'Pierre de Cagnon' text found in 1887 near the aqueduct's channel at Chagnon [2], p. 59.

Figure 28. Inscription stone 'Pierre de St. Joseph' found in 1996. Broken in four pieces and incomplete, it has a similar text as the Pierre de Chagnon [3], p. 178.

Clearly, the Chagnon detour and the continuing channel at St. Joseph were similarly treated, and they must have been equally important. So one may suppose that the detour at Chagnon was designed as such from the start. What then about the Durèze siphon? Designing a siphon and constructing it required special skills of both engineers and sufficient building knowledge, not the least because of the many lead conduits that had to be soldered together on the spot. Furthermore, an aqueduct was not constructed from source to city, but in stretches simultaneously in order to reduce the construction time needed (as for the aqueduct of Cologne [15], pp. 58–59. Once the overall design was decided upon, specialized teams would be put to the task of constructing the siphons on locations as planned, while the channel was to be connected to the siphon once the 'channel team' had made it to the header tank. The precipitously steep and 100 m deep Durèze valley not only challenged the siphon team, reducing the static pressure down in the valley by almost 2 bar with the 19 m high siphon bridge, but during operation, the siphon would be subject to pressure surges endangering the integrity of the siphon at start up, as well as when functioning because of air in the conduit and air entrained at the header tank (see for problems in operating siphons [1], pp. 47–52. Moreover, the precipitous slope going down, and to a lesser degree, going up and the considerable weight of the lead conduits may have caused the pipes to sag down and destroy the integrity of the soldered pipes even for a try-out procedure. This may have led the designers to proceed from the very start with the Chagnon detour channel and connect it to the channel at the receiving tank direction at St. Joseph. The Durèze siphon may have been built first, but bypassed as soon as severe problems of its functioning became apparent.

8. The Siphon of Garon

Downstream from Chagnon and along St. Joseph, the channel continued mostly underground, on arches and bridges or through tunnels when needed, the 825 m long and 20 m deep tunnel at Mornant being longest, before it reached a 590 m long row of 78 arches up to 16 m high crossing the Soucieux plane. The channel ended at the header tank of the siphon crossing the valley of the Garon river (Figure 29, see also Figure 22).

The wall that carried the channel and the 245–250 cm wide piers, as well as the channel walls, are covered on the outside with *opus reticulatum*, using square cut stones of granite, their sides oriented at 45 degrees from horizontal.

Apart from its aesthetic effect, this method also protected the bonding mortar from climatic influence by preventing rain water to accumulate in the joints between the stones. The channel walls on the inside were covered with the water tight *opus signinum*, a mixture of live chalk, oil, and fragments of ceramics giving it a characteristic rose/reddish color. The heavily damaged header tank of the Garon siphon, 6.2 × 3.1 m on the outside and 4.6 × 1.55 m on the inside, is still standing for a part, with 4 holes of the reconstructed 10 in situ (Figures 30 and 31). From an opening in the vaulted roof, one could enter the header tank.

The 5 m wide and partly destroyed ramp of the header tank ran down towards the 21 m high and 7.35 m wide siphon bridge that carried the 10 pressurized pipes towards the other side of the 115 m deep and 1200 m wide Garon valley.

The Garon siphon bridge counted 23 arches with a span of 6.1 m and piers 3.05 m wide, over a length of 210 m and at a level of 93.5 m below the header tank. Sections of the bridge are still standing on either side of the Garon valley (Figures 32–34) [2], p. 44. Of the receiving tank, the 6.20 × 3.40 m base and the 5 m wide ramp leading up to the tank have been preserved (Figure 35); it could be deduced that the difference in level between header tank and receiving tank was 8.8 m.

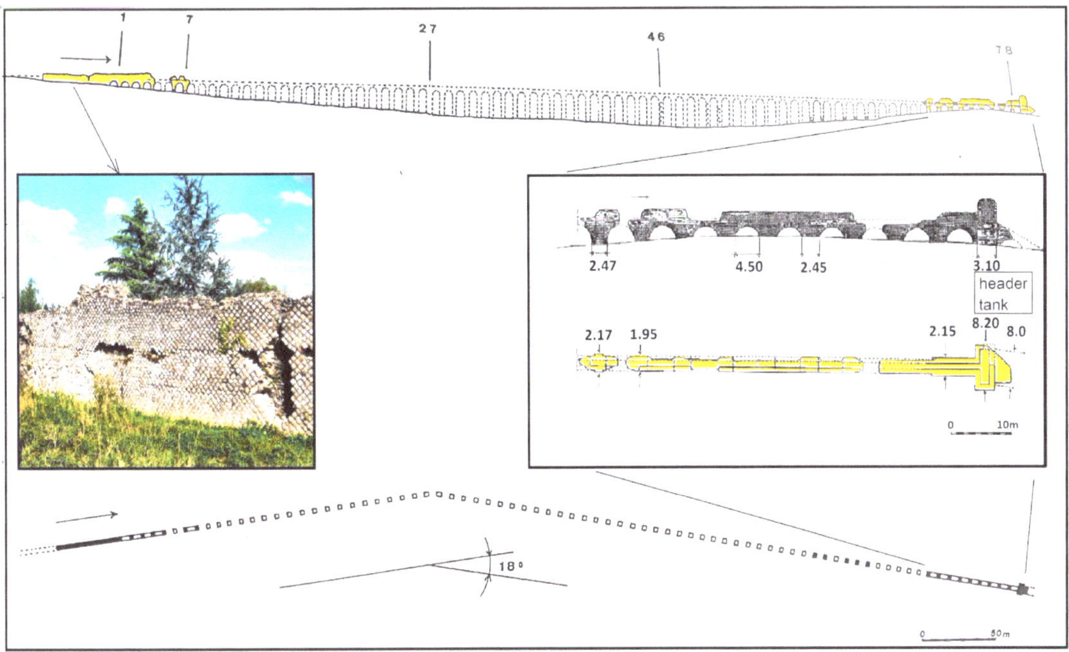

Figure 29. The 580 m long channel on arches (with exaggerated slope in the figure) crossing the Soucieux plane, ending at the header tank of the siphon crossing the Garon river. Present remains in yellow; 1: wall covered with opus reticulatum (inset), and upstream arches; 7: single arch standing; 27: change of direction, 46: highest pier (16 m); 78: final arch where channel ended at the header tank [9], pp. 137, 140; (http://archeolyon.araire.org/AqueducsLyon/Gier/GalbumU.html, accessed on 6 March 2024, adapted).

Figure 30. Remains of the header tank of the Garon siphon, with four holes for descending lead conduits (photo by author 2023).

Figure 31. Reconstruction of the header tank of the Garon siphon [9], p. 142.

Figure 32. Remains of the siphon bridge, crossing the Garon valley, upstream side (http://archeolyon.araire.org/AqueducsLyon/Gier/GalbumV3.html, accessed on 9 March 2024).

Figure 33. Remains of the siphon bridge, crossing the Garon valley, downstream side (http://archeolyon.araire.org/AqueducsLyon/Gier/GalbumV3.html, accessed on 9 March 2024).

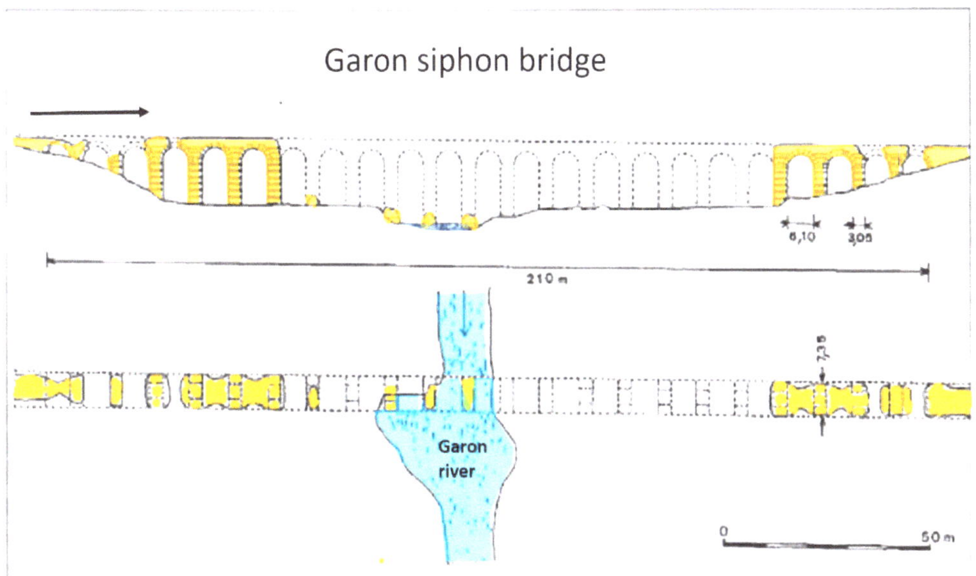

Figure 34. The 210 m Garon siphon bridge, with 23 arches of maximum height 21 m, crossing the 65 m deep Garon river valley (adapted from http://archeolyon.araire.org/AqueducsLyon/Gier/GalbumV3.html, accessed on 9 March 2024).

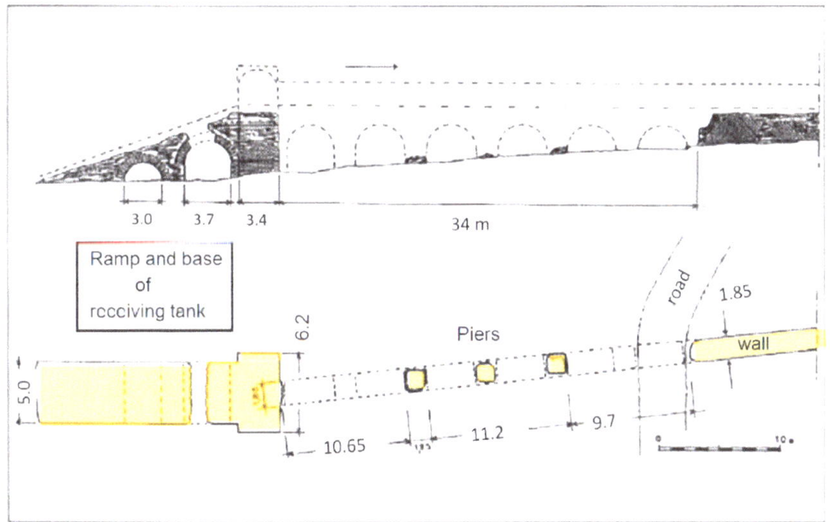

Figure 35. The Garon siphon, ramp, and base of the receiving tank, from where the channel proceeded on arches and walls. Present remains in yellow, measures in meter [9], p. 146, adapted.

9. The Yzeron Siphon

From the Garon receiving tank onwards, the channel crossed the undulating plateau of Chaponost in a northerly direction, both on arches and walls as well as underground for several hundred meters at a time. After some 4.5 km, the flat plain of *Plat de l'Air* was crossed on arches up to 15 m high until the channel reached the edge of the valley of the Yzeron river (see Figure 24 above). The channel on the row of over 90 arches terminates at the header tank situated 10.5 m above ground at a level of 313.7 m, with the start of a siphon with 11 parallel conduits crossing the 140 m deep and 3 km wide valley of the Yzeron river. In 2023, the arches plus channel and header tank were subject to restoration (Figures 36 and 37).

Figure 36. Header tank 7.1 × 3.1 m and downward ramp 5.85 m wide of the 2660 long and 122 m deep Yzeron siphon. A vaulted roof covered the header tank in which a person can stand upright (author's photo).

Figure 37. Header tank and ramp of the Yzeron siphon during restoration, November 2023 (author's photo).

The header tank and ramp were in worse condition a century ago, but now the broken upper arch of the descending ramp undergoing had repairs in the 1930s (Figures 38 and 39).

Figure 38. Header tank of the Yzeron siphon prior to restorations in the 1930s [9], p. 157.

Figure 39. Reconstruction of the header tank and ramp of the Yzeron siphon, here with nine parallel lead conduits running down the valley of the Yzeron river [16], p. 151.

The 11 lead pipes went down the valley side towards the Yzeron river that was crossed by a 270 m long, 7.35 m wide, and 18 m high 'siphon bridge' called *le Pont de Beaunant*. The bridge counted 30 arches, of which 15 still stand, 7 on the right side and 8 on the left side of the river bed. In 1875, the bridge was classified as *Monument Historique*, which, however, did not prevent the construction of a road passing it at square angles, and of several buildings blocking an open view. The 7.35 × 3.05 m piers were built with an arched opening 3.05 m wide in the 7.35 m width that for the largest piers were filled up again soon after the construction to prevent collapse, a row of five and one of three piers with unfilled openings presenting a corridor-like view (Figures 40 and 41).

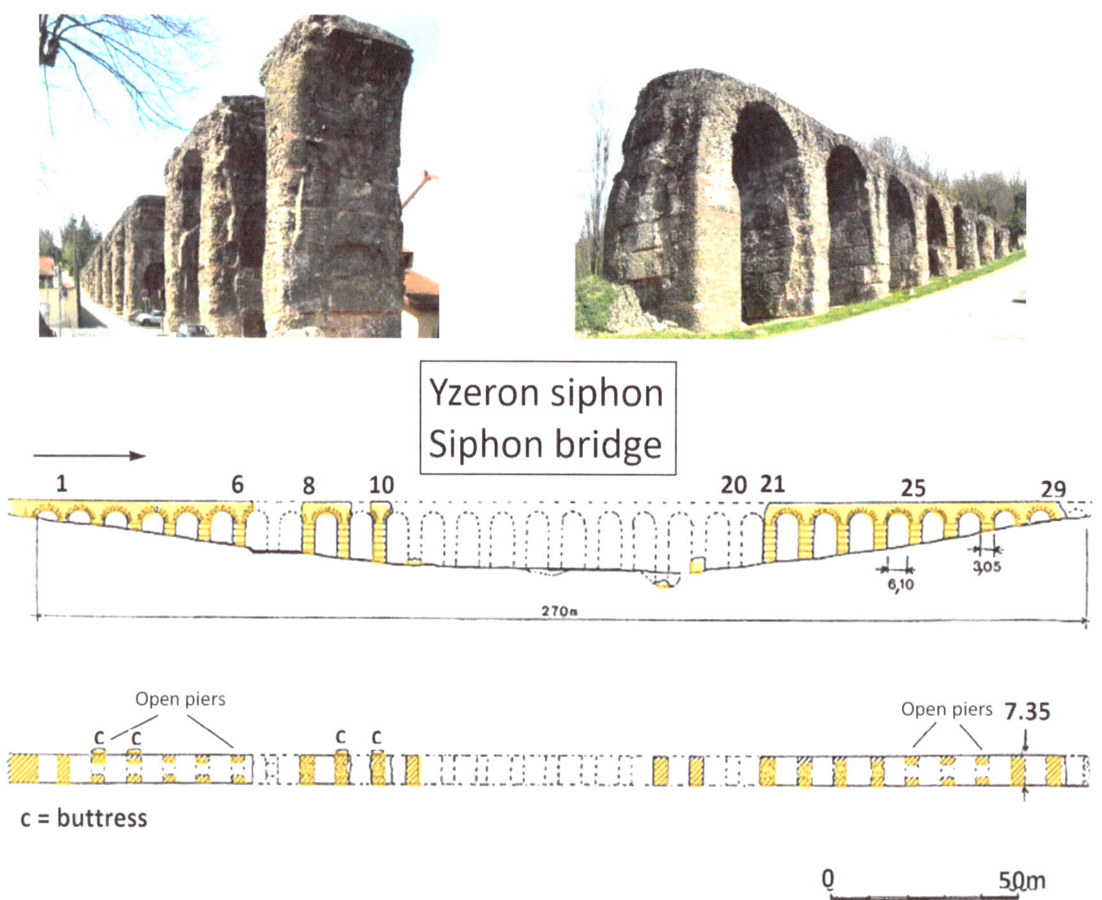

Figure 40. Pont de Beaunant, the siphon bridge of the Yzeron siphon, 16 of 30 arches standing today. Present remains in yellow. Photos above, left: seven arches nine piers, behind the three piers up front is an open pier just visible; photo right: eight arches and nine piers [9], p. 160, adapted.

In 1973, some remains were found of the receiving tank's foundation, measuring 6.75 × 1.80 m, at a level of 305.75 m, including a short section of the departing aqueduct channel 61 cm wide (Figure 42). The remains are no longer visible.

Figure 41. The Beaunant siphon bridge. Aerial photo of 1950 [9], p. 159.

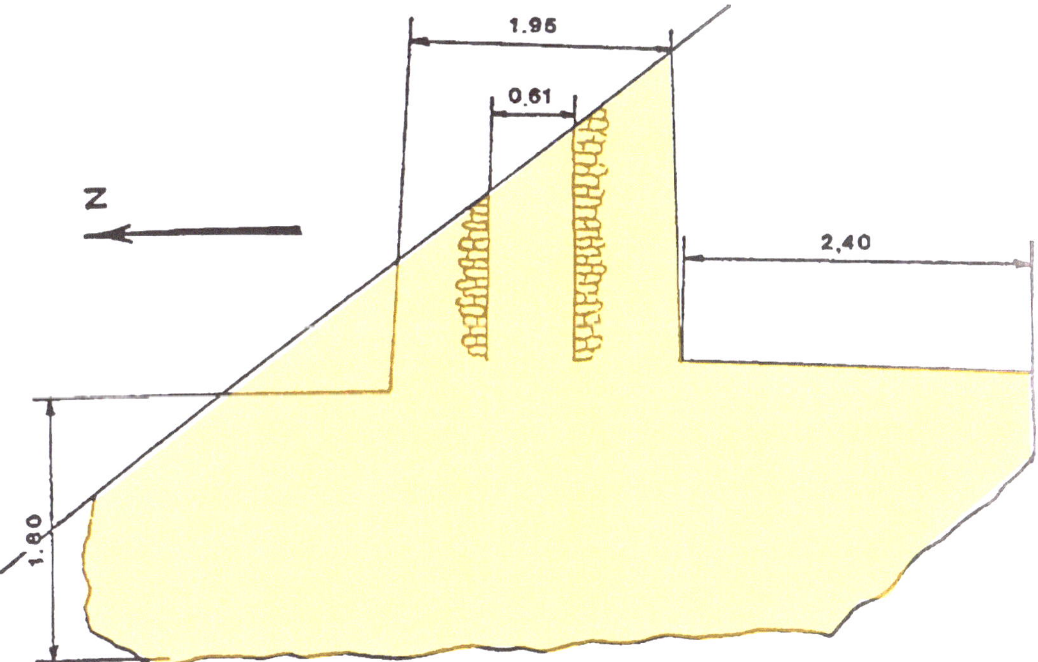

Figure 42. Remains of the receiving tank of the Yzeron siphon, measured in m [9], p. 159.

From the receiving tank, the channel of the Gier aqueduct ran in a northerly direction along the steep 100 m high edge of the Saone river on the highest 300+ m levels of the *Plateau Lyonnais*, alternatingly on arches, a bridge, walls, through tunnels, and buried segments (Figures 43 and 44). In this built-up area of Lyon, remains of piers are to be seen on some streets. The channel reaches about 3 km to the St. Irénée city quarter, the location of the header tank of the fourth siphon of the Gier aqueduct, and the Trion siphon.

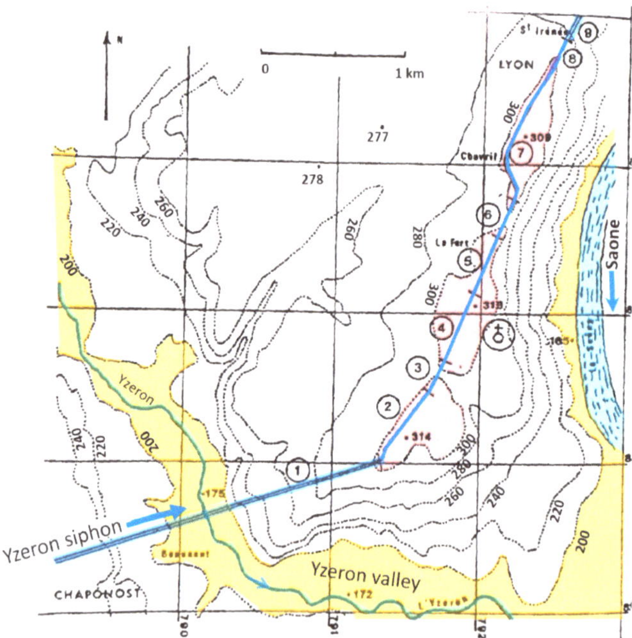

Figure 43. Plateau Lyonnais. Valley of the Yzeron river, level below 200 m. highest areas, over 300 m. 1: uphill part of Yzeron siphon towards receiving tank; 2, 5, and 7: buried channel; 4: tunnel; 3, 6, and 8: bridge/channel on arches; 9: header tank of Trion siphon [9], p. 96, adapted.

Figure 44. Piers of aqueduct bridge 160 m long on the Plateau Lyonnais, rue de Narcel (3 in Figure 41) (http://archeolyon.araire.org/AqueducsLyon/Gier/Galbum10.html, accessed on 5 April 2024).

Towards the north from *St. Irénée* (top of Figure 43), the terrain descends to the depression of *Trion* at a level 265 m. After a stretch of 500 m it goes up again to the Fourvière plateau at a level of almost 300 m (see Figure 5). At *St. Irénée*, the final stretch of the channel, towards the header tank of the siphon crossing the *Trion* depression, it was carried on arches, of which four piers, some 10 m high, still stand, one arch intact (Figures 45 and 46). In later days, the *Trion* header tank functioned as an observation post, using the opening of the arriving channel as an entrance that was accessed with a ladder, and for which windows were cut in the front and side walls of the reservoir (measuring 4.8 × 1.7 m on the inside). The window openings were bricked up again in 1995. Traces of holes in the front wall of the reservoir are indicative for a total of nine down-going lead conduits [9], p. 178.

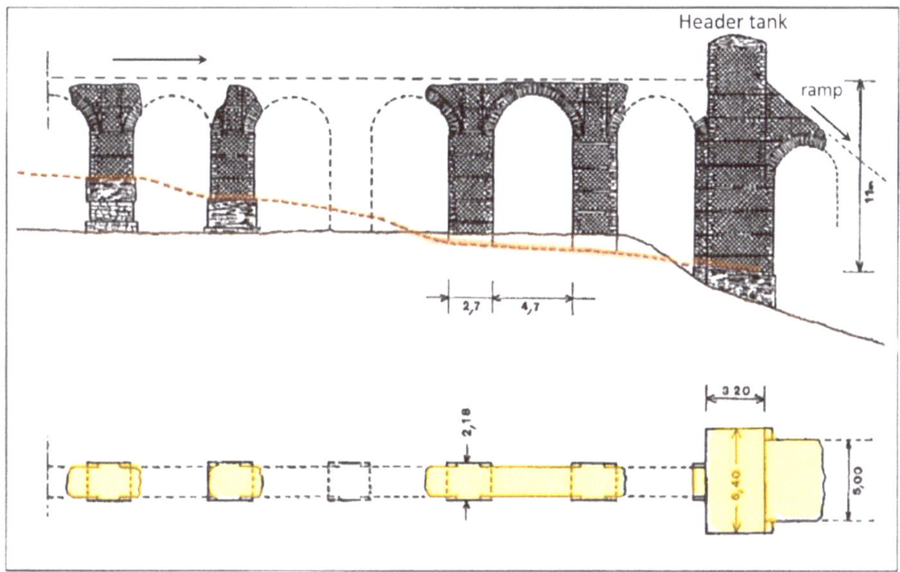

Figure 45. Four piers and one arch upstream of the header tank and remains of the ramp of the Trion siphon. Present ground level indicated, header tank 11 m above ancient ground level (bown line) [9], p. 174, adapted.

As no traces of bridge or piers have been found in the 35 m deep *Trion* depression, nor signs of a creek or river in this densely inhabited Lyon area, it must be assumed that the siphon conduits just ran on the bottom of the depression, to go up to the receiving tank on the Fourvière plateau.

No trace remains of the receiving tank whatsoever, but from 19th century publications it could be decided that its location was on the *Rue Roger Radisson* at a level of 299.6 m. The out-going channel, on arches, was connected to the right-hand wall of the receiving tank at square angles to the arriving lead conduits, presumably to avoid blocking a nearby road. Of this channel on arches, the location of 10 piers is known, of which 7 are still standing along the *Rue Roger Radisson*, 3.1 to 4.5 m wide, interdistance 5 to 7 m, with 3 not to the full height of 6 m. Interestingly, the course of the channel here makes two square angles, counting from the receiving tank at pier no. 7 to the left and at pier no. 8 to the right, resulting in a shift of the channel one arch wide (Figures 47 and 48) (such change of direction of the channel of 90 degrees is also know from the Arles aqueduct and from Anamur on the south coast of Turkey. Due to the low flow velocity of the water, neither the capacity of the aqueduct nor the integrity of the channel were disturbed, although some additional accumulation of calcareous incrustations may occur for high calcium containing water (not so for the acidic Gier waters)).

Figure 46. Remains of header tank and downward ramp of the Trion siphon (http://archeolyon.araire.org/AqueducsLyon/Gier/Galbum11.html, accessed on 5 April 2024).

Figure 47. Two piers 6 m high along the Rue Roger Radisson, with remains of two arches at 90 degrees on the left pier. To the right remains of a third pier. (http://archeolyon.araire.org/AqueducsLyon/Gier/Galbum11.html, accessed on 5 April 2024).

From the 10th pier onward the channel could not be traced. It must have ended at a *castellum divisorium* or decanted in some reservoir like the 1500 cubic meter one behind the theatre, with a floor level at 288 m (Figure 49).

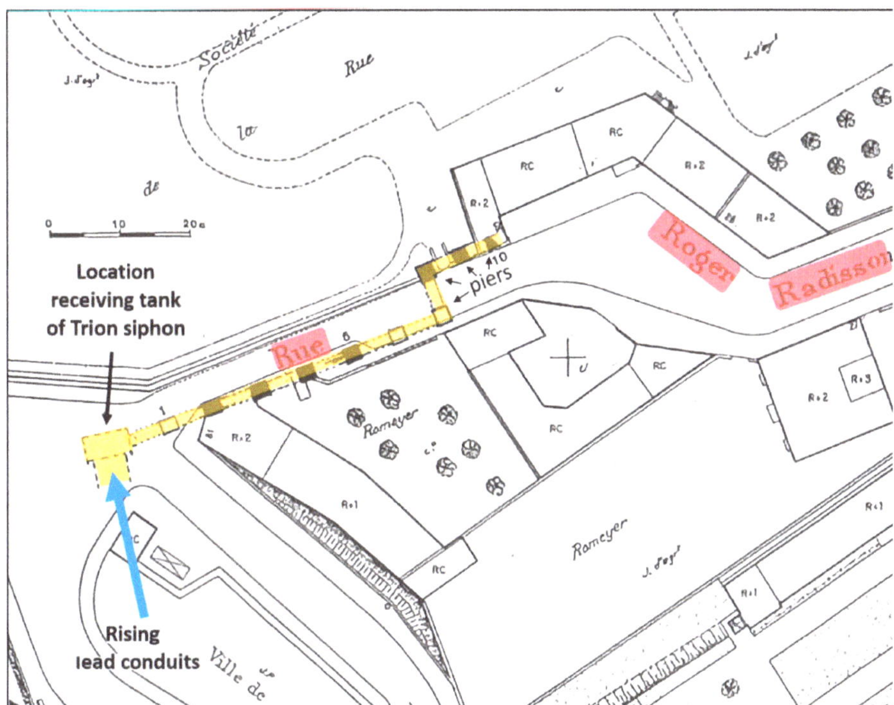

Figure 48. Location of receiving tank of the Trion siphon and 10 piers of channel on arches along the Rue Roger Radisson at Lyon Fourvière [9], p. 180, adapted.

Figure 49. Fourvière. Water reservoirs behind the Roman theatre. See also Figure 5. (http://archeolyon.araire.org/AqueducsLyon/Gier/Galbum11.html, accessed on 6 April 2024).

The overall length of the Gier aqueduct amounts to 74.5 km, and with the Chagnon detour it is 86 km. The slope of the channel varies between 0.5 and 1.2 m/km (Figure 50).

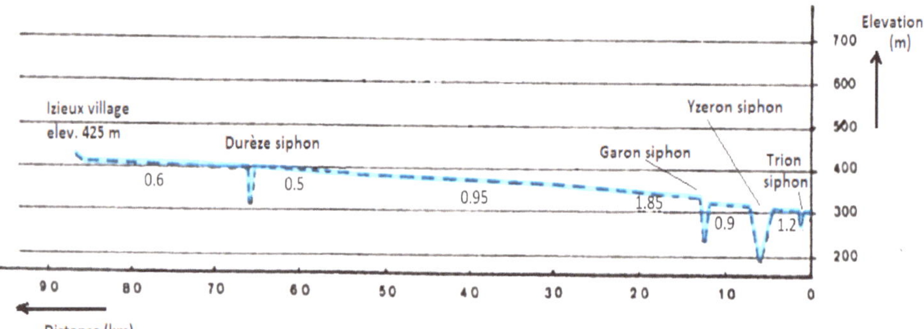

Figure 50. Gier aqueduct, distance and elevation. Length 86 km including the Chagnon detour. The numbers represent the slope of the channel in m/km [2], p. 33, adapted.

The enormous amounts of lead for the conduits of the siphons of the Gier and the other three aqueducts, as well as the choice of their size will be discussed below.

10. The Brevenne Aqueduct

The Lyon Mountains (*les Monts du Lyonnais*) range up to 933 m high and form Lyon's western skyline at 15 km from the city. The 20 km long southwest to northeast-oriented mountain range is separated by the deep valley of the Brévenne river from the Tarare mountains further west. To the east of the Brevenne river, the *Plateau Lyonnais* expands with the Yzeron river and its many tributaries.

The '*Aqueduc de la Brévenne*', probably constructed mid-1st century CE under Claudius. [2], p. 28, tapped springs at the elevation of 630 m on the right bank of the *Orjolle* stream down from *Avieze* village, springs that today supply the *Sainte-Foy-l'Argentière* village 3 km to the east (Figure 51). Of the antique water collecting provisions no trace has remained.The channel itself was found last century several hundred meters down the valley at 626.45 m. The channel down from the *Orjolle* springs near *Aveize* and further towards *Courzieu* is on the inside 54 cm wide and 150 cm high on to top of the arched vault (Figures 52 and 53. Following the contour lines in the mountainous terrain, the winding route of the channel is at times interrupted by a tunnel and by a number of drop shafts (Figure 54).

Near *Courzieu/La Verrière*, the channel descended 44 m from a level of just over 610 m to a level of 570 m, over a distance of 200 m. *La Verrière* is known for its abundant springs that today provide *Courzieu* with water. Downstream of the drop shafts, at *Sottizon*, the channel becomes wider and larger (Figures 55 and 56). This channel to Lyon may have been constructed initially, tapping the springs at *La Verrière*, while in later times the *Orjolle* waters 17 km to the south were found suitable to be added, requiring a channel of smaller dimensions.

As the channel had to descend to the level of the Forvière hill at 300 m, sections with drop shafts were incorporated in the aqueduct's course. A drop shaft consists of a vertical shaft several meters deep, into which the water falls from the channel into an ending at the shaft that is formed like a basin, and from which the water leaves the shaft again into the channelconnected to the lower end of the shaft. The purpose of (a series of) drop shafts is to reduce water velocity while substantially lowering the water level within a short distance. The departing channel is positioned above the bottom of the drop shaft so that the water falls into a water-filled basin thereby functioning as an energy dissipator preventing damage to shaft and channel (Figure 57).

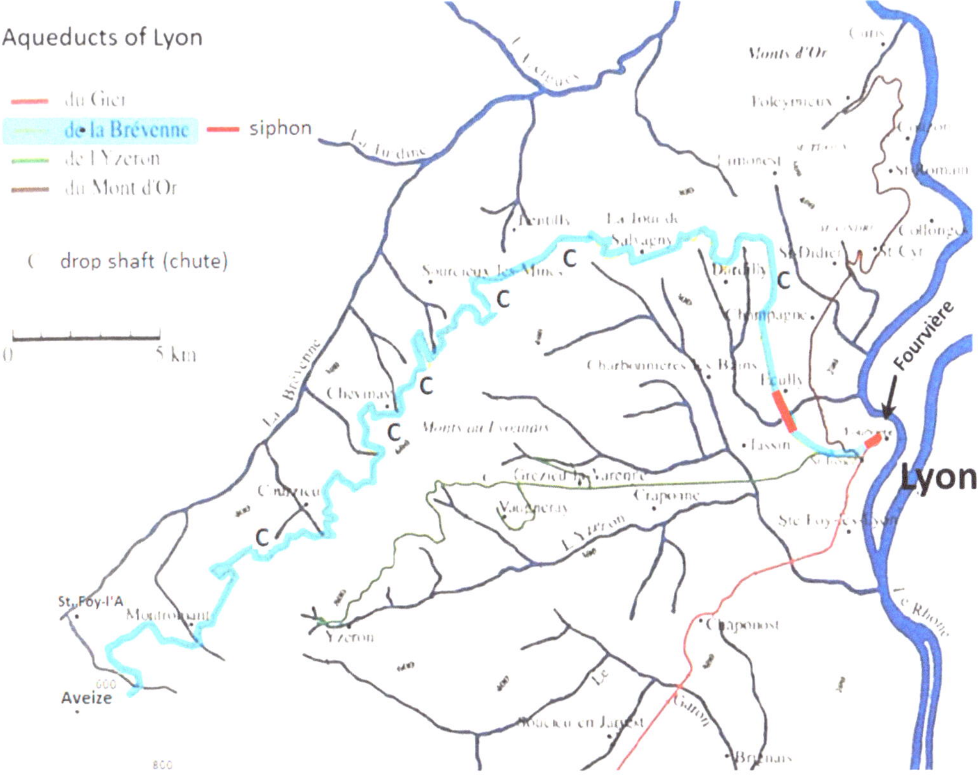

Figure 51. The 1st c. CE Brevenne aqueduct highlighted in blue, its two siphons in red. Taking its waters from a tributary of the Brevenne river at a level of 627 m near Aveizze village, it ran mainly underground in a trench. The 70 km long channel ended at 284 m close to Fourvièrre, with a number of drop shafts to overcome the steep descent of some 345 m (C on the map) [14], p. 64a, adapted.

Figure 52. Aqueduct channel downstream of the Orjolle springs, near Montromant (http://archeolyon.araire.org/AqueducsLyon/Brevenne/BalbumG.html, accessed on 10 April 2024).

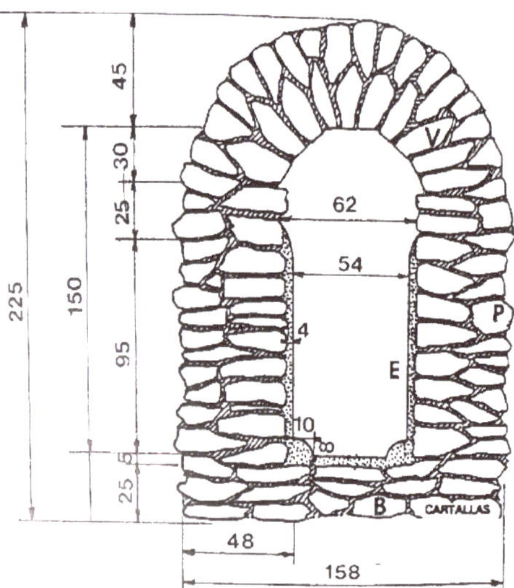

Figure 53. Channel of the Brevenne aqueduct near Courzieu, dimensions [5], p. 134.

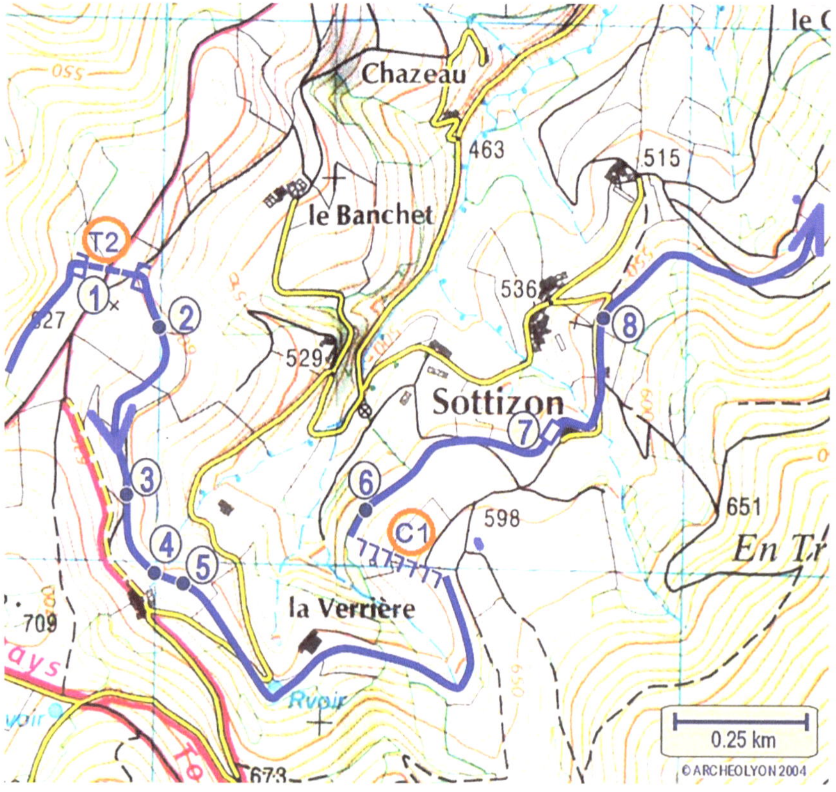

Figure 54. Winding channel route along contour lines near Courzieu/La Verrrière, with tunnel (T2) and drop shafts (C1). With the drop shafts at La Verrière (at 17.5 km from the Orjolle springs), the channel dropped 44 m over a distance of 200 m. The numbers 1 to 8 represent spots where the channel was located. (http://archeolyon.araire.org/AqueducsLyon/Brevenne/BcarteF.html, accessed on 10 April 2024).

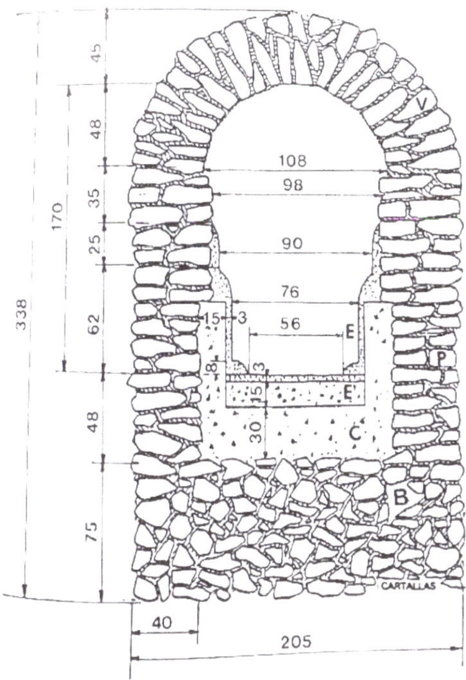

Figure 55. Dimensions of Brévenne channel downstream of Sottizon, measures in cm [5], p. 135.

Figure 56. Brevenne channel downstream of Sottizon (http://archeolyon.araire.org/AqueducsLyon/Brevenne/BalbumG.html, accessed on 10 April 2024).

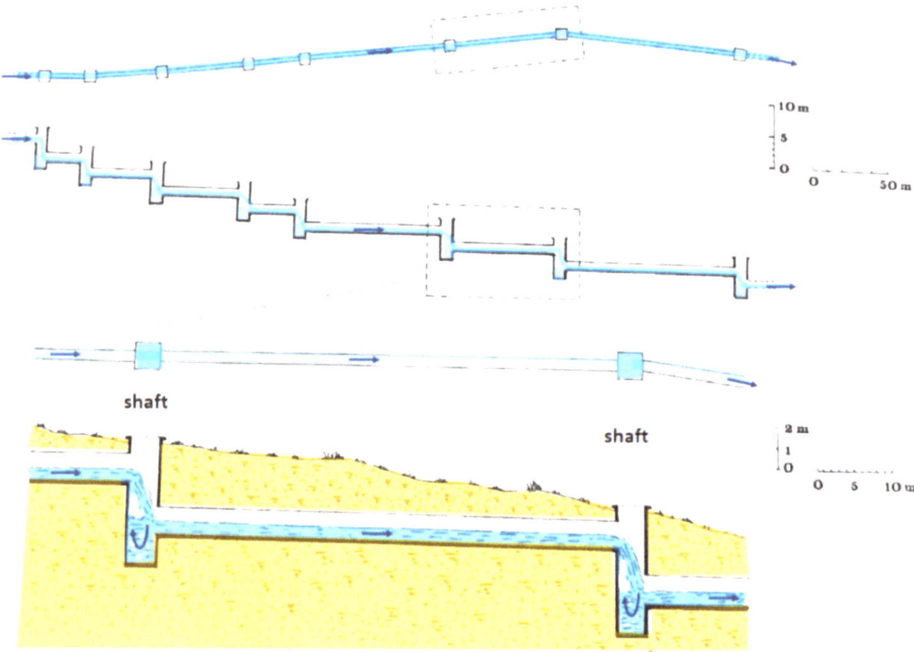

Figure 57. Technique of drop shafts. The channel ends at a vertical shaft, in which the water falls down to its water filled bottom. The connecting channel starts above the floor of the shaft and runs at low gradient towards the next drop shaft [2], p. 34.

The Brévenne aqueduct had 6 drop shaft sections, of which the one near *Chevinay* had a drop of 87 m over a distance of 300 m (C2 in Figure 58). The total drop in elevation by the six sections amounts to 240 m, which is over 70% of the difference in elevation between the Orjolle springs and Fourvière (Table 1 and Figure 59). The four tunnels, their existence assumed from landscape arguments, had modest lengths varying from 150 m to 80 m, and depths between 12 and 16 m (Table 2).

Table 1. Drop shaft sections C1 to C6 in the Brévenne aqueduct [5], p. 162.

Drop Shafts	Length (m)	Elevation Drop (m)
C1	200	44
C2	300	87
C3	150	30
C4	200	38
C5	400	33
C6	?	8
Total	1250+	240

Just nine inspection shafts, of which four of rectangular or square shape (2 × 3 feet and 3 × 3 feet) are well preserved; the remaining ones are in bad condition and have been localised for the Brévenne aqueduct, mainly in the region around *Courzieu*. This in contrast to the 63 + inspection shafts found for the Gier aqueduct [5], p. 148 to 151.

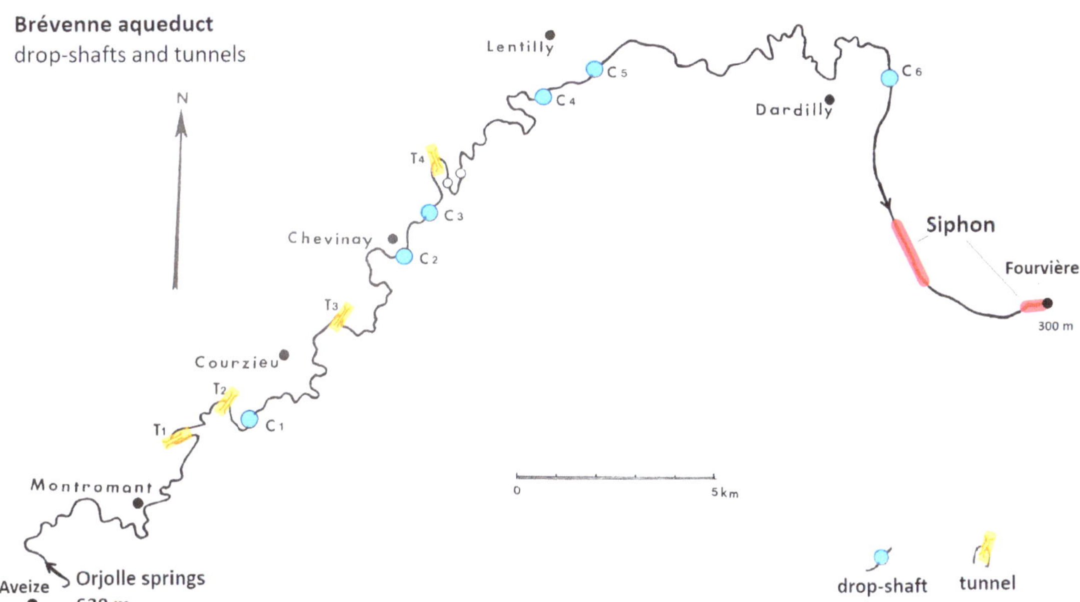

Figure 58. Route of Brévenne aqueduct, with six drop shaft sections C1 to C6, and four tunnels T1 to T4. Length of aqueduct 70 km, distance as the crow flies between Orjolle springs and Fourvière 26 km [5], p. 146, adapted.

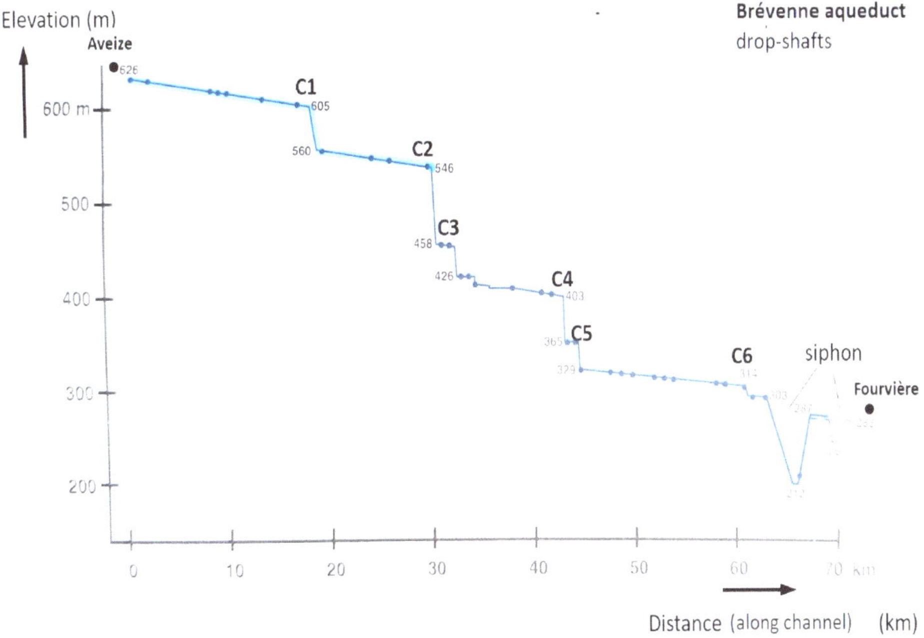

Figure 59. Profile of the Brévenne aqueduct. The six drop shaft sections account for a drop of elevation of 240 m of the total of 330 m between the Orjole springs neat Aveize at 630 m and the Fourvière hill at 300 m [5], p. 181, adapted.

Table 2. Tunnels T1 to T4 in the Brévenne aqueduct [5], p. 145.

Tunnels	Length (m)	Depth (m)
T1	150	16
T2	80	12
T3	100	12
T4	100	12

11. The Great Siphon of Ecully-Tassin

The final section of the aqueduct required the passage of a 110 m deep and 3500 m wide valley, *'le Vallon des Planches'*, which was crossed by a siphon. Of this siphon, few of the enormous piers of the 35 arches siphon bridge called the *Pont des Planches*, are in ruins, but once 22 m high, some have survived, as well as the ramp up to the non-surviving receiving tank (Figures 60–63).

Figure 60. Remains of piers of the 22 m high, 9.20 m wide, and 250–300 m long siphon bridge of the siphon crossing the Vallon des Planches [5], p. 121.

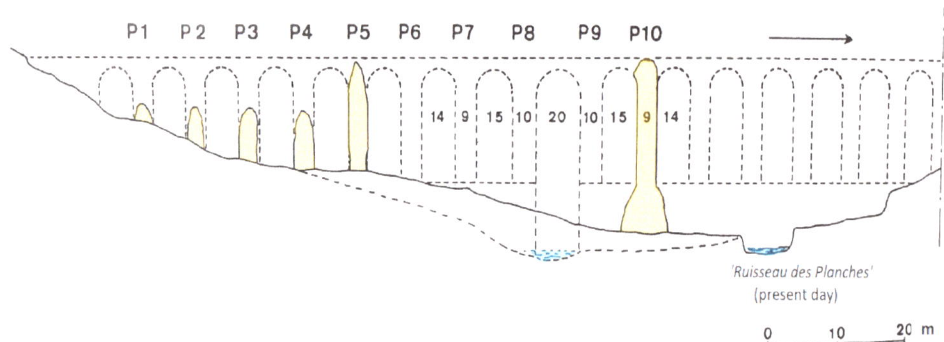

Figure 61. Reconstruction of upstream section of the 'Pont des Planches' proposed by Jean Burdy. The bed of the 'Ruisseau des Planches' is assumed to have moved in time. P1–P10: piers no. 1 to 10. Present remains in yellow. Numbers represent measures in Roman feet for width of arches and of piers [5], p. 172, adapted.

Figure 62. Ramp of the receiving tank of the siphon crossing the Vallon des Planches, photo second half 20th century [5], p. 123.

Figure 63. As Figure 10, present state. (http://archeolyon.araire.org/AqueducsLyon/Brevenne/BalbumV.html, accessed on 12 April 2024).

Taking into account historic reports on the siphon bridge from among others (*Flacheron* and *de Montauzan*) while meticulously analyzing the present remains of the piers and comparing with bridges of the Gier siphon, Jean Burdy proposed a reconstruction of the upstream section of 15 piers of the bridge (Figure 59) [5], pp. 166–174.

Of the receiving tank and its ramp, the latter has survived, but not the tank itself (Figures 62 and 63). Because the receiving tank has no remains that have survived, its reconstruction is speculative to some degree, not the least of the level of the tank where the pipes of the pressure system ended.

Comparing with similar situations for the Gier siphons where the receiving tank is not constructed on top of an arch of the ramp but on a foundation/pier of its own, a reconstruction of the receiving tank at a level of 288 m appears most probable (Figure 64).

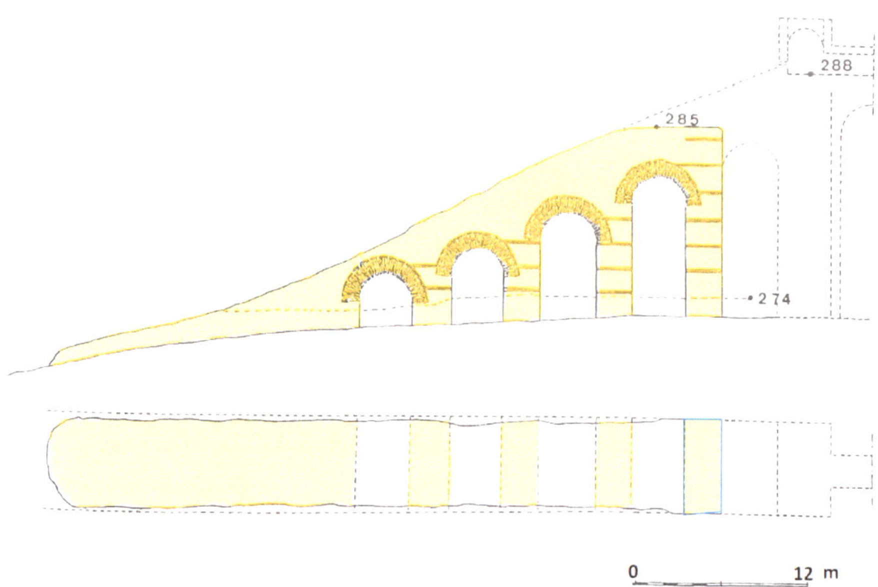

Figure 64. Reconstruction of the receiving tank and ramp, present remains in yellow. The floor level of the receiving tank is estimated at 288 m. Width of the ramp 6.3 m. Arches 12 Roman feet wide, piers 8 feet [5], p. 175.

From topographical arguments, the header tank is envisaged at a level of 302 m, the difference in level between header and receiving tank being 14 m. The siphon itself appears asymmetric, with the siphon bridge located just 800 m from the receiving tank (Figure 65). The long 2400 m upstream section shows today a minor depression at a level of 241 m rising up to 244 before it goes steeply down to the siphon bridge (Figure 65). This intermediate 'high point' in the line would interfere with the water flow because of an air pocket forming at the high point, so to avoid this, the pipes may be assumed to have been mounted in this area on a wall and then in a trench to guarantee a continuously downward slope of the conduits towards the siphon bridge [5], p. 178. Today this area is part of Lyon's suburbs *Ecully* and *Tassin-la-Demi-Lune* where, due to the modern infrastructure, no signs of such provisions have been found.

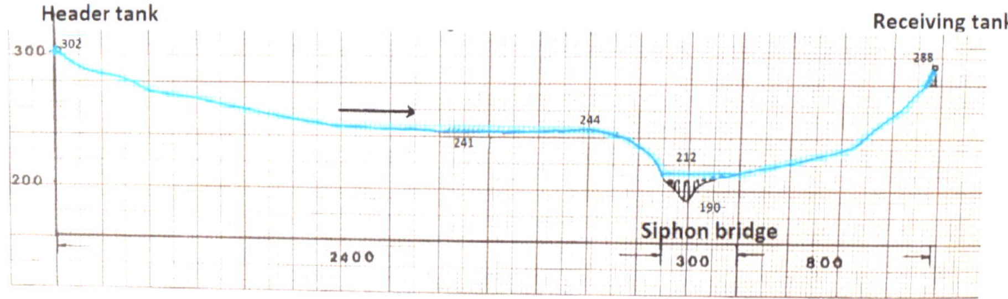

Figure 65. The 3500 m Ecully–Tassin siphon with the 22 m high siphon bridge 'Pont des Planches' ending at 800 m from the receiving tank. In the present-day terrain, the depression in the 2400 m section at 241 m followed by a rise to 244 m should have been levelled out in Roman days by wall and trench to guarantee proper functioning of the siphon [5], p. 178.

The width of 6.27 m of the ascending ramp to the receiving tank of the Great *Ecully–Tassin* siphon would allow for 13 or 14 lead conduits of 25 cm outer diameter. The theoretical capacity of the siphon with pipes of 20 cm inner diameter is estimated as 26,000 to 28,000 m^3 a day, which is, in view of available springs, channel slopes and channel dimensions, as analyzed by Burdy, reduced to an effective amount of 10,000 m^3/24 h [5], p. 185–187.

From the receiving tank onwards, the channel was carried on arches about 12 to 14 m above ground level along the *Rue des Aqueducs* in the *Massues* quarter, where the foundations (2.5 m wide, interdistance 3.65 m) of nine consecutive piers have been preserved. It is assumed that the channel at a level of 286 m was carried for 1200 m on some 300 of such arches and subsequently for another 300 m buried underground before reaching *St. Irenée* and the *Trion* depression. The 500 m wide Trion depression, lowest level at 265 m, was in its turn crossed by again a siphon towards the Fourvière hill (although a 20 m high bridge may have been possible, but no remains of piers have been found), while no bridge for a siphon was required. Header tank nor receiving tank have been preserved, and the course of the channel on the Fourvière hill is unknow [5], pp. 156–160, 185–187.

12. The Yzeron Aqueduct

Of the Lyon aqueducts, the one taking its waters near the village of *Yzeron* that gave the aqueduct its name is complex and less well known. [2], p. 21 It is dated supposedly during the reign of Augustus 27 BCE–14 CE as the aqueduct of Mont d'Or that was first to be constructed. Just west of the *Yzeron* village, springs in up to five valleys down from the tops of the *Monts du Lyonnais* which form Lyon's western skyline feed the Yzeron river (Figure 66).

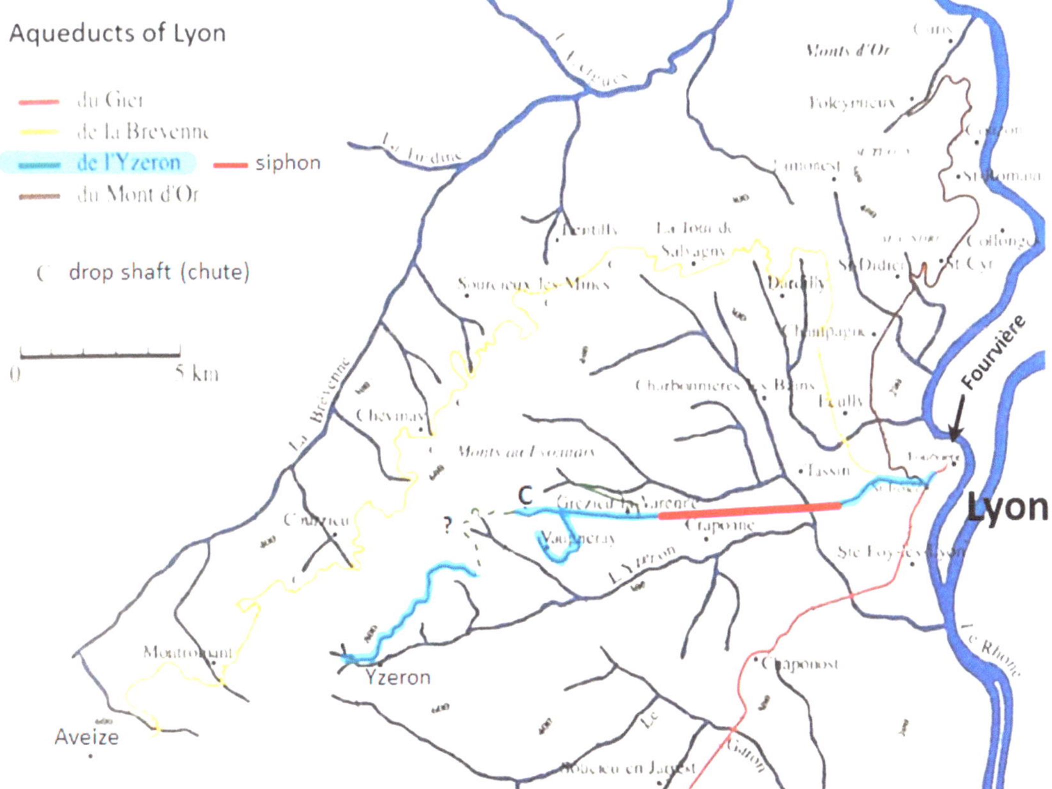

Figure 66. The Yzeron aqueduct, running east towards Fourvière. C = section of dropshafts. ? = uncomfirmed channel section [14], p. 64a, adapted.

The same springs (at an elevation of 750 to 720 m) have since long supplied local settlements as they must have also fed the aqueduct, yet no structure to this end was located. At about 1 km downhill, remains of a 40 cm wide channel have been found over a length of 365 m now called *le canal de Monteroux* (Figure 67) [8], pp. 35–38.

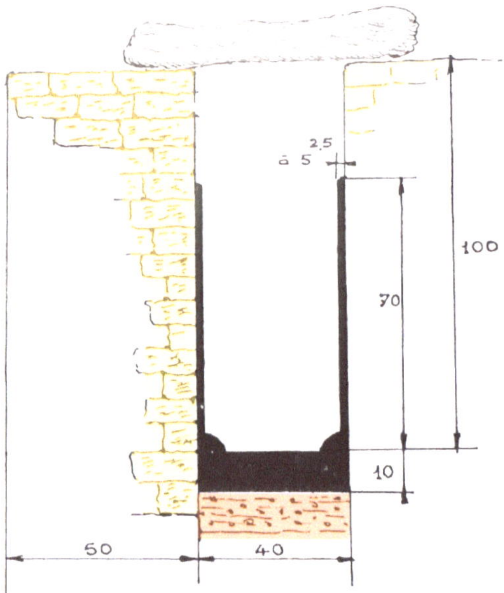

Figure 67. Canal de Monteroux, just west of Yzeron village. Measures in cm [8], p. 40.

A section of a second channel (*'le canal superieur'*) was found just along the left side of the Yzeron river bed, and it may just be that this channel also fed the aqueduct, both *le canal de Monteroux* and *le canal superieur* joining to a single channel at an unknown location uphill from Yzeron village, probably somewhere beneath the local road.

Downhill from the village to the northeast, the channel could be traced by the Burdy team at several locations, some several hundred meters apart at a settlement called *Soupat*, elevation 670 m, where in 1981 roadworks by bulldozer revealed 65 m of channel, with remnants of clay pipes inside the channel. The pipes measured some 60 cm in length and 24 cm outer/15 to 17 cm inner diameter, that were also found in this channel at several more locations (Figures 68 and 69), possibly installed to reestablish water flow after a period of neglect of the channel [8], p. 48.

To the east, at 6.7 km from the start of *le canal de Monteroux,* near the wall of a cemetery of the settlement *la Milonière,* at 655 m an section of the channel was found again, also with fragments of ceramic pipes. Further east and north no traces have been found along and around the contour lines of the terrain. A hypothetical, but not proven 4000 m long system of drop shafts may have connected to drop shafts found between elevations 506 to 458 m at *Recret* village (C in Figure 66 north of *Vaugneray* village) from where a channel departed to the east towards Lyon. Such an enormous drop shaft system could explain the fall in elevation from *la Milonière* at 655 m to 506 m at *Recret* village and onwards to 458 m, although a winding detour of the channel in the hills between *la Milonière* and *Recret* cannot be excluded (see map of Figure 66, dotted line), while it may even be possible that there was no connection at all between *la Milonière* and the drop shafts found at *Recret*. The problem remains unresolved.

Figure 68. Clay pipe fragments found in channel [8], p. 48.

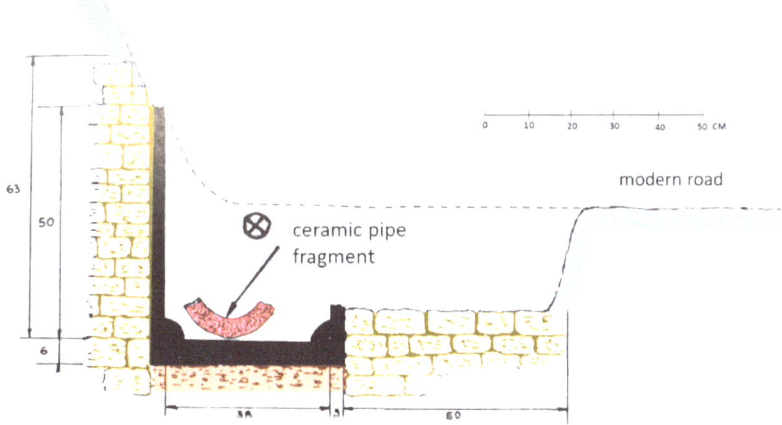

Figure 69. Channel at Soupat found alongside of modern road, with fragment of clay pipe. Measures in cm [8], p. 43, adapted.

Where the channel or channels that fed the upstream end of the section of drop shafts (a 'hydraulic stairway') at the *Recret* settlement took their water could not be established, although a number of springs are known in the hills to the north and west. Two rather intact drop shafts have been found at *Recret* 490 m apart with a level difference of 38 m (Figure 70) [17], pp. 29–44, adapted. For a drop of about 2.5 m per shaft and accounting for a shallow slope of the connecting channels, another 11 shafts may be imagined along the 490 m stretch. The terrain east of the shaft found at the lowest elevation goes down increasingly less steep towards *Grezieux-la-Varenne*, which would mean that the hydraulic stairway may have been extended some way further having longer connecting channel [17], p. 81.

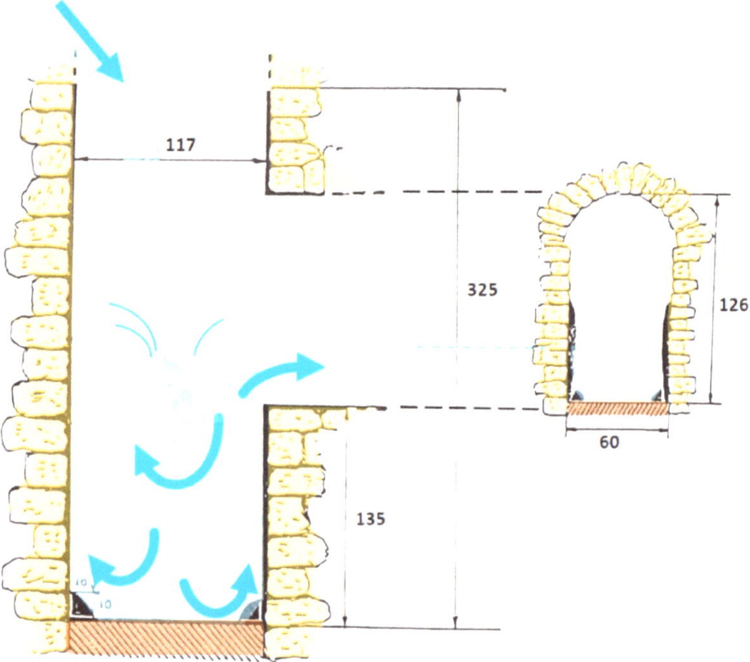

Figure 70. Downstream drop shaft of the two drop shafts found at Recret. The water fell down 2.5 m from the upper channel (not shown) into the 1.35 m deep bottom part of the shaft (117 × 117 cm), that acted as an impact cushion reducing the flow velocity of the water in the connecting channel. It must have been quite noisy in there. Measures in cm [8], p. 77, adapted.

A second channel contributing to the aqueduct originates northwest of *Vaugneray* village to the south of *Recret*, the '*branche de Vaugneray*'. During road constructions west of Vaugneray the channel came to light, showing a peculiar two floor levels 40 cm one above the other (Figure 71). In *Vaugneray* village itself, the channel passes underneath the church, where a hydraulic stairway is envisaged from level (elevation) 435 to level 420 m over a length of 250 m, with seven drops of some 2.5 m about 40 m apart [8], p. 67. About 200 m downstream of the envisaged drop shaft section the channel was traced, again showing this peculiar heightening of its floor, now by 24–28 cm [8], p. 65. Constructing a second and a higher floor level in the channel by several tens of cm while lowering the level of the channel by 15 m with drop shafts seems contradictory and remains unexplained.

After turning north, the *Vaugneray* branch joined the channel coming from *Recret*. Total length of the *Vaugneray* branch amounts to 3900 m [8], p. 68.

After the *Vaugneray* branch joined this '*Grézieu* branch', the channel ran east passing the village of *Grézieu-la-Varenne* towards an isolated hill at the area called *le Tupinier*, to feed the header tank of the siphon of the Yzeron aqueduct at a point where the terrain goes down. The header tank, no trace remaining, is thought to have been located at an elevation of 313 m [8], p. 83. The top of the hill is located at 311 m. Assuming that a header tank here would be of similar shape as the header tanks of the siphons of the Gier and the Brevenne aqueducts, Burdy assumes an extra 2 m for the level of this header tank. The total length of the Grézieu branch from Recret to le Tupinier is estimated at a bit more than 5 km.

Figure 71. Channel of the Vaugneray branch west of the village, with two channel floors [8], p. 64.

13. The Siphon of the Yzeron Aqueduct

The terrain extending towards Lyon east of *le Tupinier* goes down some 30 m to a 2 km wide plane at 280 m elevation, located between the deep valleys of the Yzeron river to the south and of the Ratier stream to the north. The terrain remounts again near the settlement of *Craponne* to a maximum elevation of 290 m. Further east the terrain again loses height rapidly over a distance of about another 2 km to the valley where the Charbonnieres stream flows at a low 195 m elevation to join the Yzeron river to the south. Then, after some 500 m, the terrain rises steeply up again to an elevation of 272 m at Fourvière (Figure 72).

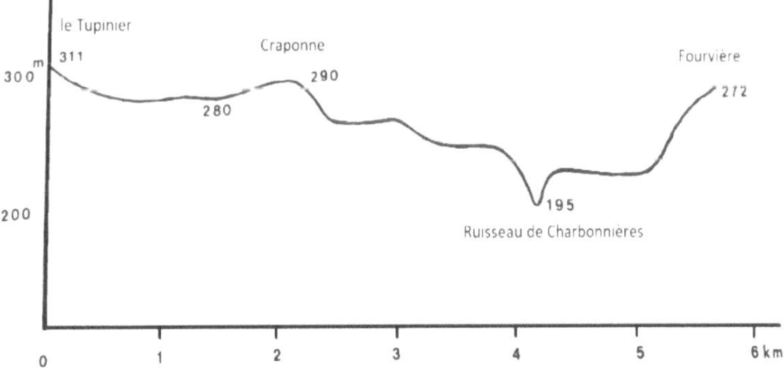

Figure 72. Profile of the terrain east of le Tupinier [8], p. 88, adapted.

On the 290 m top near *Craponne*, two 12 m high square piers stand, measuring 4.40 × 4.40 m and 4.40 × 3.77 m, 2.80 m apart (Figures 73 and 74). The two piers are since long known as '*les Tourillons de Craponne*'. First mention dates from 1599 as 'La tourray' on a medieval map [8], p. 3. The piers were restored in 2019 to prevent further degradation, during which ooccasion a study of the structure of this construction was carried out [18].

Figure 73. 'Les Tourillons de Craponne' north view early 1900 (postcard), early 2000 (author's photo), 2009 (photo by C. Frangin).

Figure 74. Les Tourillons, November 2023, after restorations. Direction Lyon is to the left. (author's photo).

Some tens of meters further east, two more remains of piers as well as of a wall rise up 1–2 m from the terrain, all arranged in an east–west line (Figure 75). The left pier in Figure 73 shows remnants of an arch on either side, but the reason for this remains unclear. Various authors presented various explanations for these massive structures, from a monumental gate to an encampment built by Julius Caesar or Marc Antony, or a triumphal arch, to a bridge and even a quay on an ancient course of the Saone river [8], p. 89.

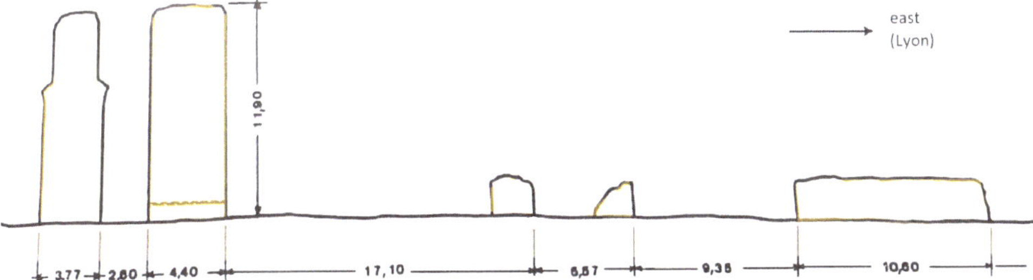

Figure 75. Present remains of 'les Tourillons' with additional remnants of two piers and a wall section. The left pier shows on either side remains of an arch at about half height [8], p. 91.

Towards the end of the 19th century, the idea that the piers were part of an aqueduct, or rather of a siphon equipped with a tower with an open reservoir on top to release air ('*ventouse*') became increasingly in vogue, supported by an 1862 report of the *XXIX Congrès archéologique De France* at Lyon of a find 60 years before of a great number lead pipes east of the piers that the local farmers had melted down on the spot into ingots [8], p. 21. Moreover, in the low plane between le Tupinier and Craponne, no remains of piers had been found, so that the continuation of the aqueduct channel at le Tupinier on a bridge towards les Tourillons was not deemed probable, a bridge that had to be over 2 km long and 30 m high.

Various reconstructions were proposed, among others by Germain de Montauzan with sloping ramps on both sides, supported by further remains at ground level west and east of the piers, carrying lead conduits up to an open reservoir on top [8], pp. 92–96.

The most recent reconstruction is by Burdy, with an open reservoir on top of the 4.40 × 4.40 m pier at an elevation of 316 m (Figure 76).

Figure 76. Reconstruction by Burdy, with open reservoir/tank on top of the largest pier [3], p. 138.

Once the idea of a siphon was accepted, the over 3 km long stretch towards Fourvière, with the *Ruisseau de Charbonnières* flowing at 195 m elevation, had to be covered as well. In the 1880s during the construction of a new traffic bridge (*Pont d'Alaï*), massive foundations of a Roman structure including large limestone blocks with connecting iron cramps sealed with lead and parts of arches came to light, which were interpreted as the remains of the bridge on behalf of the siphon crossing the *Ruisseau de Charbonnières*. The findings were confirmed again in 1969 during works in relation with a large sewer system beneath the *Pont d'Alaï*. The bridge is thought to have been 20 m high, the siphon passing the *Ruisseau de Charbonnières* at an elevation of 215 m (Figures 77 and 78) [8], pp. 97–99, adapted.

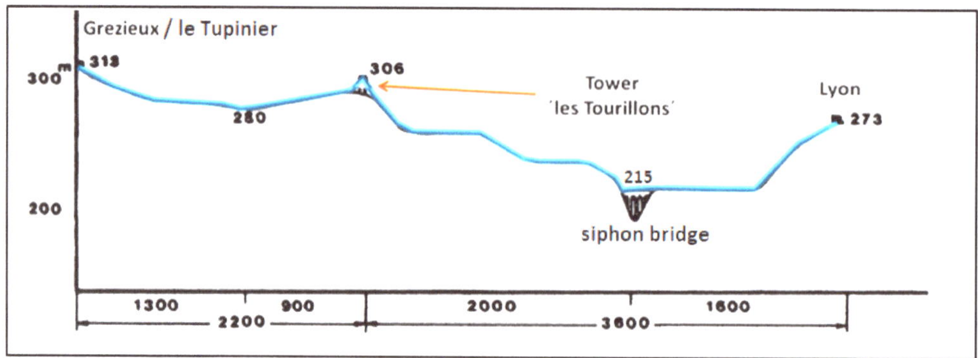

Figure 77. Profile of the double siphon of the aqueduct of Yzeron [8], p. 101, adapted.

Figure 78. Foundation bloc of the siphon bridge that crossed the Ruisseau de Charbonnières near the modern Pont d'Alaï, Lyon (http://archeolyon.araire.org/AqueducsLyon/Yzeron/YalbumK.html#, acessed on 15 April 2024).

After passing the *Pont d'Alaï*, the siphon, first turning a bit to the right and proceeding horizontally for some 600 m, had to go up the steep east side of the valley of the *Ruisseau de Charbonnières* towards Lyon, where the receiving tank is envisaged at an elevation of 273 m (see below). With the open reservoir on top of the tower at Craponne, the siphon is split up into two, with two siphons connected end to end, the open tank in between serving both as receiving tank and as header tank, a double siphon. Of the remains of the siphon bridge, little may be seen today, just one block of the foundations near the modern traffic bridge, the *Pont d'Alaï*, on the left side of the *Ruisseau de Charbonnières* (Figure 78).

Until the 1990s, it was unclear where the siphon ended, whether there was a continuing channel, and where it went. However, in June 1990 during ground works on behalf of new edifices, a channel was found at no. 53 of the '*Rue Joliot-Curie*' 120 cm below ground level, of which 20 m was already removed before archaeological inspection could be performed on a subsequent 20 m section, after which this part was destroyed as well with no remains preserved. The 62 cm wide channel had walls 40–45 cm wide, the floor of the channel being at an elevation of 271.8 m. This elevation conforms to the proposed elevation of the receiving tank at 273 m.

The discovery of the channel at the *Rue Joliot-Curie* allowed for a reconstruction of the course of the channel in this area called *'Plat du Jour'* just south-west of the Fourvière hill and the Trion depression where the siphons of the Brevenne and the Gier aqueduct pass to Fourvière. Following the irregularities of the terrain the channel would have a sinuous course and may have crossed the channel of the Mont d'Or aqueduct some 8 m above it, ending just south of the Baths and the Roman Theatre/Odeon at elevation of 268 m (see Figure 5). Length of channel from receiving tank to end point about 3.85 km.

14. Les Tourillons de Craponne

One wonders what the reason was for this 5.8 km long siphon to construct an open tank 2.2 km downstream from the header tank on a tower 16 m high with sloping ramps on either side, and that on top of a hill at an elevation of 290 m?

About 80 siphons are known from classical times, the majority of them of the 'simple siphon' type, as shown in Figure 14.

Few are equipped with a hydraulic tower such as 'Tourillons de Craponne'. These hydraulic towers were often installed on an intermediary hill in the path of the siphon, but there were also siphons with intermediary hills and no hydraulic towers, and siphons with hydraulic towers but no intermediary hills (Table 3). The conduits of the siphons were made of prefabricated lead pipes, connected together on site, with soldered joints as was the case for the siphons of the Lyon aqueducts, or of stone/terracotta pipes with weak male/female joints that had been made watertight with a mixture of quicklime and oil that expands when in contact with water but that has low tensile strength. The hydraulic towers are thought to release air from the closed conduit of the siphon.

Table 3. Siphon typology. Madradag siphon: [19]. Cadiz and Aspendos: [1].

Simple Siphon	The Majority of Classical Siphons
Siphon with intermediary hills but no hydraulic towers	Madradag siphon of Pergamon, length 3250 m (Turkey)
Siphon with intermediary hills + hydraulic tower(s) on top	Les Tourillons de Craponne, Yzeron aqueduct Los Arquillos siphon (3500 m), aqueduct of Cadiz (Spain)
Siphon with hydraulic towers but without intermediary hills	The siphon of Aspendos (1650 m) Aspendos aqueduct (Turkey)

City water distribution systems as in Pompeii and Herculaneum (lead conduits) and also in the east (e.g., Laodikeia, (Turkey, stone and ceramic conduits) applied low pressure siphons which included hydraulic towers to keep the pressure elevated. Compared to siphons in aqueducts, this rather low static pressure was sufficient to serve second stories in houses, as may be seen in Pompeii. This 'Pompeian' type water distribution system with hydraulic towers was common practice in Europe until the late 18th, early 19th century, and even later, as in Palermo, Sicily, until in the 1960 1970s [20].

But how to start a siphon? It is not difficult for a simple siphon: fill it slowly with water from the header tank, and when the water reaches the receiving tank, slowly increase the water flow to the maximum quantity. Filling a siphon with an intermediate hill, a 'high point', is another matter. When the conduit that crosses an intermediate hill is filled with water, an air pocket will form downstream of the hill top because air tends to rise up in water, and as such, the air pocket cannot escape.

In Figure 79, a siphon with two intermediate hills (two high points in the line) is shown schematically. When filling this siphon, air pockets will form downstream of the high points. If the combined vertical heights of the air pockets is greater than the difference in height between header tank and receiving tank, the water will not reach the receiving tank and the water will not flow: an airlock has been created, and the siphon does not start. On the other hand, the air pockets are partly compressed by the water pressure, so that this effect is reduced. In such case, the siphon may start and operate at reduced capacity. When starting at reduced capacity, air entrainment at the header tank may subsequently enlarge the air pocket leading to an incurable standstill.

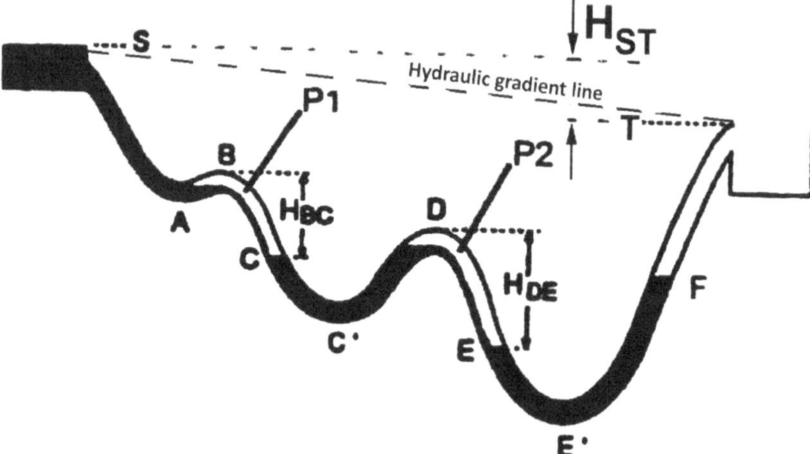

If $H_{BC} + H_{DE} > H_{ST}$ the siphon will not start

Figure 79. Starting a siphon made of a closed conduit with two high points. When filling the siphon air pockets will form at the high points. If H1 + H2 > H the siphon will not start. If H1 + H2 < H the siphon will start at reduced capacity. Because of water pressure, the air pockets may be reduced in size. The problem may be solved by releasing the air, which may be done by constructing an open reservoir at the high points, which, however, must be positioned near the overall hydraulic gradient line (but below the level of the header tank) [1], p. 39.

The 16 m high tower of les Tourillons with the open tank on top to release air must have been built on purpose, but why? The profile of the Yzeron siphon as shown in Figure 70 has a high point at 290 m. Could this high point have been avoided? Figure 80 shows a map with the siphon's course. It may be seen that from the header tank, the siphon extends east between the deep valleys of the Ratier river to the north and the Yzeron river to the south, and finally crossing the Charbonnières stream by means of a 20 m high bridge where today the *Pont d'Alaï* serves the modern traffic as discussed above. The siphon has three horizontal bends (not counting the *les Tourillons* tower at Craponne, which for the lead conduits with strong joints between pipe elements may not be problematic (see for the various aspects of classical siphons and their problems, [1], pp. 12–59.

The tower of *les Tourillons* was built on a top of a hill at 290 m (inset), which must have been deliberately chosen for its construction. Just west of the 290 m high point, a lesser hill at 283 m was avoided by the siphon. The dotted line in the inset represents an alternative course of the siphon which, however, would incorporate this lower hill. The shallow depression downstream of this hill could have been leveled out with a wall to avoid a second high point just downstream. Whatever the case and whatever the course, the Yzeron siphon had to deal with a high point, either at 290 m positioned at 2200 m from the header tank, or at 283 m at 1500 m from the header tank, the plane downstream of the header tank being at a level/elevation of 280 m. This means that in either case, an open tank was obligatory for the siphon to avoid reduced capacity at start up with even the possibility of an incurable standstill due to air entrainment at the header tank (Air entrainment at the header tanks of classical siphons was common due to the insufficient depth of the intake below the water level in the header tank. See [1], p. 47; for the les Tourillons case: [22], in French.

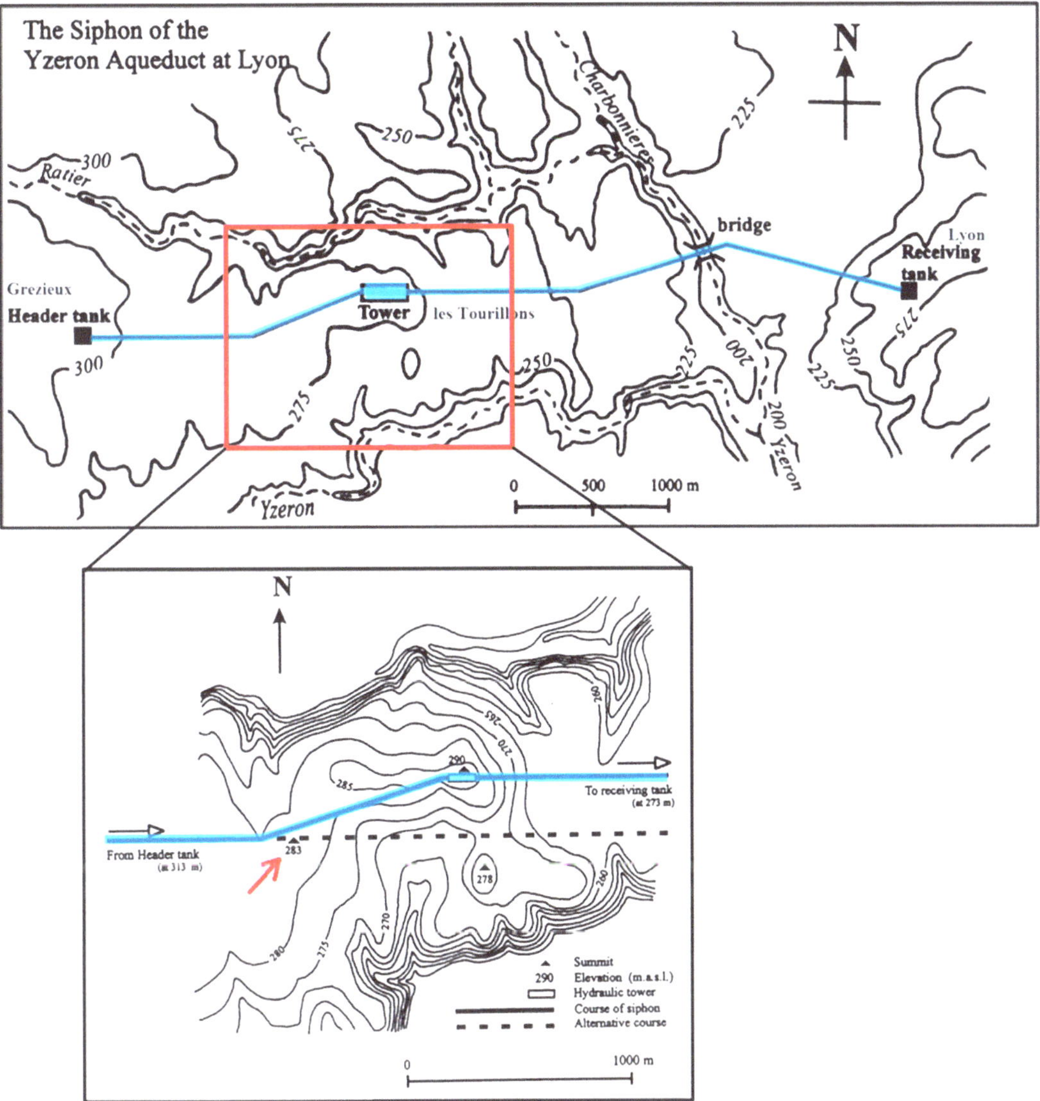

Figure 80. The Grezieux/le Tupinier–Craponne–Lyon siphon of the Yzeron aqueduct, plan. Inset: the 16 m high tower of les Tourillons is located on top of a hill at 290 m. The dotted line represents an alternative course of the siphon, which incorporates a high point at 283 m (red arrow), after [21], p. 269.

So if an open tank at a high point, either at 290 m or at 283 m (the receiving tank being at 273 m, see above) would guarantee the water flow, why built a 16 m high tower with sloping ramps on top of the 290 m hill to have an open tank at an elevation of 306 m? Let us compare the two situations: an open tank at 306 m elevation on top of the 16 m tower, and an open tank 16 m on top of the 283 hill.

The 4.4 × 4.4 m open tank on top of the *les Tourillons* tower is estimated by Burdy to have walls 60–80 cm thick, the reservoir being 2.80–3.20 m wide on the inside. This leaves room for six or seven lead pipes (at the most), pipes with an external diameter of 25 or maybe 23 cm as Jean Burdy proposes (internal diameter 20 cm or 18 cm), similar as for the siphons of the Gier aqueduct. Such a reservoir may also be imagined on top of the 283 m hill. We have to consider two situations, each with two consecutive siphons that differ with respect to the position of the intermediate open tank, and thus with the driving force for each section, assuming for each case lead conduits of similar dimension.

Burdy estimates that the open tank on top of the tower of *les Tourillons* could accommodate six or seven lead pipes of 25 cm outer diameter, and calculated that the total discharge Q, with a level difference ΔH of 7 m between header tank and the *les Tourillons* tank. With 20 cm inner diameter of the pipes, the total discharge Q would be about 13,000 m^3 per day. As a consequence, the 13,000 m^3 per day must be considered the average capacity of the Yzeron aqueduct. Burdy also estimated that because of the greater ΔH of 33 m between the tank of *les Tourillons* and the receiving tank, only four pipes would suffice for the longer stretch of 3600 m, [8], pp. 122–123.

The average velocity V of water in a pipe of length L can be estimated with the Darcy-Weisbach formula (See http://www.pipeflow.co.uk/public/articles/Darcy_Weisbach_Formula, accessed 10 April 2024):

$$V^2 = (8 \cdot g \cdot \Delta H \cdot Rh)/(\lambda \cdot L)$$

with g = 9.81 m/s^2; ΔH = driving force (m); Rh = hydraulic radius, for fully filled pipe Rh = D/4; D = inside diameter of pipe (m); λ = factor related to pipe wall roughness; L = pipe length (m).

For identical pipes, we may write $V^2 = ((8 \cdot g \cdot Rh)/\lambda) \cdot \Delta H/L = C \cdot \Delta H/L$ with C = 8·g·Rh)/λ. This means that the velocity V of the water in identical pipes (C has a fixed value) depends only on the square root of ΔH/L. In the expression V (:) $(\Delta H/L)^{1/2}$ the water transport Q (volume per time) in a pipe of inner cross-section O is V·O. As V (:) $(\Delta H/L)^{1/2}$, then also Q (:) $(\Delta H/L)^{1/2}$. The ΔH between header tank and intermediate open tank for the situation I is larger than for situation II (with *les Tourillons*), see Table 4 above, so that pipes in the upstream section in situation I would transport more water; that is, fewer pipes were needed here for the same discharge of the arriving aqueduct.

Table 4. An obligatory open tank: in situation I on the 283 hill, and in situation II on top of the tower at Craponne. For each situation, the distance between the header tank and the open tank, and the distance between open tank and receiving tank, plus their difference in height ΔH is given.

	Situation I Open Tank at 283 m		Situation II les Tourillons de Craponne	
	L	ΔH	L	ΔH
Section length upstream	1500 m	30 m	2200 m	7 m
Section length downstream	4300 m	10 m	3600 m	33 m

What are the consequences? Any pipe in the 1500 m section towards the open tank at 283 m would transport more water per pipe than in the 2200 m section case of *les Tourillons*, the water transport Q being related to $(\Delta H/L)^{1/2}$ as discussed above. Table 5 gives the value of $(\Delta H/L)^{1/2}$ for each section in both situations.

Table 5. Calculated value of $(\Delta H/L)^{1/2}$ for each section and for both situations.

	Situation I Open Tank at 283 m			Situation II Les Tourillons de Craponne		
	L (m)	ΔH (m)	$(\Delta H/L)^{1/2}$	L (m)	ΔH (m)	$(\Delta H/L)^{1/2}$
Section length upstream	1500	30	0.1414	2200	7	0.0564
Section length downstream	4300	10	0.0482	3600	33	0.0957

Each pipe in in the 1500 m section delivers more water than any pipe in the 2200 m section with *les Tourillons*, that is, 0.1414/0.0564 = 2.507, say, 2.5 times more. Fewer pipes are needed in situation I to have the water, that arrives at the header tank, to be transported to the open tank at 283 m, or 7/2.5 = 2.8, say, three pipes. Moreover, from Table 5, it may be estimated how many pipes would suffice for the 3600 section of situation II: each pipe in this section delivers a factor 0.0957 compared to 0.0564 for the 2200 m section, that is, 0.0957/0.0564 = 1.70 times more water. For the seven pipes of the 2200 m section, only 7/1.70 = 4.12 (say four) pipes are needed for the 3600 m section, conforming to the estimate of Burdy mentioned above.

On the other hand, the downstream section needed in situation I may be estimated as well. The three pipes needed in the upstream section with capacity Q (:) 0.1414 must be compared with the pipes in the downstream section with Q (:) 0.0482, that is, the latter pipes have a capacity of 0.0482/0.1414 = 0.34 times that of the pipes in the upstream section. Thus, for the 2.8 (say, three) pipes in the upstream section there are needed 2.8/0.34 = 8.2, say, eight pipes, in the 4300 m downstream section.

The total length of the lead conduits needed for each situation may now be calculated, see Table 6. It will be seen that situation I, with an open tank on top of a 16 m high tower on top of a 290 m hill, required fewer long conduits than for situation II, with an open tank on top of the 283 m hill. For both situations, an open tank was obligatory for the functioning of the siphon. However, the building of a 16 m high tower on top of the 290 m high hill saved 9 km of lead conduit. While lead metal was an expensive commodity, in the immediate environment sufficient building stones for constructing *les Tourillons* were available.

Table 6. Total length of lead conduits needed for Situation I and Situation II.

	Situation I Open Tank at 283 m			Situation II Les Tourillons de Craponne		
	L (m)	Number of Pipes	Total Length (m)	L (m)	Number of Pipes	Total Length (m)
Section length upstream	1500	3	4500	2200	7	15,400
Section length downstream	4300	8	34,400	3600	4	14,400
Total (m)	5800		38,900	5800		29,800

This means that the construction of the 16 m high tower with open tank/reservoir on top on a hill at 290 m compared with an open tank on top of a hill at 283 m saved 9 km of lead piping of 23 or 20 cm outer diameter and 20 or 18 cm inner diameter. That is almost 1800 tons of lead. The decision by the Roman engineers to construct *les Tourilons de Craponne* for the siphon of the Yzeron aqueduct was a rational one: it was less expensive The overall profile of the double siphon of the Yzeron aqueduct is shown in Figure 81.

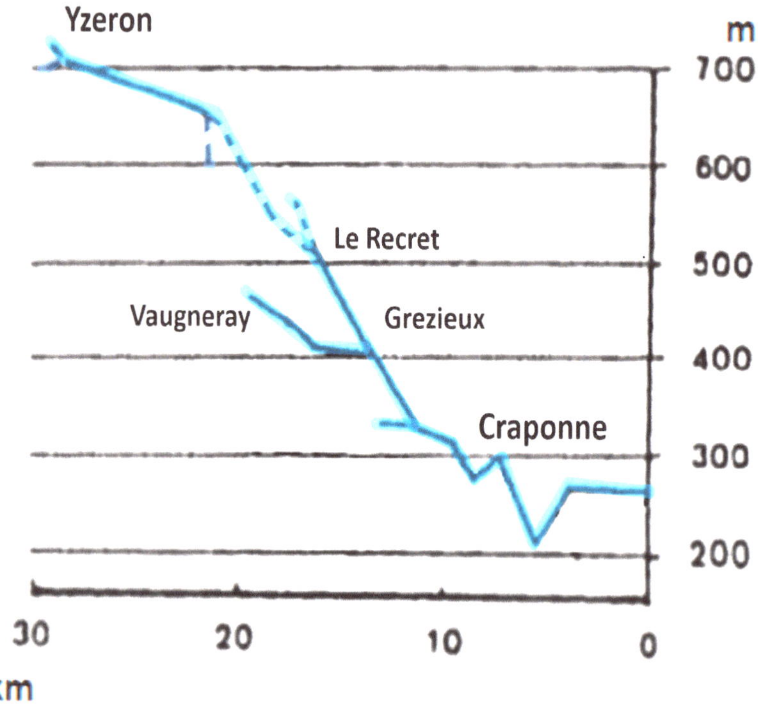

Figure 81. Profile of the Yzeron aqueduct, with a possible but not found connection between the channel coming from Yzeron village and the channel towards Lyon at Le Recret. A double siphon with the hydraulic tower at Craponne [8], p. 12, adapted.

15. Discussion

The four aqueducts of Lyon were built in a time span of one and a half century, from about 20 BCE (Mont d'Or aqueduct) to 140 CE (Gier aqueduct). Only the Gier aqueduct provided the entire city on the Fourvière promontory, the other three aqueducts reaching lower sections of the town including some public fountains (Figures 82–84).

By the end of the 3rd century CE under Diocletian, Lugdunum loses her title as capital of the Gaules, after which decline set in. In the 4th century, the city was deserted and the aqueducts lost their function, the lead of their siphons becoming subject to pillaging [2], p. 28. The enormous amount of lead for the four Gier siphons alone is estimated to have been 10,000 tons, assuming lead pipes of 23 cm outer diameter and a wall thickness of 2.5 cm [9], pp. 269–275. Where did all the lead come from?

Pliny the Elder mentions that lead is to be found in Spain and Gallia, but especially in Britannia in great amounts: Pliny the Elder XXXIV ch 164 [24]: 'Nigro plumbo ad fistulas lamnasque utimur, laboriosus in Hispania eruto totasque per Gallias, sed in Brittannia summo terrae corio adeo large, ut lex ultro dicatur, ne plus certo modo fiat'. Translation: Black lead is used for water pipes and sheets, and is laboriously delved in Spain as well as in Gallia, but in Brittannia it is in the upper soil in such great quantities, that a law is said to have been issued, that one can only take certain amoun (Translations of Brittannia may vary from England to Bretagne).

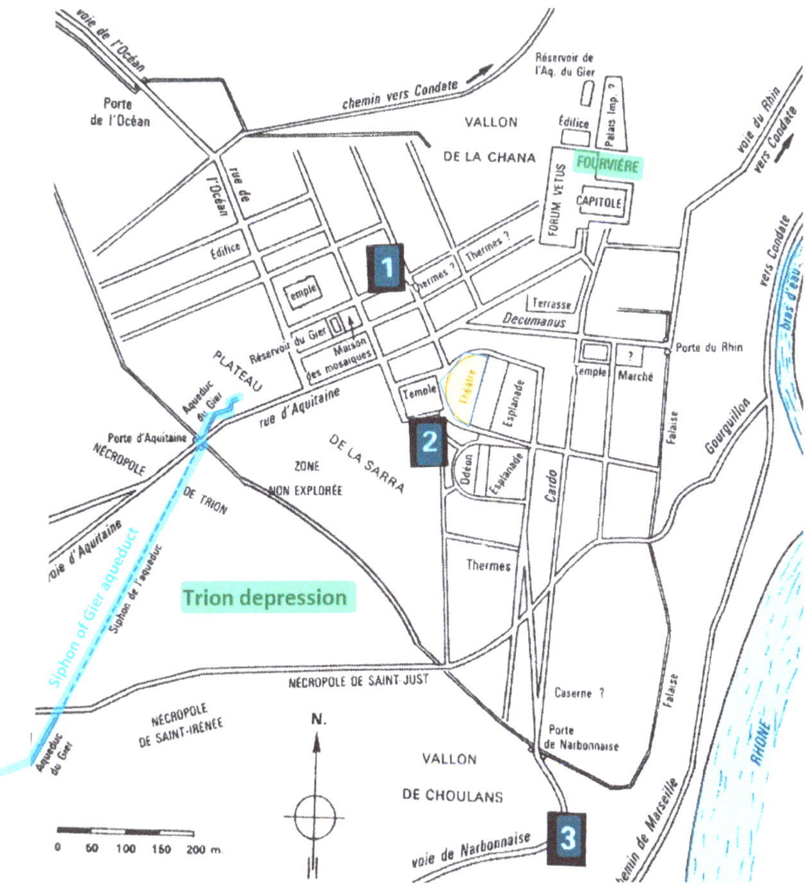

Figure 82. Fourvière, and the Trion depression to the south crossed by the siphon of the Gier aqueduct. Yellow: theatre. Location of found public fountains 1: fountain 'du Verbe Incarné; 2: Odeon fountain; 3: fountain of 'Choulans' [23], p. 83.

Figure 83. Fountain 'du Verbe Incarné, as found in situ [23], p. 84.

Figure 84. Fountain 'du Verbe Incarné, after restorations, in 1991 installed on the Place de Trion [2], p. 53.

The question is, why were multiple pipelines chosen for the Lyon siphons? No multiple-conduit siphons are known from classical times except at Lyon. The two parallel stone siphons at Laodikeia a/L had each a separate header tank and receiving tank [21], p. 152. Could not a single conduit have sufficed to transport the water to Lyon?

Lead conduits of siphons have been found, for instance at Arles and at Vienne, to cross the Rhône on the river bed [25,26]. Diameter and wall thickness of these conduits are, respectively, 10 cm/1.2 cm and 32 cm/1.6 cm. For the Lyon siphons, lead conduits of 23 cm outer diameter and wall thickness 2.5 cm will be taken as 'standard' in the discussion below, although a wall thickness half as wide may be regarded as an option.

To see whether a single conduit siphon would do for Lyon, one has to regard the water transport by means of say 10 conduits with internal diameter, e.g., 18 cm (outer diameter 23 cm wall/wall thickness 2.5 cm), and see what diameter pipe would transport the same amount of water for the same siphon stretch.

The velocity V of water in a pipe may be estimated with the Darcy–Weissbach formula as mentioned above:

$V^2 = (8 \cdot g \cdot \Delta H \cdot Rh)/(\lambda \cdot L)$, with $Rh = D/4$ for full pipe flow, $D = 2R$ = internal pipe diameter, R = internal pipe radius.

For the same siphon length L and the same difference in level between header tank and receiving tank ΔH, as well as for the same material of the pipe (wall roughness λ), the flow veleocity only depends on Rh, that is, on the internal pipe diameter, V^2 depends only on D, that is, V^2 depends only on R so V depends only on $R^{\frac{1}{2}}$.

The water transport Q through a pipe of internal radius R is equal to the water velocity V in the pipe times the cross-section area O of the pipe: $Q = V \cdot O$, where $O = \pi \cdot R^2$. So, in this case Q depends only on $R^{\frac{1}{2}} \cdot R^2$ or $R^{5/2}$.

Thus, the water transport by 10 conduits of 18 cm inner diameter (with radius 9 cm = R1) would relate to $10 \cdot (R1)^{5/2}$, while the transport of a single conduit of inner radius R2 relates to $(R2)^{5/2}$. To have both transports equal: $(R2)^{5/2} = 10 \cdot (R1)^{5/2}$, or $(R2)^5 = 100 \cdot (R1)^5$.

That is R2 = 2.512·R1, (say, 2.5·R1). With R1 = 9 cm, it would mean R2 = 22.5 cm, or inner diameter 45 cm.

A single conduit of 45 cm inner diameter would transport the same amount of water as 10 conduits of 18 cm inner diameter; surprisingly, only 2.5 times as wide, and with only 6.25 times larger cross-section.

This is due to the lesser resistance from the larger conduit wall, which covers less surface area than for the 10 smaller conduits, by which the water will have larger velocity in the larger conduit.

What about the required wall thickness for the 45 cm conduit? In the case that the inside pressure in a pipe becomes too large, a pipe will burst lengthwise. The burst pressure P(l) relates to the tensile strength τ of the pipe wall material, to the thickness d of the pipe wall, and to the inner diameter D = 2R (R = inner radius): the thicker the pipe wall and the higher the tensile strength of the pipe wall material, the better the pipe is resistant to failure, but the wider the pipe, the less it is resistant it will be [1], p. 22. That is,

$$P(l) = \tau \cdot (d/2R).$$

The 45 cm conduit made of the same material as the smaller 18 cm one (with 2.5 cm wall thickness) will burst at the same pressure only if its wall thickness d is 45/18 = 2.5 times larger, that is, 6.25 cm. Thus, the pipes of the single conduit siphon will have an outer diameter of 45 + 2 × 6.25 = 57.5 cm. These are big pipes!

The 33 lead pipes that have been dredged up from the Rhône river bed at Arles over a period of many centuries belonged to the siphon that crossed the river between Trinquetaille and Arles. They are on exposition in the Arles museum (Figures 85 and 86). In 1988 the pipes were investigated by Jørgen Hansen from Denmark [25]. The conduit of the siphon was composed of 3 m prefab lead pipes that were soldered together at the ends. According to Hansen, the connections were as strong as the pipes themselves (*'waren nicht das schwache Glied der Kette'*). The quality of the lead was 99.97% pure [25], p. 485.

Figure 85. Arles museum, depot. About 3 m long lead pipes 10–13 cm outer diameter, wall thickness 10–12 mm, weight 100–150 kg. The pipes were soldered together by pushing one end into the other, driving a nail through both ends, and covering both ends with lead soldered onto the pipes (see wider pipe ends on left side (photo author).

Figure 86. Arles museum, pipes from Rhône river on exhibition (https://thefogwatch.com/roman-baths-thermae-arles/_mg_2361-2/, accessed on 12 April 2024).

The soldering of the joints in those days must have been done on the spot to make the long conduits. The samples that Hansen took were not indicative for the use of soldering alloys, nowadays tin–lead alloys like 60% tin/40% lead, having a lower melting temperature than lead, in combination of solder flux are used.

The soldering in those days may have been performed by scraping the oxidized surfaces of the lead part to be soldered with a knife to a shiny surface, cover this with bee's wax or candle wax to prevent renewed oxidising of the lead surface while heating the lead slowly, and when the lead becomes locally mallable but not melting, push the pipes together. Such procedure would then make the lead of the two surfaces join (this is an old technique called lead-welding dating back to 1000 B.C. It does not require solder flux nor solder alloys, but it is a rather time-consuming procedure (https://silk-leadwork.co.uk/20 21/07/21/lead-welding-what-is-it-and-what-is-it-used-for/, accessed on 2May 2024).

Hansen, on the other hand, mentions that the pipe ends were shoved one into the other for a few centimeters, then an iron nail or two was driven through both ends, and that subsequently lead was cast around the connection. For this, some mould may have been required and the problem of oxidised surfaces solved (Figures 87 and 88). How exactly the Romans managed to solder the pipes together remains an open question.

Figure 87. Soldered joint of Arles pipe, the right-hand pipe cut off next to the joint [25], p. 505.

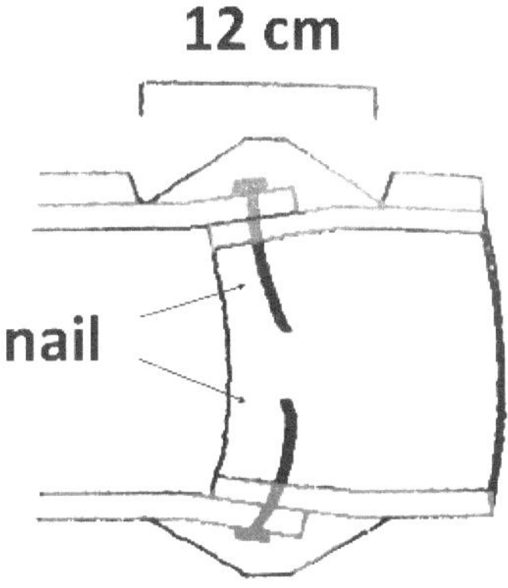

Figure 88. Fixating nails [25], p. 517.

Whatever the case, strong lead conduits could be and were produced in these days. For the siphons of Lyon, a similar technique as at Arles may be envisaged, with 3 m prefab pipes soldered together on the spot to make the long conduits. For the envisaged Lyon siphon conduits with an outer diameter of 23 cm and wall thickness 2.5 cm, such 3 m pipes would weigh about 550 kg, requiring at least 10 men to lift and carry 50–60 kg per person. With today's labour regulations for lifting weights not over 20–25 kg, it would even require 20 persons, let alone to manipulate such pipes (https://en.wikipedia.org/wiki/Manual_ handling_of_loads, accessed on 2 May 2024). With half a wall thickness of 1.25 cm, the load would be halved, 250 kg—much better to work with yet still heavy (the safety factor against bursting of the conduits of the Lyon siphons was put at a factor 4, which may be in view of this exaggerated, a lesser factor of 2 may be more realistic). For the theoretical single conduit siphon as discussed above, the weight of a 3 m pipe (outer diameter 58 cm wall thickness 6.25 cm) would be an impossible 3400 kg (safety factor 4) or 1600 kg (safety factor 2), by which this possibility as an alternative for the 10 parallel 23 cm pipes may be excluded.

Now that it is evident that a single lead conduit instead of the 10 parallel smaller ones is not realistic, one may conclude that a siphon with one conduit of 18 cm inner diameter was not deemed suitable to transport sufficient volumes of water to Fourvière. We do know that the difference in level of the Lyon siphons was rather small, while the siphons themselves were rather long, see Table 7 and Figure 4 for the siphons that crossed the great valley just west of the Roman Fourvière settlement.

Table 7. The siphons of the four Lyon aqueducts crossing the great valley west of the Fourvière promontory and associated heights depicted in Figure 4. Discharge Q per lead conduit of 18 cm inner diameter. For the data of the siphon of the Gier aqueduct see [9], p. 291.

Aqueduct	Siphon	Lk(m)	ΔH (m)	ΔH/L (m/km)	Q for Each Lead Conduit 18 cm Inner Diameter (m^3/day)
Mont d'Or	Limonest	3.5	11	3.1	1351
Yzeron	Grezieux-Craponne	2.2	7	3.2	1337
	Craponne-Lyon	3.6	33	9.2	2329
Brevenne	Ecully-TAssin	3.3	14	4.2	1574
Gier	Yzeron	2.6	7.9	3	1330

For any siphon, the maximum water quantity that is transported is set by difference in level between header tank and receiving tank, and by the length of the conduit (and roughness of the pipe wall), independent of the amount of water that the incoming channel would deliver to the header tank that regularly is equipped with an overflow. In view of the numbers in Table 7, each lead conduit of the Lyon siphons would be sufficient for some 1300 persons at a 100 liters a day, an amount that today is asssumed a minimum (https://cdn.who.int/media/docs/default-source/wash-documents/who-tn-09-how-much-water-is-needed.pdf, Accessed on 2 May 2024). The Roman engineers must have been aware of the problem, which they solved by installing a number of parallel conduits for each siphon instead of siphons of larger diameter—this an inventive idea, like *les Tourillons* at Craponne. According to Burdy, the numbers would for the Gier aqueduct amount to 10,000 m^3/day, and for all aqueducts some 40,000–45,000/day (which, at the time that the Gier aqueduct was running, may not have functioned simultaneously [9], p. 292.

16. Epilogue

The four aqueducts of Lyon represent an extraordinary example of Roman hydraulic expertise, which has come to us by the work of authors like Delorme and de Montauzan, and especially, in the second half of last century, by Jean Burdy from Lyon and his colleagues. Constructed in the time span from 20 BCE to 140 CE, to bring waters to the settlement on the Fourvière hill overlooking the Rhône and Saone rivers to the east, the Roman engineers had to cope with mountainous areas to the north and west, as well as with wide plains and deep valleys to the west and south. Ranging in length from 26 to 86 km, three of the four aqueducts incorporated 'hydraulic stairways' to trespass steep terrains, while all aqueducts were equipped with pressurized conduits, siphons, to pass deep valleys, the longest aqueduct (of Gier) even having four siphons. These at times kilometers long siphons were laid out with ten or more parallel-laid-out lead pipes, a unique feature contrasting the usual single conduit siphons as known from classical times. This may be regarded as an inventive step to guarantee sufficient water flow in terrains of gentle sloping. For the longest, 5.8 km, siphon (of the Yzeron aqueduct) an obligatory 'hydraulic tower' was built, splitting the siphon into two, and that on top of a hill that could be avoided, but which reduced the number of lead pipes required for the siphon, a sign of great expertise and inventiveness. Moreover, gravity driven siphons have remained until our days (Figure 89).

Figure 89. Steel 900 m siphon transporting waters from the Rhue river to the lake of the Barrage de Bort, Corrèze, France (postcard).

Funding: This research received no external funding.

Data Availability Statement: Data are contained within the article.

Conflicts of Interest: The authors declare no conflict of interest.

References

1. Kessener, H.P.M. Roman Water Transport: Pressure Lines. *Water* **2022**, *14*, 28. [CrossRef]
2. Burdy, J. *L'Alimentation en eau de Lugdunum*; Digitprim: Lyon, France, 2017.
3. Burdy, J. *Les Aqueducs romains de Lyon*; University Press of Lyon: Lyon, France, 2002.
4. Burdy, J. *Guillaume Marie Delorme 1700–1782*; l'Araire: Lyon, France, 2015.
5. Burdy, J. *L'Aqueduc Romain de la Brevenne*; Departement du Rhône: Lyon, France, 1993.
6. Germain de Montauzan, C. *Les Aqueducs Antiques de Lyon. Etude Comparée d'archéologie Romaine*; E. Leroux: Paris, France, 1909.
7. Burdy, J. *L'Aqueduc Romain du Mont d'Or*; Departement du Rhône: Lyon, France, 1987.
8. Burdy, J. *L'Aqueduc Romain de l'Yzeron*; Departement du Rhône: Lyon, France, 1991.
9. Burdy, J. *L'Aqueduc Romain du Gier*; Departement du Rhône: Lyon, France, 1996.
10. Kessener, H.P. Moderne Persleidingen en Romeinse Hydraulische Technieken. *Rioleringswetenschap en-Techniek* **2004**, *15*, 10–44
11. Passchier, C.W.; Sürmelihindi, G.; Spötl, C. A High-resolution palaeoenvironmental record from carbonate deposits in the Roman Aqueduct of Patara, SW Turkey, from the time of Nero. *Sci. Rep.* **2016**, *6*, 28704. [CrossRef] [PubMed]
12. Sürmelihindi, G.; Passchier, C.; Crow, J.; Spötl, C. Carbonates from the ancient world's longest aqueduct: A testament of Byzantine water management. *Geoarchaeology* **2021**, *36*, 643–659. [CrossRef]
13. Hodge, A.T. *Roman Aqueducts and Water Supply*; Duckworth: London, UK, 1992.
14. Burdy, J. *Les Aqueducs romains de Lyon*; l'Araire: Lyon, France, 2008.
15. Grewe, K. *Aquädukte, Wasser für Rom's Städte*; Regionalia Verlag: Euskirchen, Germany, 2014.
16. Haberey, W. *Die Römischen Wasserleitungen Nach Köln*; Rheinland Verlag: Bonn, Germany, 1971.
17. Burdy, J. Some directions of Future Research for the aqueducts of Lugdunum (Lyon). In *Future Currents in Aqueduct Studies*; Hodge, A.T., Ed.; Francis Cairns Ltd.: Leeds, UK, 1991.
18. Baldassari, D. L'Aqueduc de l'Yzeron, nouvellles donnessur le double siphon de Craponne (Métropole de Lyon). *Gallia* **2023**, *80-1*, 199–215. [CrossRef]
19. Garbrecht, G. Altertümer von Pergamon, Band 1—Teil 4. In *Die Wasserversorgung von Pergamon*; DAI: Berlin, Germany; New York, NY, USA, 2001.
20. Kessener, H.P.M. A Pompeiian-type Water System in Modern Times. In *Historische Wasserleitungen, Gestern—Heute—Morgen*; Wiplinger, G., Ed.; BABesch Supplements: Leiden, The Netherlands, 2013; Volume 24, pp. 229–240.
21. Kessener, H.P.M. *Roman Water Distribution and Inverted Siphons—Until Our Days*; Ipskamp: Nijmegen, The Netherlands, 2017.
22. Kessener, H.P.M. Sur le siphon romain et ses problèmes: Les Tourillons de Craponne. In Proceedings of the Premières Assises du Patrimoine Hydraulique 2023, Caluire & Cuire, France, 23–24 November 2024.
23. Burdy, J. Lyon: Les Fontaines publiques de Lugdunum. Caesarodunum XXXV–XXXVI, 2001, pp. 77–87.
24. Plinius, C. *Secundi, Naturalis Historiae, Libri XXXVII Libri XXXIV*; Latin–Deutsch, Artemis Verlag: Munich, Germany, 1989.

25. Hansen, J. *Die antike Wasserleitung unter der Rhône bei Arles*; Mitt. des Leichtweiss-Institut für Wasserbau, Heft: Braunschweig, Germany, 1992; Volume 117, pp. 470–531.
26. Burdy, J.; Cochet, A. Une Date Consulaire (213 après J.-C.) sur un Tuyau de Plomb Viennois. *Gallia* **1992**, *49*, 89–97. [CrossRef]

water

Article

A Summary of 25 Years of Research on Water Supplies of the Ancestral Pueblo People

Kenneth R. Wright [1,2]

1   Wright Water Engineers, Inc., 2490 W. 26th Ave., Denver, CO 80211, USA; krw@wrightwater.com
2   Wright Paleohydrological Institute, 1440 High St., Boulder, CO 80304, USA

Abstract: Six ancestral Pueblo community water supply sources were investigated by a team of engineers, scientists, archeologists, and other specialists affiliated with Wright Water Engineers, Inc. (WWE), and the Wright Paleohydrological Institute (WPI) from 1996 to 2021. The team members applied their various technical backgrounds and research methods to gain more insight into the water available to the ancestral Pueblo people living in the Four Corners area of the United States between 750 and 1280 CE, and how these indigenous people managed the water. Using lab analyses, field research, surveys, and analyses of sediment layers, the WWE/WPI team determined that four mounded areas discovered at Mesa Verde National Park had been ancestral Pueblo reservoirs. Through climate research, lab analyses, and investigations at these and two other sites, the team learned that water in this region was limited, and the community had to work diligently to harvest this water and maintain access to it. In the case of the four reservoirs studied, for example, the runoff used as water supply carried a high volume of sediment that required the water storage basins to be frequently dredged to maintain adequate capacity. These and other examples indicate that the ancestral Pueblo people were resourceful, hardworking, and organized water harvesters.

Keywords: ancestral Pueblo; paleohydrology; paleohistory; Mesa Verde; Four Corners; sdendrochronology; water supply; reservoirs; water storage

Citation: Wright, K.R. A Summary of 25 Years of Research on Water Supplies of the Ancestral Pueblo People. *Water* **2024**, *16*, 2162. https://doi.org/10.3390/w16172462

Academic Editors: Bommanna Krishnappan and Charles R. Ortloff

Received: 17 July 2024
Revised: 14 August 2024
Accepted: 22 August 2024
Published: 30 August 2024

1. Introduction

As early as 7500 BCE, nomadic Paleo-Indians chose the area now known as the Four Corners region of the United States to be their home (Figure 1). The Four Corners area, where Utah, Colorado, Arizona, and New Mexico meet, accommodated six eras of ancestral Pueblo people: Basketmaker I, Basketmaker II, Basketmaker III, Pueblo I, Pueblo II, and Pueblo III. For about 750 years, the ancestral Pueblo people lived at Mesa Verde, first as nomadic hunters, later as farmers living in pit houses (Figure 2), and eventually as builders of remarkable cliff dwellings in the sheltered alcoves of the canyon walls (Figure 3) [1].

Water scarcity was a defining factor in the later ancestral Pueblo inhabitation of the Four Corners area. In the best of times, the climate was arid and water was a precious resource to be carefully managed. The worst of times brought a major drought that lasted from 1135 to 1180 CE. This was about the time that the ancestral Pueblo people moved from pleasant valley bottoms and mesa tops to cliff dwellings that could be better defended. Beginning around 1140 CE, the residents of the Four Corners area experienced famine, water uncertainty, and raiding parties. People began leaving Mesa Verde around 1250 CE to resettle along the Rio Grande, in southern Arizona and in Old Mexico. From 1275 to 1300 CE, there was another major drought; it was during this time that most of the remaining ancestral Pueblo people either abandoned the region or were killed by hungry raiders [2].

Figure 1. Four Corners area of the United States.

Figure 2. Ancestral Pueblo pit house remains at Mesa Verde National Park (MVNP).

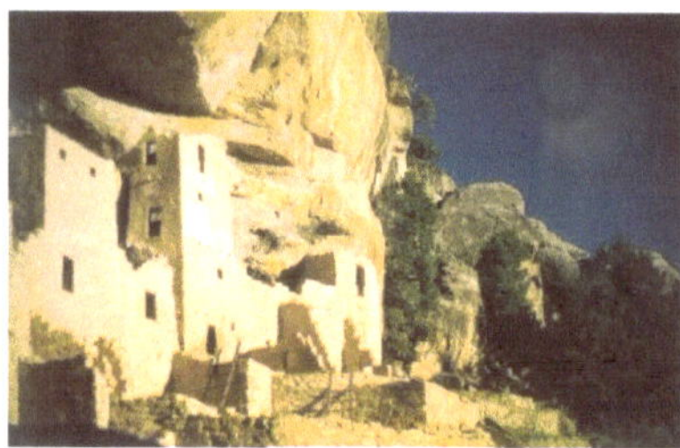

Figure 3. Ancestral Pueblo cliff dwelling remains at MVNP.

Mesa Verde National Park (MVNP), established in 1906 by Congress as a national park and ranked by the National Geographic Society in 1999 as number six of all world wonders, encompasses 210 square kilometers of the Four Corners area. The Pueblo I, Pueblo II, and Pueblo III people who lived in what is now MVNP and the surrounding area proved themselves to be successful in implementing water works and civil engineering projects that made life possible in the dry climate of southwestern Colorado. Wright Water Engineers, Inc. (WWE), and the Wright Paleohydrological Institute (WPI) collaborated on the study of water supply development throughout these three ancestral Pueblo eras. The multi-disciplinary WWE/WPI team explored reservoirs at MVNP that were later distinguished collectively as one of five American Society of Civil Engineers National Historic Civil Engineering Landmarks in Colorado, plus two other ancestral Pueblo water supply artifacts in the region—a cistern and a spring. The four reservoirs studied are Morefield, Far View (aka Mummy Lake), Sagebrush, and Box Elder, all described in detail in The Water Mysteries of Mesa Verde [3]. The WWE/WPI team also performed water supply research at Mug House cistern (MVNP) and Goodman Spring (Hovenweep National Monument). The data and analyses developed through this research were catalogued and made available to the public at the Norlin Library at the University of Colorado in Boulder, Colorado, in 2024.

The purpose of this manuscript is to provide a "one-stop" summary of WWE/WPI's water research in the southwestern United States and the resultant findings. No other paper by the WWE/WPI team provides the overview and synthesis towards which this manuscript strives. This manuscript is intended to make other researchers aware of the studies conducted at these six sites to date, what was learned, and where further data can be gleaned. A larger purpose of this research is to look at the prehistorical water handling practices of a culture—the ancestral Pueblo people—to understand the importance of water as a resource and its role in shaping history. In the face of climate change, shifting resources, and evolving technology, learning about the extensive effort made by the ancestral Pueblo people to secure limited amounts of water should be both inspiring and prescient. While modern problems and solutions differ from those of the ancestral Pueblo people, their resourcefulness and determination can serve as a positive example for coming generations.

2. Materials and Methods

Webster's Third International Dictionary defines paleohydrology as "the study of ancient use and handling of water" [4]. Research by the WWE/WPI team in the Four Corners area has been aimed at that definition. During a 25-year period, this team of water resource engineers, hydrologists, scientists, geologists, archeologists, palynologists, and anthropologists sought to learn about the water-harvesting practices of the ancestral Pueblo people through the six studies described herein. This work was intended to harness the expertise of veteran scientists from various fields to address a specific query—how did the ancestral Pueblo people of the southwestern United States obtain water and what can we learn about their practices and results?

The team obtained permits from the National Park Service (NPS) and followed NPS's strict field protocols for preservation and protection of these sites. In some cases, the team collaborated to combine resources and data with archeologists who were already conducting excavations or studies.

The analyses typically applied at each site included procedures that, when integrated, would validate the purpose and function of each site and provide evidence of how the water structure was developed, how and when it was used, and the likely characteristics of operation and maintenance. These analyses included the following:

- Field topographic surveys using traditional civil engineering technology with theodolites, level instruments, measuring tapes, and planetables. The team defined and mapped the topography so that shapes of features, heights of mounds, and lengths of canal routes could be calculated (Figure 4).
- Using soil augers to recover soil profile data from the archeological sites so sediment layering could be defined with specificity, along with charcoal evidence. Soil profile data also helped define the physical extents of reservoir and canal building where natural, undisturbed soil layers were confirmed (Figure 5).
- Surface infiltration tests for site-specific data on present-day infiltration rates of bare and vegetated soils and typical permeability. These data informed estimates of rainfall–runoff relationships (Figure 6).
- Geologic descriptions for an understanding of basin characteristics and the source of eroded materials. The team defined bedrock outcrops and faults.
- Geomorphological analyses to define the character of canyon bottoms.
- Ceramic analyses performed by archeologists to estimate the period of human activity for reservoir dating.
- Paleo climate evaluations using dendrochronology (tree rings) for estimates of precipitation and temperature going back to about 500 CE. Overall, the long-term precipi-

tation in the Four Corners area has been roughly equal to modern times, i.e., about 46 cm per year [5].

- Rainfall–runoff determinations to verify that a surface water supply existed.
- Pollen analyses of various sediments to define prehistoric vegetation, the type and location of agricultural practices, and the presence or absence of wetland plants. Forest successional periods and the presence or absence of sagebrush and medicinal plants were also evaluated.
- Laboratory analyses of reservoir sediments using hydrometer and sieve analyses of soils and sediments to define sand, silt, and clay content to estimate sediment transport and depositional character. The team looked for redoximorphic features that indicate long-term soil saturation.
- Carbon dating, where feasible, to determine the age of prehistoric structures and artifacts, such as a deer antler, to supplement pottery analyses.
- Groundwater evaluations to estimate water sources available and whether reservoir inflow might have originated via aquifers.
- Archeological studies of the reservoir structures, canals, watersheds, and adjacent villages, allowing the engineering data and studies to be properly placed in their anthropological context.
- Aerial, standard black and white, color and infrared photographs to define the ground conditions, wetlands, and water tables, and assist with the evaluation of geomorphological processes.

All or a combination of the above activities were conducted for the six sites summarized in this paper. The following sections highlight new information gleaned from the study of each site.

Figure 4. Field topographic surveying at Far View Reservoir.

Figure 5. Use of a soil auger at Far View Reservoir.

Figure 6. Surface infiltration testing at Box Elder Reservoir.

3. Discussion and Results

3.1. Morefield Reservoir

Morefield Reservoir, on the floor of Morefield Canyon, was the first ancestral Pueblo artifact studied by the WWE/WPI team after a MVNP superintendent asked for the team's opinion on what this relic mound might have been. Several theories were in play, including that the mound could have been a ceremonial dance platform or perhaps a former reservoir (Figure 7). Through the processes of excavation and examination of mound layering (Figure 8), as well as soil testing, the team determined that the mysterious mound had originally been a 1.22-meter-deep, 15-m-diameter storage system for runoff and groundwater circa 750 CE. The evidence of this former use was fulsome: (1) the mound was comprised of sediment layer patterns consistent with the frequent dredging of waterlogged deposits that had been washed into the basin from the silty terrain upstream; (2) core samples extracted using soil augers indicated the site had been a tadpole-shaped indentation into the earth up to 6.5 m deeper than the highest part of the mound (Figure 9); (3) the sediment layers contained a concentration of shards from broken water vessels dating between 750 CE (at the deepest) and 1100 CE (at the shallowest) (Figure 10); (4) the soil analyzed from the excavation had redoximorphic features, indicating it had been submerged in water for long periods; (5) adequate runoff would have existed to occasionally fill this reservoir; and (6) the sediment contained high concentrations of maize pollen that were likely deposited by runoff from uphill agricultural fields.

This hand-dug basin captured runoff via an inlet canal (the tail of the tadpole shown on Figure 9) established by the ancestral Pueblo people around 750 CE. The team found evidence that the canal had been lined and relined with flat rocks to reduce erosion. Its length was extended over time as the erosive, silty soil that had washed in required continual dredging, creating rising side berms where the dredged material was thrown. WWE/WPI analyses determined that the silty sediment washed into the reservoir from the basin above would have had to have been dredged almost continually. Despite the cleaning operations, the valley-bottom reservoir and canal rose in elevation about 1.8 cm per year. After the reservoir was abandoned, the silt in the reservoir accumulated within the raised berms until the feature became the mound we see today.

Figure 7. The origin of the Morefield mound was a source of curiosity.

Figure 8. Morefield mound excavation from above.

The excavated mound layers contained more detailed information on how the reservoir developed. Strata of hardpacked sand indicated long-ago wind direction (Figure 11), while mangled layers indicated a berm failure in a section that was dug at too deep a slope. Maize pollen accumulation in the sediment layers indicated that the runoff filling the reservoir flowed through upstream corn crops. Carbon deposits in the sediment layers demonstrated that 14 forest fires had occurred during the reservoir's 350-year period of use.

The Morefield Reservoir would have filled sporadically due to large rainfalls and increased runoff from periodic forest fires. Additional WWE/WPI team analyses of capacity and weather patterns suggested the Morefield Reservoir stored up to 450,000 L of water about five times per year, or about 2.25 million liters of water per year. While the WWE/WPI team did not specialize in anthropology, they were curious about the rough amount of water available per person in the Ancestral Pueblo era. According to Wilshusen [6], the average population of Pueblo I communities was about two hundred people. Dividing the estimated 2.25 million liters of water stored per year at Morefield Reservoir by 200 people, divided by 365 days per year, indicates that the residents of Morefield Reservoir might have had available approximately 30 L of water per person per day for drinking, cleaning, livestock watering, and other uses. This likely would have been a best- case scenario because Morefield Reservoir was the largest Four Corners water storage feature studied by the team. For comparison, Americans in 2015 used an average of 310 L of water per day [7].

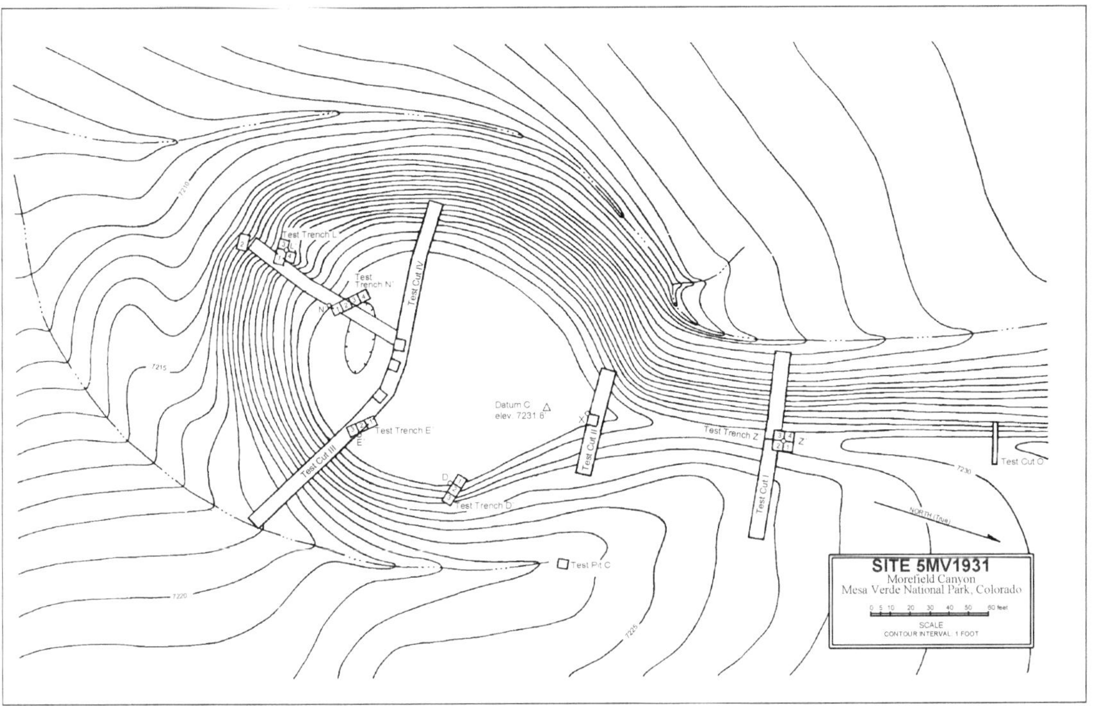

Figure 9. Morefield Reservoir plan view showing the location of excavations. Test pits of 1.52 m by 1.52 m, indicated by squares, were excavated first and later expanded into trenches where warranted.

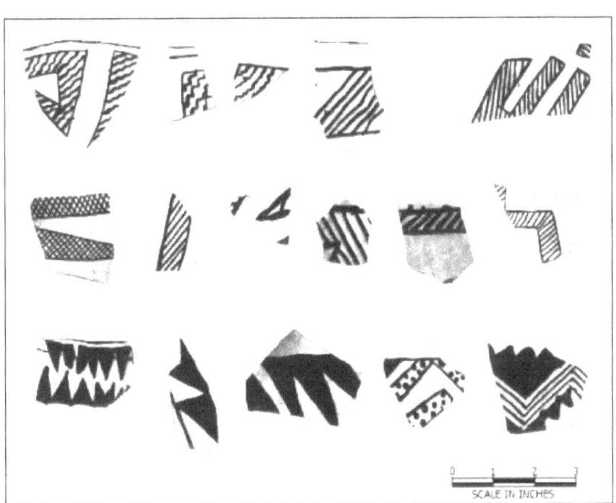

Figure 10. Broken pieces of ceramic material (potsherds) collected from a Morefield Reservoir wall, indicated the reservoir was used during the Pueblo II era.

Figure 11. A compacted sand layer shows ripples caused by the wind blowing from the southwest during the 9th century.

The main thing that analysis of the Morefield mound layers taught the WWE/WPI team was that the ancestral Pueblo people were willing to engage in near-constant maintenance of this reservoir. Clearly, the water must have been regarded as a precious resource to merit this investment of extensive time and effort. Based on their various specialties, the WWE/WPI team members also concluded the ancestral Pueblo people must have been well organized and communal to successfully execute these reservoir-building and maintenance operations.

3.2. Far View Reservoir

The next reservoir studied by the WWE/WPI team was Mummy Lake, later renamed Far View Reservoir. In contrast to Morefield Reservoir's valley-bottom location, Far View Reservoir is located on the top of Chapin Mesa (Figure 12). Far View Reservoir is one of about three dozen MVNP archeological sites with public access and is therefore quite popular. The former use of the site was particularly enigmatic because it is comprised of stone walls that make the structure appear significant, and perhaps ceremonial. In response to archeologists' questions on whether there had been a water supply at Far View Reservoir, the WWE/WPI team conducted hydrological and scientific studies to analyze water supply feasibility and possible reservoir operations. The reservoir structure shown in Figure 12 is about 28 m in diameter and would have contained a maximum depth of 1.4 m, representing maximum storage of about 300,000 L of water. Sand and silt in the reservoir excavation were clearly deposited by water.

Palynological analyses of soil samples taken upstream of the reservoir found high concentrations of maize pollen. This suggested to the team that agricultural fields would have existed above the reservoir, allowing for greater runoff than would have flowed from undeveloped land. The team analyzed the soil inside of the stone walls outside of the water storage area and found higher levels of maize pollen there. The team developed a theory that if the reservoir stored water, like Morefield Reservoir, there would be a great deal of sediment running into it. The logical place for the dredged sediment to have been disposed would have been within the perimeter stone walls. The presence of high levels of maize pollen in this area suggested to the team that runoff from the corn fields above would have flushed maize pollen along with the sediment, which would have been thrown into the walled ring during dredging. This theory was borne out by archeologist David Breternitz, who analyzed sediment layering and pottery sherds to determine that the reservoir began as a simple hole in the ground with the walls added later, likely to hold the dredged material. A study of the material within the walls provided further evidence that Far View was a water storage feature, in the form of potsherds and the presence of pollen that indicated water-loving plants had grown there.

Figure 12. The remains of Far View Reservoir from above.

Surveying by the WWE/WPI team found that Far View Reservoir lies about 95 m in elevation below the highest point of the ridge. By analyzing weather patterns and typical runoff rates for agricultural and hard-packed ground conditions, the WWE/WPI team concluded that the site was indeed a reservoir that could have received runoff from interceptor ditches along the ridgeline, foot-packed public areas, and farmland, above. The reservoir has been known to collect water during modern times in much the same way it did during ancestral Pueblo times (Figure 13). The reservoir would not have received any groundwater and would have held surface water infrequently.

Figure 13. This National Park Service file photograph shows cowboys watering their horses at Far View Reservoir in the early 1900s.

3.3. Sagebrush Reservoir

The WWE/WPI team also performed research at Sagebrush Reservoir (Figure 14), which resides 1.6 km west of Morefield Reservoir on an unnamed mesa. Sagebrush Reservoir is another mesa-top water storage facility, utilized by the ancestral Pueblo people from 950 to 1100 CE. A plan of the site can be seen in Figure 15. A WPI colleague, Dr. Jack Smith, had excavated Sagebrush Reservoir in the 1970s and found that it began as a dug-out water storage facility, without walls, and would have held approximately 340,000 L of water. Due to about 1.5 cubic yards of sediment accumulation per year, the capacity of the reservoir decreased over time and walls were constructed to retain the dredged material.

The reservoir's capacity was ultimately reduced to roughly 150,000 L by the time it ceased to be used as a water storage facility in about 1100 CE. Based on early water-borne deposits in the reservoir, followed by a period of eolian deposits (windblown sand, loess, and long-range-transported dust), the WWE/WPI team determined that the reservoir had been used by the ancestral Pueblo people for other purposes during its final stage of occupation. This conclusion was further confirmed by evidence of human activity, such as a metate fragment, found in the upper layer of deposits.

Figure 14. The faint cross in the center of this photograph shows the path of Sagebrush Reservoir excavations conducted in the 1970s.

No water could have flowed to Sagebrush Reservoir without human intervention. Evidence was found on the ridge of several interceptor ditches beginning at higher elevations that routed runoff to the reservoir. As was the case with Far View Reservoir, Sagebrush Reservoir relied on hard-packed sand and clay terrain above it to produce runoff. The WWE/WPI team determined that about $\frac{1}{2}$-acre of this water-tight ground surface could produce runoff during periods of about 1.3 cm or more of rain per hour. Based on precipitation data, the reservoir would have stored water about five or six times per year. A small spring that is still producing water exists at the bottom of East Fork Navajo Canyon, next to Chapin Mesa. This spring would likely have been a distant secondary water supply to which those who lived on this mesa had to trek when Sagebrush Reservoir was dry.

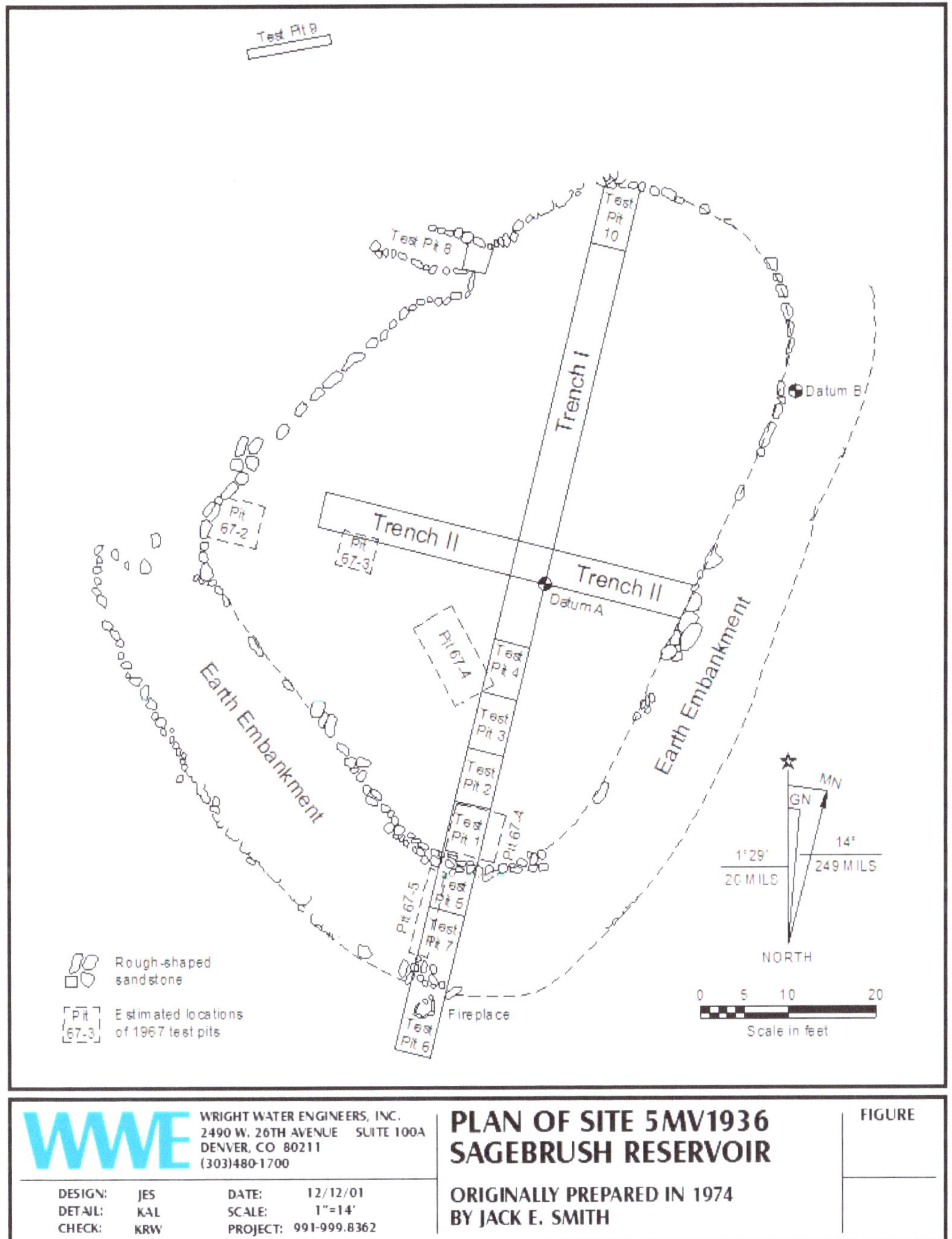

Figure 15. Plan view of 1974 Sagebrush Reservoir excavations.

3.4. Box Elder Reservoir

The final reservoir investigated by the WWE/WPI team was Box Elder Reservoir, a valley-bottom water storage site in Prater Canyon about 1.2 km west of Morefield Canyon. Box Elder Reservoir strongly resembles Morefield Reservoir as they are both on an east–

west latitude at the bottom of canyons, they were similarly constructed, and they both collected groundwater and runoff from nearby agricultural fields. Box Elder Reservoir was unknown to MVNP staff until the Bircher Fire of 2000 consumed the brush and trees that covered the mound left by the reservoir (Figure 16). A park ranger noticed the mound and saw it had many of the hallmarks of Morefield Reservoir. The WWE/WPI team conducted auguring and other studies that found water-borne sediment, layers of dredging evidence, and potsherds, confirming the Box Elder site had been a reservoir.

Figure 16. The Bircher Fire exposed a mound that was the remains of the Box Elder Reservoir.

WWE/WPI team research suggests that the Box Elder Reservoir was initiated by Pueblo I people in 800 CE or earlier. Based on precipitation data, the reservoir would have stored water about five times per year. Eventually the reservoir rose so high in elevation due to sediment accumulation and dredging that a diversion canal and increased maintenance became necessary. Team member Gregory Hobbs searched the bed of the Prater Canyon Creek for remnants of an ancient canal diversion that would have fed the reservoir and found none. However, while tracking the likely route of the canal between the creek bed and the reservoir area, he found an erosional remnant of a canal with typical canal liner stones. Interspersed with the stones were clay potsherds that Dr. Breternitz identified as Pueblo I era. This helped verify our hypotheses that this was the ancient diversion canal that carried water to Box Elder Reservoir.

By about 950 CE, the sediment from the canal caused the Box Elder Reservoir to reach six meters in elevation, become unmanageable, and no longer be viable for water storage. Park archeologists and the WWE/WPI team surmised that Morefield Canyon residents transferred technical reservoir-building knowledge to the residents of Prater Canyon around 800 CE (Figure 17) based on the many similarities of the two storage structures, the earlier evolution of the Morefield Reservoir, and the proximity of the two sites to each other.

Figure 17 shows that both reservoir sites had nearby springs that could also have been used for water supply, although springs were far more abundant in and near Morefield Canyon than Prater Canyon. This is congruent with an estimate by park archeologists that Prater Canyon had about 60% of the population of Morefield Canyon.

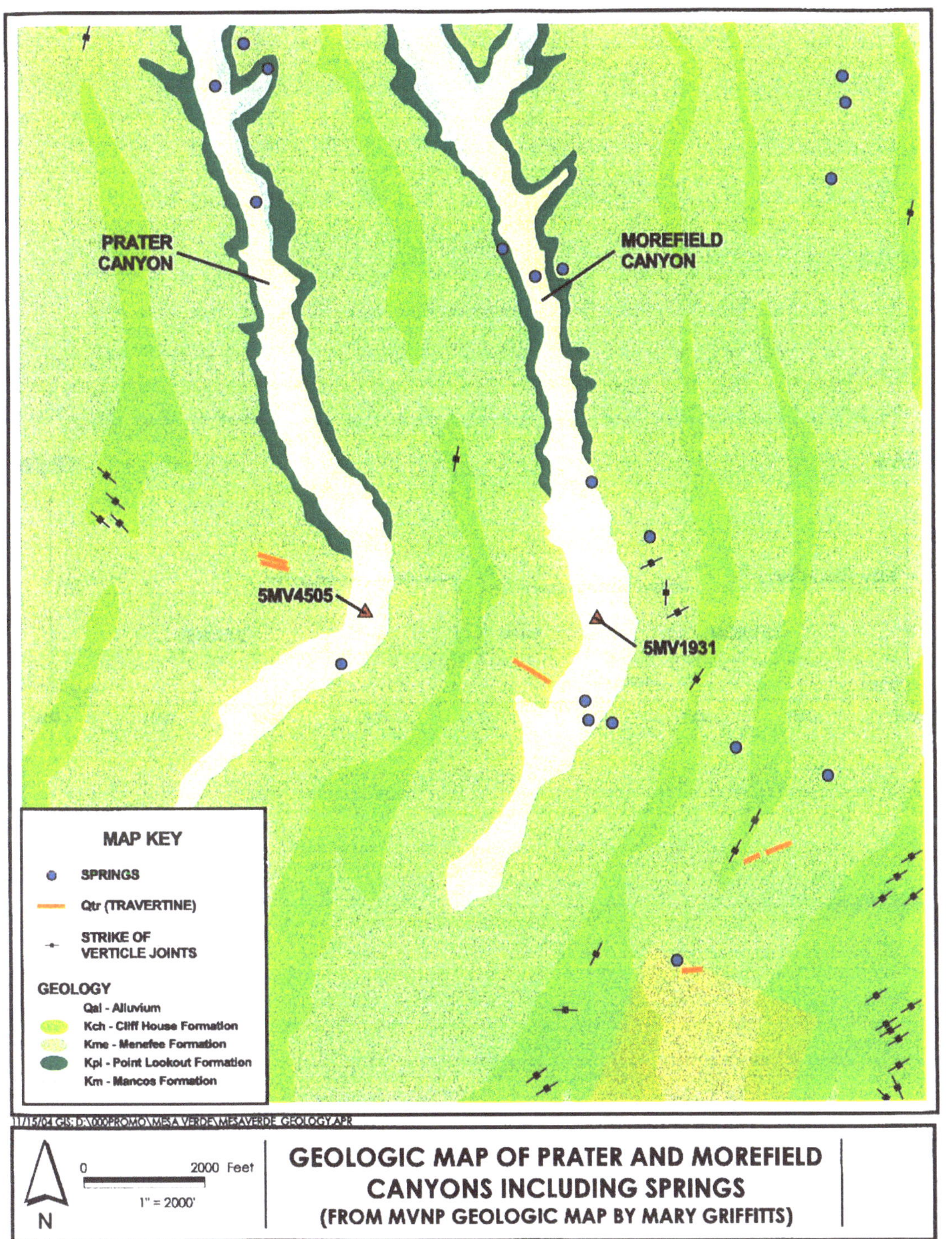

Figure 17. Geologic map of Prater and Morefield Canyons.

3.5. Mug House Cistern

After the reservoir research, the WWE/WPI team obtained a permit to study a water collection cistern located near the Mug House Ruin (Figure 18). This cliff dwelling on Wetherill Mesa was home to about 80–100 Pueblo III people during the 13th century. As conditions became drier and competition for resources surged, the ancestral Pueblo people began to develop cliff dwellings in locations that were more protected and defensible than the pit houses and kivas. Mug House was abandoned abruptly around 1280 CE, likely due to encroaching raiders [8].

Figure 18. The Mug House cliff dwelling was named for pottery mugs found hanging at the site when it was discovered by modern explorers.

The Mug House cistern (Figure 19) is a 15,000-L water storage vessel at the base of a 26-m-high cliff. At the top of the cliff, there is a deep notch that serves as a collection point for a six-acre forested drainage basin. A WWE/WPI team of about 20 professionals performed topographical surveys, inspected the drainage basin, analyzed the cistern structure, collected soil samples for soil gradation testing, performed hydrological analyses, and collected pollen samples to determine pre-historic vegetation.

The team conducted eight flow tests, discharging water at different flow rates at the collection notch to simulate ancient runoff as a rappeler recorded video of the hydraulic jet formed by the notch (Figure 20). The team learned that the hydraulics of the notch were functional and delivered water to the cistern at a range of flows. The notch was formed, either by nature or through manipulation, to jet the water away from the cliff wall so that little water was wasted during the drop.

Figure 19. Author Ken Wright and former Colorado Supreme Court Justice Gregory Hobbs preparing to take measurements of the Mug House cistern.

The cistern is roughly rectangular with dimensions of 2.5 by 7 m, with the longer axis of the cistern running parallel to the cliff face. The maximum depth for water storage before overtopping is one meter. On the west side of the cistern, the ancient engineers incorporated an overflow port allowing excess water to safely exit the structure. At the bottom of the north end of the cistern are two large sandstone splash blocks. Splash blocks minimize erosion due to impact from the high-velocity water falling from the cliff above, keeping the cistern floor intact and reducing turbidity levels that affect water quality. The blocks were also well-placed to be used as a jug-filling platform so that a water gatherer could avoid stepping onto the cistern bottom and raising clouds of sediment when the water level was low.

The ancestral Puebloans capitalized on 230 square meters of bare sandstone rock above the notch that shed rapid and immediate runoff when it rained. The team noted that the ancient people of Mesa Verde were likely aware of the need to control channel erosion. In the drainage basin above the Mug House cistern, the ancestral Pueblo people built sandstone block erosion control checks in two natural channels that convey water from east to west to the cliff edge. These ancient structures helped stabilize the channels by preventing downward channel cutting.

Hydrological analyses indicate that the Mug House cistern, under average conditions, would fill to capacity (15,000 L) from a single rainfall event about once each year. About 3400 L would have been generated four times per year, and twice per year, it would receive 4500 L. About 40 times per year, the cistern would receive 760 L of water, or more, per rainfall event. In a typical year, the cistern would have received approximately 380,000 L of water supply for useful storage, which could have supplied most of the domestic water needs of the Mug House community during approximately the April through October period. The team judged that the Mug House water gatherers would have had to trek much farther to a local spring for water during other times of the year [9].

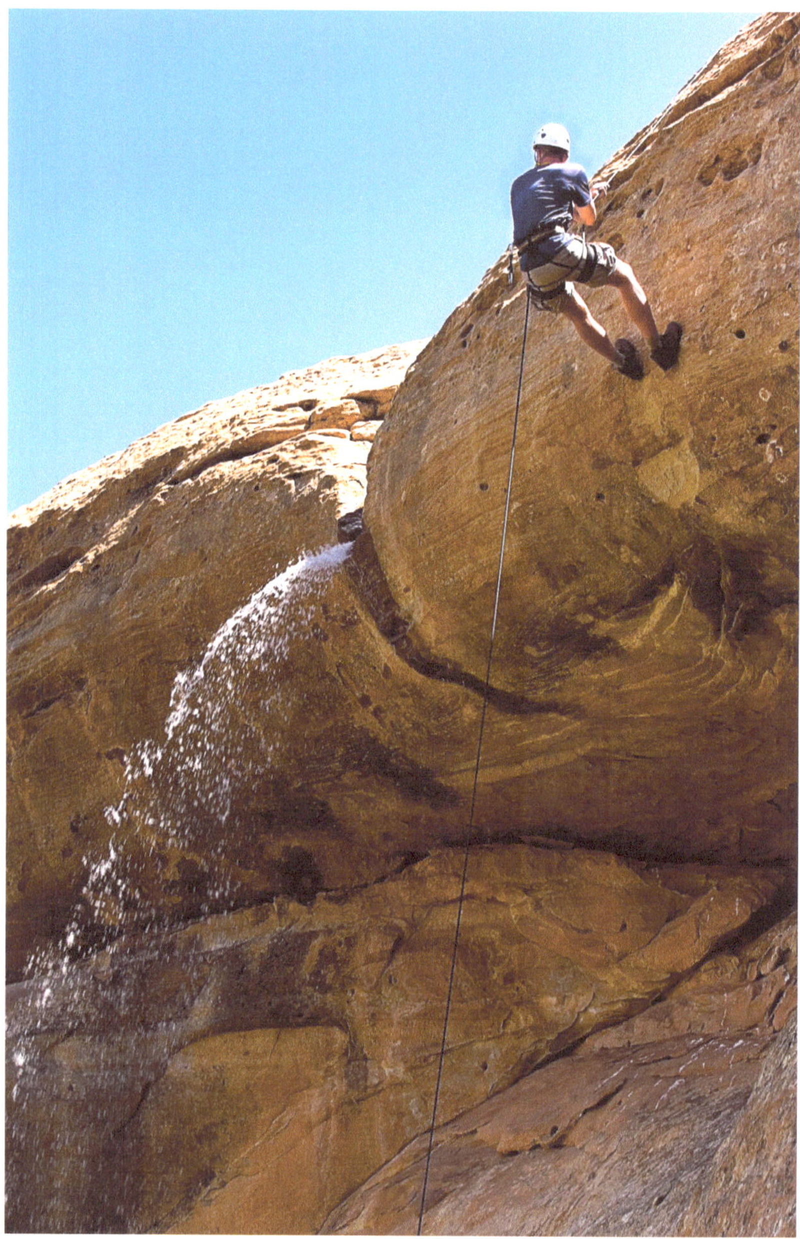

Figure 20. A rappeler collected video footage of various levels of flow through the notch above the cistern to test the notch's ability to jet water at different flow rates and the approximate amount of water that would have reached the cistern during a range of precipitation events.

3.6. Goodman Point Spring

The WWE/WPI team's final ancestral Pueblo water supply study site was Goodman Point, the one site described in this paper that is part of Hovenweep National Monument rather than MVNP. Goodman Point, approximately 18 km northwest of Cortez, Colorado, was occupied during roughly the same period as Mug House. The settlement was a walled community of about 500 to 800 people who had earlier lived in scattered family farmsteads but gathered at Goodman Point, likely for increased security, around 1260 CE. Like Mug House, Goodman Point was abandoned abruptly around 1280 CE [10].

Team studies focused on Juárez Spring (Figure 21, which was within the walled settlement and would have been the main source of water for the Goodman Point community. The team also researched other water sources that would have existed at the time: an ancient reservoir, now called Goodman Lake, and Mona Spring. Both Goodman Lake and Mona Spring were about one kilometer south-southwest of the Goodman Point Pueblo. Juárez Spring water issues from the Dakota Sandstone at two distinct points designated "Juárez Spring No. 1" and "Juárez Spring No. 2". The distance between the two springs is about 24 m. WWE/WPI studied geology, soils, spring location, water quality, and ground infiltration rates to assess the water supply provided by Juárez Spring. The team greatly benefitted from prior archeological studies performed at Goodman Point by the Crow Canyon Archaeological Center.

Figure 21. A team member installs a weir at Juárez Spring to measure flow.

Juárez Springs Nos 1 and 2 were developed on the sides of the slopes where the water issued from the bedrock. These locations would allow for the direct filling of jars. Below each spring outlet were pools that collected spring water. Because Juárez Spring has a low flow rate of about 10,600 L per day, it would have been necessary for Goodman Point residents to use both water directly from the spring and the pooled water. There would have been enough water to maintain the turkeys that were a major food source, but not for irrigation of crops. Luckily, dryland farming was, and still is, possible in the Four Corners area [11].

Mona Spring issues from a sandstone outcrop near the bottom of a shallow gully. Here, the likely headworks was a gully-bottom pool. Goodman Lake, with a sporadic water supply, provided stored water from direct rainfall-runoff (Figure 22). Stepping stones in the lake bottom would have allowed access to the pooled water for the filling of jars. The ancestral Pueblo people could have walked the kilometer to Goodman Lake or Mona Spring to fill their jars if necessary, helping ensure an adequate supply.

Figure 22. Goodman Lake during the winter.

The team determined that the quality of the Juárez Spring water was good for its purposes, even with rather high total dissolved solids levels. The water was essentially free of harmful organisms, such as bacteria. However, the pooled water adjacent to the springs and in Goodman Lake was susceptible to contamination [12].

4. Conclusions

The six ancestral Pueblo community water supply sources investigated by the WWE/WPI team provide much insight into these people and their circumstances. Through these artifacts, the ancestral Pueblo people demonstrated an impressive application of water harvesting knowledge and skill. They showed good hydrological understanding through their water storage and transportation activities, and also displayed organizational abilities through their collaborative creation and maintenance of reservoirs over many generations. The climate of this period was harsh, but the ancestral Pueblo people pulled together to create effective community systems that allowed them to survive with limited resources.

Funding: The work described herein was funded by Wright Water Engineers, Inc., Wright Paleohydrological Institute, and research grants from the Colorado Historical Society. The Colorado Historical Society was not involved in the study design, collection, analysis, interpretation of data, the writing of this article or the decision to submit it for publication.

Data Availability Statement: The data presented in this study are openly available in Wright Paleohydrological Institute records, COU:5090, Rare and Distinctive Collections, University of Colorado Boulder Libraries.

Acknowledgments: The author of this paper is indebted to two groups of researchers and experts who helped conduct the studies described herein. The first group is made up of past and present archeologists who have helped make Mesa Verde understandable and relevant to the public (those archeologists who collaborated with the WWE/WPI team are denoted by an asterisk after their name): David Breternitz,* Frederick H. Chapin, Mona Charles, Susan Collins, Calvin Cummings, Jesse W. Fewkes, Harold S. Gladwin, Alden Hayes, Bonnie Hildebrand, James Kleidon,* Kristin Kuckelman,* James A. Lancaster, Sean Larmore, Robert H. Lister, George McLellan, Earl Morris, Kara Naber, Larry Nordby,* Gustaf Nordenskiold, Arthur Rohn, Jeri Smalley, Jack Smith,* Linda Towle,* Guy Stewart, Joe Ben Wheat, Cynthia Williams, Richard Woodbury, Don G. Wyckoff, and Ezra Zubrow. The second group that requires acknowledgement consists of the professionals and technicians who donated their time and expertise to ancestral Pueblo water supply research and study over the 25 years WWE/WPI conducted this work. Some of these individuals participated in a single project, while some were involved in all six projects: Jason Alexander, David Baysinger, Eric Bikis, Aurora Bouchier, Joel Brisbin, Chris Brown, Ted Brown, Melissa Churchill, Chris Crowley, Linda Scott Cummings, Owen Davis, Eric DeLony, Andrew Earles, John Ewy, Elizabeth Fassman,

David Foss, Peter Foster, Michael Frachetti, Matt Gavin, Mary Gillam, Brad Hagen, Bobbie Hobbs, Gregory Hobbs, Richard Holloway, Robert Houghtalen, Robert Jarrett, Lisa Klapper, Sally Kribs, Bastiaan Lammers, Tom Langan, Brendon Langenhuizen, Peter Laux, Charles Lawler, Kurt Loptien, William Lorah, Wayne Lorenz, Rita Lovato, McKim Malville, Scott Marshall, Gordon McEwan, David Mehan, Grosvenor Merle-Smith, Peter Monkmeyer, John O'Brien, Lisa O'Connor, Ernie Pemberton, Ben Peterson, Patricia Pinson, Rachel Pittinger, Maria Prokop, Douglas Ramsey, Derek Rapp, Shannon Richardson, Warren Rider, John Rold, Lynette Shaper, Janice Sheftel, Terri Shelefontiuk, Chad Taylor, Donald Tucker, Jean Tucker, Ryan Unterreiner, Linda VanDamme, Robin VerSchneider, Rose Wallick, Kim Warhoe, Dwight Warren, Robert Weiner, Kyle Westendorf, Neil Williams, Richard Wiltshire, Gary Witt, and Ruth Wright.

Conflicts of Interest: Author Kenneth R. Wright was employed by the company Wright Water Engineers, Inc. The author declares no conflicts of interest.

References

1. Charles, M. The First Pueblonians: Hunters, Foragers, and First Farmers. In *The Mesa Verde World: Explorations in Ancestral Puebloan Archaeology*; Nobel, D.G., Ed.; School of American Research Press: Santa Fe, NM, USA, 2006; pp. 8–17. ISBN 978-1-930618-75-6.
2. Lekson, S.H. *The Chaco Meridian: One Thousand Years of Political and Religious Power in the Ancient Southwest*, 2nd ed.; Rowman and Littlefield: Lanham, MD, USA, 2015; ISBN 978-1-4422-4645-4.
3. Wright, K.R. *The Water Mysteries of Mesa Verde*; Johnson Books: Boulder, CO, USA, 2006.
4. Gove, P.B. *Webster's Third New International Dictionary of the English Language*; Unabridged Gove, P.B., Ed.; Merriam-Webster: Springfield, MA, USA, 1993; pp. 1902–1972.
5. Dean, J.S.; Robinson, W.J. *Climatic Variability in the American Southwest, AD 680 to 1970*; Laboratory of Tree-Ring Research: Tucson, AZ, USA, 1977.
6. Wilshusen, R.H. The Genesis of Pueblos: Innovations between 500 and 900 CE. In *The Mesa Verde World: Explorations in Ancestral Puebloan Archaeology*; Nobel, D.G., Ed.; School of American Research Press: Santa Fe, NM, USA, 2006; pp. 18–27. ISBN 978-1-930618-75-6.
7. United States Geological Survey. *Estimated Use of Water in the United States in 2015*; Dieter, C.A., Maupin, M.A., Caldwell, R.R., Harris, M.A., Ivahnenko, T.I., Lovelace, J.K., Barber, N.L., Linse, K.S., Eds.; Circular 1441; Water Availability and Use Science Program: Reston, VA, USA, 2017.
8. Rohn, A.H. *Wetherill Mesa Excavations Mug House Mesa Verde National Park-Colorado*; National Park Service: Washington, DC, USA, 1971.
9. Wright Paleohydrological Institute and Wright Water Engineers, Inc. *Mug House Cistern Hydrology: Site 5MV1586 Mesa Verde National Park*; NPS Permit #MEVE-2007-SCI-005; Wright Water Engineers, Inc.: Denver, CO, USA; Wright Paleohydrological Institute: Boulder, CO, USA, 2008.
10. Kuckelman, K.A.; Coffey, G.D.; Copeland, S.R. *Interim Descriptive Report of Research at Goodman Point Pueblo (Site 5MT604), Montezuma County, Colorado, 2005–2008*; Crow Canyon Archaeological Center: Cortez, CO, USA, 2009.
11. Creswell, R.; Martin, F.W. Dry land Farming Crops and Techniques for Arid Regions ECHO, 17391 Durance Rd., North Ft. Myers FL 33917, U.S.A. 1998. Available online: http://www.echonet.org/ (accessed on 15 December 2010).
12. Wright Paleohydrological Institute and Wright Water Engineers, Inc. *Goodman Point Palaeohydrology: Hovenweep National Monument*; Wright Water Engineers, Inc.: Denver, CO, USA; Wright Paleohydrological Institute: Boulder, CO, USA, 2011.

water

Article

Investigation of Improved Energy Dissipation in Stepped Spillways Applying Bubble Image Velocimetry

Lars Marius Mikalsen [1], Kasper Haugaard Thorsen [2], Aslı Bor [3,4,*] and Leif Lia [3]

1 Dr.techn. Olav Olsen AS, Vollsveien 17A, 1366 Lysaker, Norway; lmm@olavolsen.no
2 Multiconsult Norge AS, Nedre Skøyen vei 2, 0276 Oslo, Norway; kasperhaugaard.thorsen@multiconsult.no
3 Department of Civil and Environmental Engineering, Norwegian University of Science and Technology (NTNU), S.P. Andersens veg 5, 7491 Trondheim, Norway; leif.lia@ntnu.no
4 Department of Civil Engineering, Izmir University of Economics, 35330 İzmir, Turkey
* Correspondence: asli.b.turkben@ntnu.no

Abstract: This study investigates skimming flow regimes, two-phase air–water flow conditions, and simple measures to improve energy dissipation in stepped spillways. Experiments were conducted using two different scale physical models, 1:50 and 1:17, within separate rectangular flumes to define scale effects. Flow patterns were analyzed using the Bubble Image Velocimetry (BIV) technique, which tracks air bubbles. The introduction of splitters resulted in a 7% increase in relative energy dissipation. Additionally, the length of inception was reduced to $L_i/k_s = 10$, thereby decreasing the potential for subsequent cavitation. Beyond the BIV experiments, two experiments were conducted on the large-scale model using Acoustic Doppler Velocimetry (ADV), with and without splitters, to examine the impact of splitters on the velocity profile above the crest. In the experiment with splitters, the vertical velocity vector (v) contributed to turbulence by changing direction, thereby reducing average velocities both in front of and behind the ogee crest. This led to a reduction in energy on the downstream side of the spillway. Although the small-scale model appears unsuitable for studying two-phase flow, the change in relative energy dissipation from the baseline to the splitter configuration was practically identical for both scale models, thereby supporting the findings of the large-scale model.

Keywords: stepped spillways; energy dissipation; inception point; skimming flow; air-entrainment; ADV

Citation: Mikalsen, L.M.; Thorsen, K.H.; Bor, A.; Lia, L. Investigation of Improved Energy Dissipation in Stepped Spillways Applying Bubble Image Velocimetry. *Water* **2024**, *16*, 2432. https://doi.org/10.3390/w16172432

Academic Editors: Helena M. Ramos and Charles R. Ortloff

Received: 12 July 2024
Revised: 11 August 2024
Accepted: 20 August 2024
Published: 28 August 2024

1. Introduction

Stepped spillways, a type of energy dissipator, have garnered increased interest since the advent of roller-compacted concrete (RCC). These structures have been studied extensively through both physical and numerical models, with particular emphasis on the transitions of flow regimes and modifications of step geometry (Chanson [1], Kokpinar [2], Zhang & Chanson [3], Sánchez-Juny et al. [4] and Kramer [5]). The earliest laboratory experiments in the literature utilized electrical resistance probes (Rajaratnam [6]). Kramer [5] and Kokpinar [2] measured local values of air concentration, air bubble frequency, and average beam length in the air–water flow region using a fiber optic instrumentation system, while Murillo [7] employed hot film anemometry. Additional methods used in experiments to investigate the interaction of air–water two-phase flows include Laser Doppler Anemometry (LDA) and Particle Image Velocimetry (PIV). Ohtsu & Yasuda [8] measured velocities using a one-dimensional fiber laser-Doppler (LDA) velocity meter. Amador et al. [9] characterized the evolving flow using Particle Image Velocimetry (PIV), although their study was limited to a narrow range of laboratory flow conditions. They found PIV to be particularly useful in the non-aerated region.

Bubble Image Velocimetry (BIV) has been applied to visualize the complex flow patterns over stepped spillways. Bung & Valero [10] used BIV to study the air–water flow structure in stepped spillways, revealing intricate details of turbulence and flow separation. Their findings demonstrated that BIV could effectively capture the interaction between water and air phases, providing valuable insights into the energy dissipation process. Kramer & Chanson [11] and Sánchez-Juny et al. [4] explored the use of BIV to measure velocity fields in stepped spillways, obtaining high-resolution velocity profiles. These studies showed that BIV can accurately measure the distribution of velocities in both horizontal and vertical directions. The data obtained from BIV were compared with results from traditional phase-detection probes, highlighting the advantages of BIV in terms of data density and spatial resolution. Investigations into the scale effects on energy dissipation using BIV have been conducted to ensure the validity of laboratory findings. Zhang & Chanson [3] examined the influence of scale on the energy dissipation characteristics of stepped spillways, using BIV to compare flow patterns and velocity fields at different scales. Their results indicated that BIV could reliably capture scale-dependent variations in flow behavior, supporting its use in both small-scale models and full-scale prototypes. The application of BIV in optimizing the design of stepped spillways has been a focus of recent research. Studies by Amador et al. [9] and Kramer [5] utilized BIV to assess the impact of design modifications, such as the inclusion of crest splitters, on energy dissipation. These investigations demonstrated that crest splitters could enhance energy dissipation by increasing turbulence and reducing flow separation, as evidenced by detailed BIV measurements. Other studies that have utilized the BIV technique include those by [Lopes et al. [12], Leandro et al. [13] and Kramer & Chanson [11]. Despite the advantages of BIV, certain challenges remain. Issues such as image distortion, bubble size consistency, and lighting conditions can affect the accuracy of BIV measurements. Future research should aim to refine BIV techniques and integrate them with other measurement methods, such as Particle Image Velocimetry (PIV) and Laser Doppler Anemometry (LDA), to improve data reliability. Additionally, the development of advanced image processing algorithms could enhance the capability of BIV to capture flow dynamics in more complex hydraulic environments.

The hydrodynamics of flow over stepped cavities have been classified into nappe, transition, and skimming flow regimes, each presenting complex analytical challenges. This study focuses on the skimming flow regime (Figure 1), two-phase flow conditions, and straightforward measures for improving energy dissipation. These were investigated using physical hydraulic models at two different scales (1:50 and 1:17). Crest splitters were installed to further enhance energy dissipation, and the hydraulic jump method was applied to measure the residual energy level. This allowed for a more detailed study of energy dissipation and provided additional insights into the flow patterns using high-speed cameras and state-of-the-art photogrammetry technology. The flow pattern during the experiments was measured using the Bubble Image Velocimetry (BIV) method, a modified version of Particle Image Velocimetry (PIV) that uses air bubbles in the two-phase air–water flow as tracers. Data obtained from this method are extremely detailed and information-intensive, making it ideal for studying highly turbulent biphasic flows. As expected, scale effects were observed in both air entrainment and energy dissipation. Flow patterns in the spillway ogee crest were investigated using Acoustic Doppler Velocimetry (ADV) measurements, both with and without crest splitters. The analysis focused on the effect of the splitters on the crest.

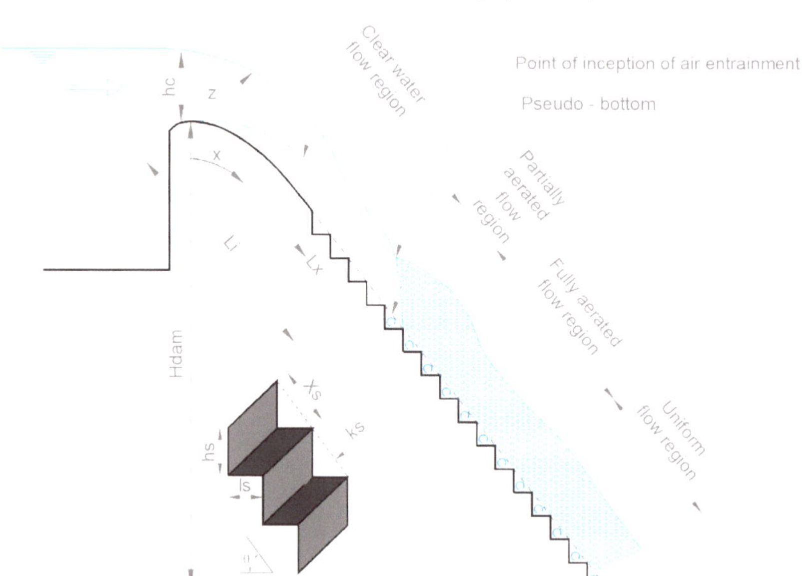

Figure 1. Definition sketch of stepped spillways in skimming flow conditions.

2. Materials and Methods

2.1. Experimental Set-Up and Procedure

Experiments were conducted at the Norwegian Hydraulic Laboratory at NTNU using two different scaled stepped spillway models (C-Model and D-Model) in separate flumes, adhering to the Froude similarity law. These models are based on a hypothetical prototype RCC dam with a height of 20 m, corresponding to a slope angle of $\theta = 51°$, featuring a vertical water edge and a downstream slope of 1:0.8 (V:H). The prototype consists of 11 steps (N_s), each with a step height h_s of 1.5 m. The scaled models, built to Froude similitude, were constructed in rectangular flumes to achieve the desired skimming flow regime within the limitations of flume size and available discharge. This design mirrors typical stepped spillways on existing gravity dams in the lower height range (Wright & Cameron-Ellis [14] and Chanson H. [15]). The test results were compared with data from the Hinze Dam Stage 3 (Chanson H. [15]) on the Gold Coast in Australia, which has similar slopes and step heights, with a stepped spillway approximately 35 m high. The C-Model, scaled at 1:50 (Figure 2a), was built in a 20-m-long C-flume using horizontal layers of marine plywood, analogous to the RCC construction method, with a crest of extruded polystyrene (XPS). The D-Model, scaled at 1:17 (Figure 2b), was constructed using planks and marine plywood, also with a crest of XPS, and situated in a 25-m-long, 2-m-high, and 1-m-wide D-flume. Originally, the D-Model had a step of approximately 0.37 m located 1.2 m downstream from the toe, but this plateau was extended to 4 m to allow a hydraulic jump to develop undisturbed from the bed level difference. The use of two different scales facilitated the investigation of scale effects. The size of the D-Model closely matches the 1:15 scale recommended by Boes & Hager [16], though slightly smaller due to flume limitations.

A standard ogee crest was designed following USBR (1987) [17] guidelines, with two upstream radii, $R_1 = 0.5H_0$ and $R_2 = 0.2H_0$, where $H_0 = 4.89$ m in the prototype, corresponding to a unit discharge of $q_w = 24$ m^2/s. For this design, the downstream curvature is defined by $y = -0.5H_0(x/H_0)^{1.87}$, y is the vertical distance from the top of the weir, and x the horizontal distance. The height of the curved crest is 2 m, resulting in a slope angle of approximately 40° before the first step. Extending the curvature to an angle of $\theta = 51°$ would reduce the number of steps, which was not desirable for the research objectives. Flow regulators were used in both channels to minimize turbulence and create a calm water surface with minimal cross-flow in the flumes. An overview of the C-Model and D-Model is summarized in Table 1.

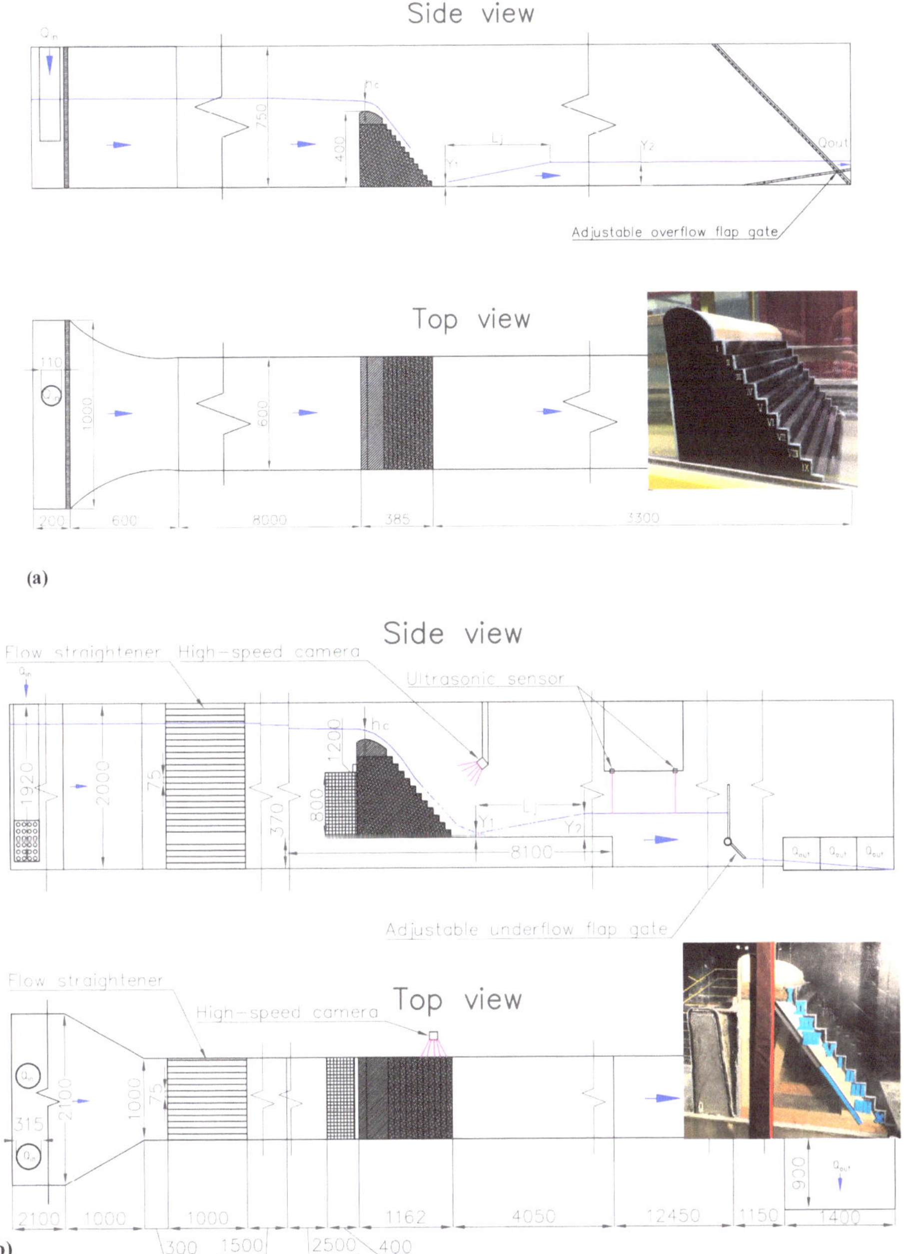

Figure 2. Experimental setup for (**a**) C-flume and (**b**) D-flume. Flow direction from left to right and dimensions in millimeters (Mikalsen & Thorsen [18]).

Table 1. Overview of model and prototype dimensions.

Name	Scale	H (m)	h_s (m)	H_f (m)	B_f (m)	Q_{max} (m^3/s)
Prototype	01:01	20	1.5	-	-	-
C-Model	01:50	0.4	0.03	0.75	0.6	0.058
D-Model	01:17	1.2	0.09	2	1	0.51

Note: H = local energy head, H_f = height of flume, B_f = flume width, Q_{max} = maximum water discharge.

The discharge was monitored using a Siemens Sitrans FM MAG 5100 W (München, Germany), an electromagnetic flow sensor mounted on the inlet pipes. The transmitter was a Siemens Sitrans FM MAG 5000, with an accuracy of 0.2% ± 2.5 mm/s. In the D-flume, the downstream sequent depth of the hydraulic jump was measured by two ultrasonic sensors, Microsonic mic + 130/iu/tc (Dortmund, Germany) with a stated accuracy of ±1%. These sensors were positioned in the center of the flume, with an internal distance of 0.8 m between them, and the upstream sensor was placed 4 m from the spillway toe. In the C-flume, the sequent depth of the hydraulic jump was measured manually using a ruler. The ultrasonic sensors measured depths within a specified range, and preliminary experiments were conducted to determine the relevant range of depths for the different discharges used.

2.2. Crest Splitters

Experiments were conducted both with and without energy-reducing crest splitters. The crest splitters were included to induce additional turbulence near the crest without decreasing the spillway capacity. These crest splitters were inspired by the design criteria outlined by Robert [19], with some modifications to facilitate their implementation in new and existing stepped spillways. The splitters were constructed from wood, as shown in Table 2 and Figure 3. In this study, the experimental results were labeled as "base case" (without splitters) and "with splitters" (with splitters).

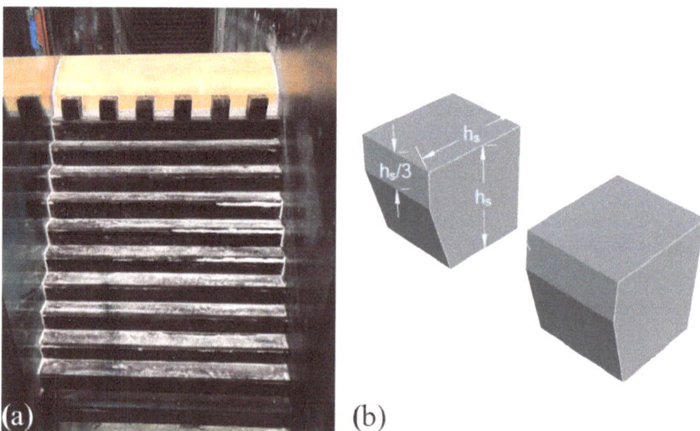

Figure 3. Crest splitters positioning: (**a**) downstream face of the D-Model with crest splitters and (**b**) design of splitters (Mikalsen & Thorsen [18]).

Table 2. Crest splitters design parameters, prototype scale.

Dimension	Length (m)	Relative Size
Height	1.5	h_s
Top length	1.5	h_s
Bottom length	1.2	l_s
Width	1.2	l_s
Gap	1.5	h_s
Vertical face	0.5	$h_s/3$

2.3. Procedure for BIV

In this study, flow characteristics and velocity profiles on the steps were developed from side and top video recordings. Videos of bubbles illuminated with intense white light were captured using a SONY DSC-RX0M2G camera with a resolution of 1920×1080 pixels, a 16:9 aspect ratio, and a frame rate of 1000 frames per second. The recordings were then analyzed using PIVlab, an open-source toolbox in MATLAB R2023a. Strong white light sources were utilized during the experiments to increase the concentration of air within the flow, which darkened the areas farthest from the flow and made the air bubbles more distinct. Using the classical Particle Image Velocimetry (PIV) technique, PIVlab tracks the tracer particles to obtain the flow velocity field through image processing (William & Stamhuis [20]). There are three stages in BIV analysis: image pre-processing, evaluation, and post-processing. Figure 4 illustrates the result of the pre-processing stage and the consequent improvement in the photograph quality.

Figure 4. One representative frame from the video recordings before and after pre-processing and masking in PIVlab—flow direction from left to right: (**a**) original frame; (**b**) pre-processed and masked frame (The arrow indicates the direction of flow) (Mikalsen & Thorsen [18]).

The red areas in the photo were then masked, and particle movements in these regions were omitted from the analysis to save processing time. This image evaluation phase leaves only the air bubbles, which the program identifies as white particles on a black background, facilitating frame-to-frame analysis. The results from each particle movement between frames were averaged over 2042 ms, and default settings were used for post-processing. To ensure accurate velocity extraction, a known measured reference length was used for calibration.

2.4. Experiment Parameters

The experiments in the C- and D-flumes were conducted with an initial normalized critical depth of $h_c/h_s = 0.8$. The discharge was controlled by manual valves and measured using an electromagnetic flow sensor. All models were run with increased discharges from a depth of $h_c/h_s = 0.8$ to approximately $h_c/h_s = 3.3$ with intervals of $h_c/h_s = 0.1$. This corresponds to prototype unit discharges ranging from $q_w = 4 \text{ m}^2/\text{s}$ to $q_w = 34 \text{ m}^2/\text{s}$. Table 3 shows the range of Reynolds numbers and Weber numbers, along with the range of discharges for the experiments. The Weber number in this study was calculated in the upstream cross-section of the hydraulic jump at the dam toe, as measuring the depth-averaged mixture velocity $\overline{u_m}$ is not possible without intrusive probes. To reduce uncertainty in the results, the experimental program was repeated three times for each of the four models.

Table 3. Summary of parameters in experiments.

Model	λ_f (-)	Q (L/s)	h_c/h_s (-)	Re (-)	W (-)
C-Model	50	7.2–58	0.6–3.2	1.2×10^4–9.5×10^4	32–62
D-Model	17	61–510	0.6–3.3	6.1×10^4–5.1×10^5	106–179

Note: λ_f = length scale factor in Froude similitude, R_e = Reynolds number, W = Weber number.

Three-component instantaneous flow velocities (u, v, w) on the ogee crest were measured with a Nortek ADV (Bologna, Italy) (10 MHz) device at a sampling frequency of 100 Hz for the D-Model with a discharge of $Q = 110 \; L/s$. Velocity measurements were recorded for 3 min at each measurement point, resulting in the collection of approximately 18,000 data points per location. This sample size was considered appropriate given the turbulence in the downstream region of the full crest point (Bor [21]). The collected velocity component data were processed and cleaned using the phase-space threshold (PST) filter algorithm, which was employed to remove outliers from the velocity time series based on the physical principle that maximum particle acceleration should not exceed a certain value (Goring & Nikora [22]). In the initial data processing, the ExploreV Pro program was used to discard points with an average correlation below 70% or an SNR below 15 dB. Subsequent data processing and time-averaging of the velocity measurements were performed using MATLAB code, which was developed specifically for ADV data post-processing and consists of several independent functions. In Figure 5, the measurement mesh points are shown both in plan and vertical views: 495 mesh points were used, including 6 profiles in the x-direction and 10 profiles with 5 cm vertical spacing in the y-direction.

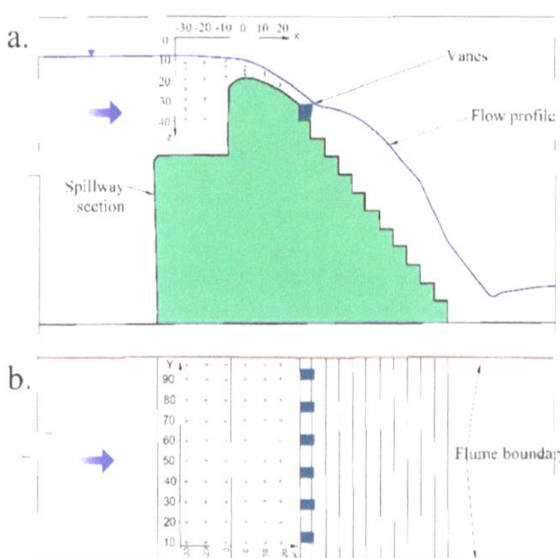

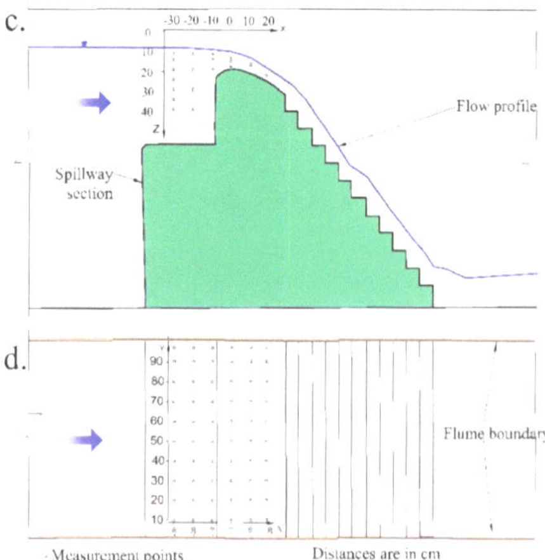

Figure 5. The measurement velocity grids in (**a**) base case cross-section and (**b**) base case plan, with (**c**) with splitters cross-section and (**d**) with splitters plan (units are in centimeters).

3. Experimental Results

3.1. Post-Processing of the Flow Characteristics

Velocity values were measured for a highly aerated flow over a stepped spillway using the BIV method, which employs air bubbles trapped in the flow as tracers. In Figure 6a, velocity vectors from the BIV analysis can be observed at the maximum discharge of the D-Model base case. No free surface aeration occurred in either C-Model or D-Model under maximum discharge conditions; therefore, bubbles were introduced at the first step to serve as BIV tracers. Figure 6a,b illustrate the phenomenon of circulating vortices in the step cavities. In the BIV analysis, the density of vectors does not represent the density of information in the velocity field; rather, the length of the vector represents the magnitude of the velocity.

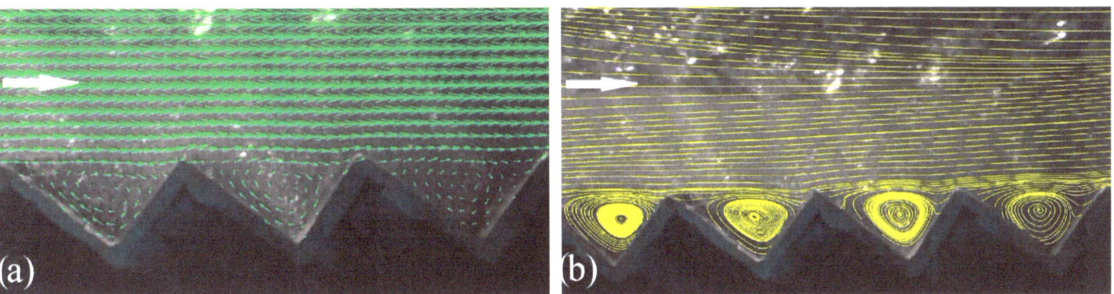

Figure 6. (**a**) Velocity vectors after post-processing; (**b**) streamlines in D-Model obtained from BIV analysis (flow direction from left to right).

Figure 7a displays the skimming flow conditions in the stepped spillway models. The point of inception was observed at $L_x/X_s = 5$ under the $h_c/h_s = 1.2$ flow condition in the D-Model. Figure 7b shows the C-Model under maximum flow conditions. For very low discharges, up to approximately $h_c/h_s = 0.6$ in the C-flume and slightly less in the D-flume, the water formed a deflection nappe from the first step, which descended as a free-falling nappe before approaching the spillway further downstream. Figure 7c illustrates the nappe deflecting out from step I and landing on the stepped chute further downstream at step IV. For even smaller discharges, the nappe was observed hitting step VIII/IX.

Figure 7. (**a**) Skimming flow for D-Model base case $h_c/h_s = 1.2$; (**b**) skimming flow for C-Model splitters $h_c/h_s = 3.3$; (**c**) a nappe is formed for very low discharges, C-Model base case $h_c/h_s = 0.6$. (Both of the models' steps are labelled with Roman symbols).

Figure 8 shows a comparison of the C-Model and D-Models for the transition of flow regimes in base case and splitters configuration experiments under $h_c/h_s = 1.2$ flow conditions. It can be observed that the splitters can alter the skimming flow regime at intermediate discharges. For discharges of $h_c/h_s = 1.2$ in the D-Model and $h_c/h_s < 1.5$ in the C-Model, the splitters caused a water jet to deflect off the spillway, resulting in a free-falling jet that descended to a lower level of the stepped spillway invert.

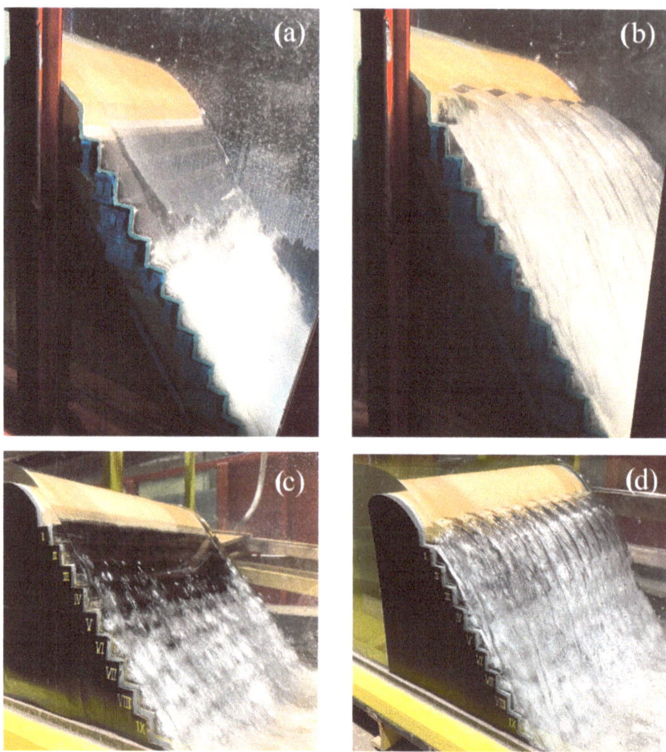

Figure 8. Comparison of flow regimes in the (**a**) D-Model base case (**b**) D-Model splitters (**c**) C-Model base case (**d**) C-Model splitters, at the same normalized critical depth $h_c/h_s = 1.2$. (Both of the models' steps are labelled with Roman symbols).

In addition to this deflecting jet, it is possible to observe air pockets inside the cavities, which reduce the traditional step-cavity circulation of the skimming flow regime, as shown in Figure 9. The regional drowning of the splitters is rather unstable and appears to be influenced by upstream turbulence and transverse flow.

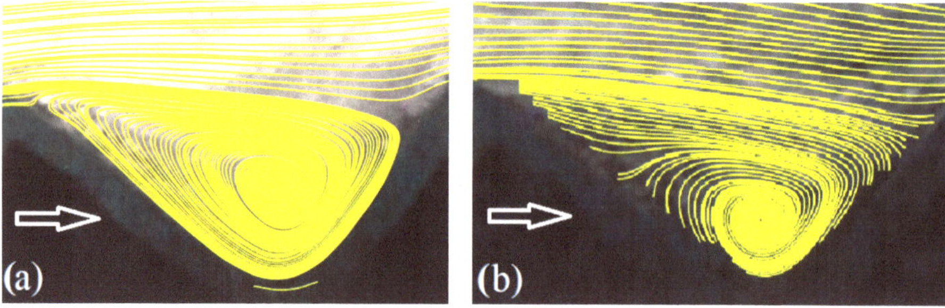

Figure 9. Comparison of streamlines obtained from BIV analysis of (**a**) base case and (**b**) splitters configuration in D-Model $h_c/h_s = 1.2$ and $7 < L_x/x_s < 8$ flow direction from left to right.

This new flow regime observed at low to intermediate discharges, combined with the splitters' configuration, exhibits considerable aeration and lacks a distinct pseudo-bottom, resembling a transitional flow regime, although it appears more stable. Figure 10 displays the unstable transitions occurring at flow rates of about $1.8 \leq h_c/h_s \leq 2.0$ in the D-flume.

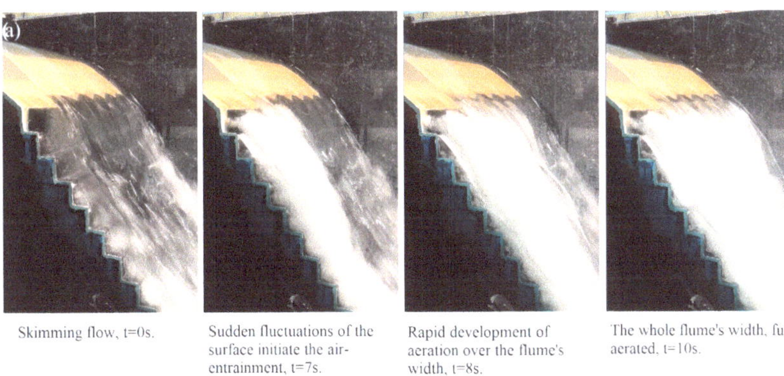

Skimming flow, t=0s.

Sudden fluctuations of the surface initiate the air-entrainment, t=7s.

Rapid development of aeration over the flume's width, t=8s.

The whole flume's width, fully aerated, t=10s.

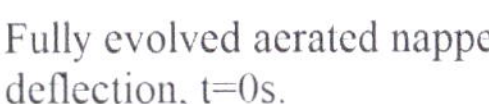

Fully evolved aerated nappe deflection, t=0s.

Step cavities get filled with water, t=5s.

Splitters and cavities fully submerged, t=8s.

Figure 10. Video frames displaying unstable transition from (**a**) a skimming regime to a deflecting nappe regime and (**b**) a deflecting nappe regime to a skimming regime over approximately 10 s, $h_c/h_s = 1.8$. (Both of the models' steps are labelled with Roman symbols)

3.2. Inception of Free-Surface Aeration

The classical definition of the inception of free-surface aeration is the point at which the turbulent boundary layer reaches the free surface in stepped chutes (Figure 1). To maintain consistency with the literature, the studies by Hunt & Kadavy [23] and Hunt et al. [24] were used to guide the introduction of the inception of free-surface aeration. In these studies, L_i s defined as the downstream distance from the downstream edge of the spillway crest to the point where "white water" first appears across the entire width of the stepped chute's free surface—in other words, where the turbulent boundary layer reaches the surface. Wood [25] derived an empirical relationship for defining the inception of free-surface aeration L_i for smooth chutes, as follows:

$$\frac{L_i}{k_s} = 13.6(sin\theta)^{0.0796} F_*^{0.713} \tag{1}$$

where $\theta = 51.3°$ is the spillway slope for this study; k_s is the step roughness (normal height of the step in this study); F_* is the roughness Froude number and given as:

$$F_* = \frac{q_w}{\sqrt{gsin(\theta)k_s^3}} \tag{2}$$

where q_w is the discharge per unit width and g is the gravitational acceleration. Validating with several independent studies, Boes & Hager [16] stated the following free-surface inception point relationships for skimming flow conditions as the point with 0.01% air for slopes of $26^0 \leq \theta \leq 55^0$,

$$\frac{L_i}{k_s} = \frac{5.90(cos\theta)^{1/5}}{sin\theta} F_*^{4/5} \tag{3}$$

In their study, they defined the depth-averaged air concentration at the point of origin as follows:

$$\overline{C}_i = 1.2 \times 10^{-3}(240 - \theta) \tag{4}$$

where $\overline{C}_i$ is the depth-averaged air concentration at the point of inception. Matos [26] determined the point of inception as where the boundary layer reached the free surface and determined that the average value of $C_{meani} \approx 0.2$ for steep slopes.

$$\frac{L_i}{k_s} = 6.289 F_*^{0.734} \tag{5}$$

$$\overline{C}_i = 0.163 F_*^{0.154} \tag{6}$$

In this study, the inception of free-surface aeration was determined visually by analyzing digital images for each experiment. The dimensionless inception point distance L_i/k_s was observed with respect to the free-surface inception point L_i and surface roughness for vertical face steps k_s. Figure 11 presents a comparison of the experimental measurements with the empirical relationships proposed by Wood [25], Boes & Hager [16] and Matos [26] alongside the prototype data from the Hinze Dam (Chanson H. [15]) used for validation.

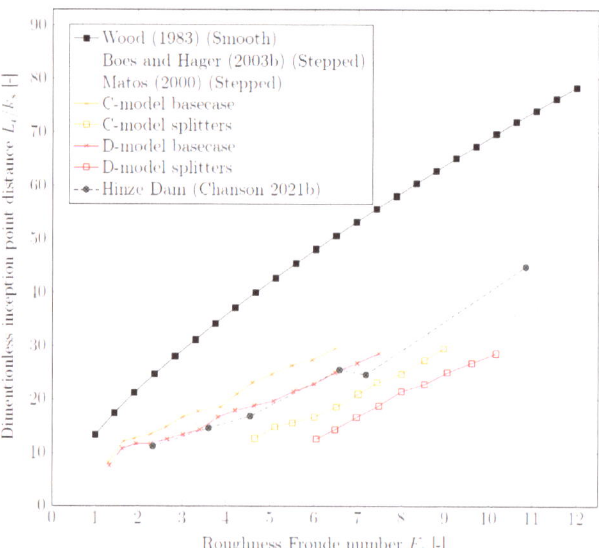

Figure 11. Comparison of experimental data, prototype data, and empirical relations for the location of the inception point (Chanson H. [15], Boes & Hager [16], Wood [25] and Matos [26]).

3.3. Velocity Profiles over Step Edges

For flow in both the aerated and non-aerated regions, the velocity distribution is best described by a power law (Chanson H. [1], Amador et al. [27] and Meireles et al. [28]):

$$\frac{u_m(z)}{u_{90}} = D\left(\frac{z}{z_{90}}\right)^{1/n} \tag{7}$$

$$\frac{u(z)}{u_{FS}} = \left(\frac{z}{\delta}\right)^{1/n} \tag{8}$$

where $\frac{u_m(z)}{u_{90}}$ is the dimensionless mixture velocity with $u_m(z)$ being the mixture velocity in the normal distance z over the pseudo-bottom. u_{90} is the mixture flow velocity at the characteristic flow depth z_{90} with the local air concentration of 90%, often defined as the surface of the air–water mixture flow. $\frac{z}{z_{90}}$ is the dimensionless mixture flow depth. For non-aerated flows, $\frac{u_z}{u_{FS}}$ is the dimensionless velocity, u_{FS} is the free-stream velocity, and δ is the boundary layer thickness, defined as the normal distance z, where the velocity reached 99% of the maximum value (Zhang & Chanson [3], Kramer [5] and Amador et al. [9]). The coefficients D and n are selected to fit the data obtained from the experiments, with n often exhibiting a wider range of values than D, which is typically set to 1. The coefficient n ranges from 3.0 to 5.4 in the non-aerated zone and from 6.6 to 14 in the aerated region. Some studies have reported for which values of $H_{90} = \frac{z}{z_{90}}$ the flow follows the power law. Above the boundary layer in both non-aerated and aerated flows, the velocity remains relatively constant up to the surface. An empirical relation between chute slope, normalized discharge, and the power-law coefficient n was proposed by Takahashi & Ohtsu [29] and later utilized by Kramer [5] as follows:

$$n = 14\theta^{-0.65}\frac{h_s}{h_c}\left(\frac{100}{\theta}\frac{h_s}{h_c} - 1\right) - 0.041\theta + 6.27 \quad \text{for} \quad 19^\circ \leq \theta \leq 55^\circ \tag{9}$$

where θ is the angle of the chute slope, h_s is the height of the step, and h_c is the critical depth. Researchers have proposed other models for describing flow velocities over triangular cavities. Kramer [5] combined four different models in his study, creating a multilayered velocity model that corresponded well with measurement data. The number of parameters required for such a model is higher than for simpler models using the power law. The model proposed for the mixing layer by Kramer [5] is as follows:

$$\overline{u}_{ML} = \left(\overline{u}_{if} - \overline{u}_{min}\right)\left(1 + tanh\frac{z - z_{if}}{L_e}\right) + \overline{u}_{min}, \quad \text{if } z < \delta \tag{10}$$

where $\overline{u}_{ML}$ corresponds to the mixing layer velocity, $\overline{u}_{if}$ is the velocity at the infection point, $\overline{u}_{min}$ is the minimum velocity in the mixing layer, and L_e is the characteristic length scale of the mixing layer. z_{if} denotes the elevation of the inflection point above the pseudo-bottom, under the conditions outlined below::

$$\frac{\overline{u} - \overline{u}_{min}}{\overline{u}_{FS} - \overline{u}_{min}} = 0.5 \tag{11}$$

where $\overline{u}$ is the time-averaged streamwise velocity. In this study, the velocity profiles were calculated from time-averaged BIV measurements without validation. Due to the absence of instrumentation for measuring air concentration, the velocity distribution equation was applied in both the aerated and non-aerated zones. Figure 12 shows a comparison of the theoretical power law velocity distribution with the velocity profiles measured by BIV over step edges at locations $L_x/x_s = 4, 5, 6, 7, 8, 9$ with $h_c/h_s = 1.2$ for the D-Model base case.

Figure 13 presents a comparison of velocity profiles measured with BIV at locations $L_x/x_s = 5.5$ and $L_x/x_s = 6.5$ with $h_c/h_s = 1.2$ in the D-Model for both the base case and the splitters configurations.

Figure 14 presents a comparison of surface velocities measured with BIV from a top view at $h_c/h_s = 1.2$ and $h_c/h_s = 1.6$ in the C-Model and D-Model for the base case configuration. "IP" corresponds to the inception point, or the location where free-surface aeration begins.

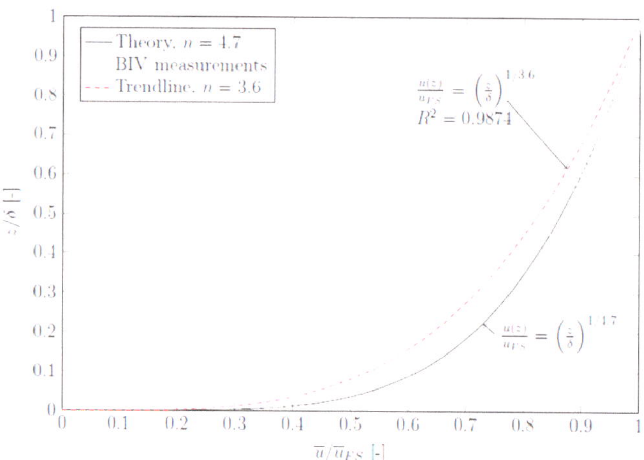

Figure 12. Dimensionless velocity profiles measured with BIV over step edges in D-Model base case compared with power-law approach $h_c/h_s = 1.2$, $z < \delta$ and $L_x/x_s = 4,5,6,7,8,9$.

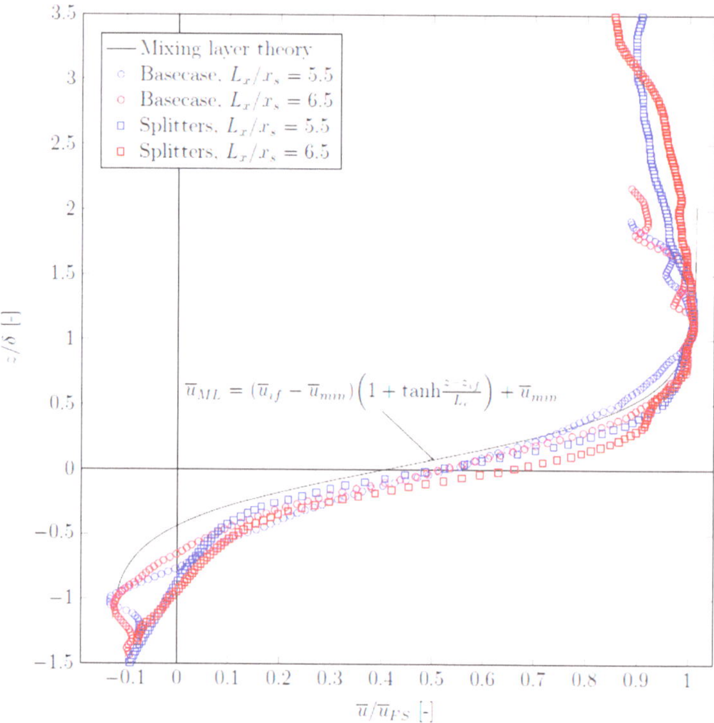

Figure 13. Velocity profiles in the D-Model measured with BIV compared with mixing layer theory $h_c/h_s = 1.2$, $z < \delta$, $L_x/x_s = 5.5$ and $L_x/x_s = 6.5$.

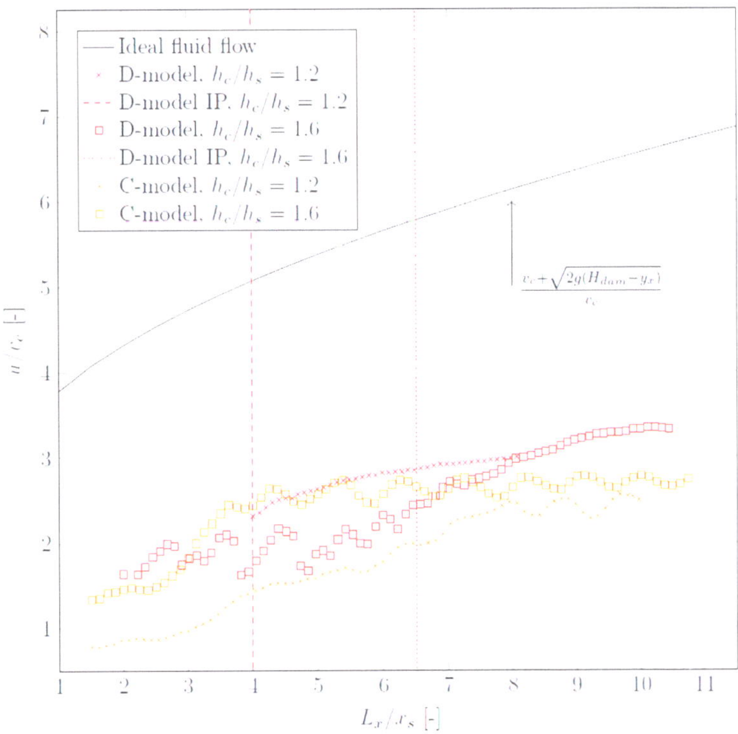

Figure 14. Surface velocities in the C-Model and D-Model base case measured with BIV from top-view compared with ideal fluid flow.

3.4. Flow Patterns on Ogee Crest

For each data point, the time-averaged components of the velocity vectors were measured and normalized by the average flow velocity (u_0). The figures were generated using dimensionless lengths, where h represents the ogee height. The contours of the normalized horizontal velocity components u/u_0 and v/u_0 are shown in Figure 15 for both the base case and splitters configurations. Figure 16 presents contour plots of $U_{xy} = \sqrt{(u/u_0)^2 + (v/u_0)^2}$, along with horizontal velocity vector plots of u and v the flow surface, displayed on a horizontal plane with a color scheme for both the base case and splitters configurations.

According to ADV measurements, the normalized vertical velocity vectors w/u_0 along to the channel axis are plotted for stream-wise locations $x/h = 0.55$, 0.2, 0, -0.25, -0.5, -0.75, and -1.0 as presented in Figure 17.

In Figure 17, the normalized vertical one-dimensional profiles w/u_0 taken along these streamwise locations are compared for both the base case and splitters configurations. As shown in Figure 17, the profiles become nearly identical after the streamwise location $x/h = -0.25$.

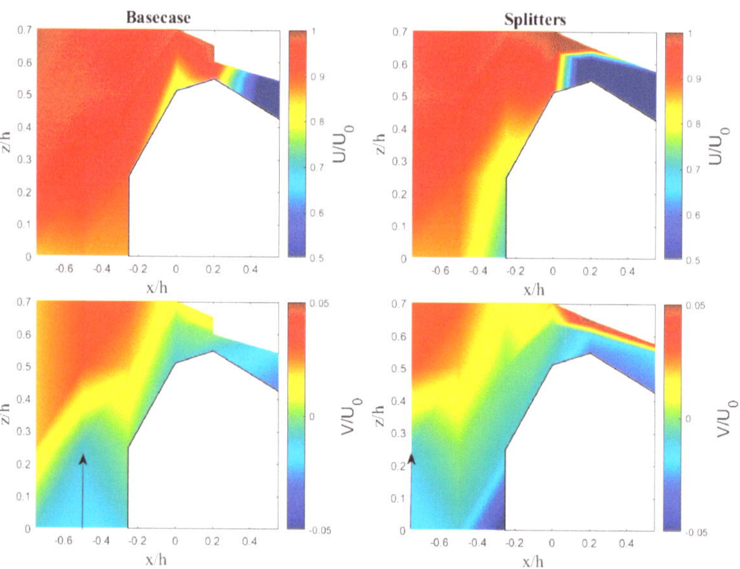

Figure 15. Contour plots of the normalized horizontal velocities u/u_0 and v/u_0 for the D-Model.

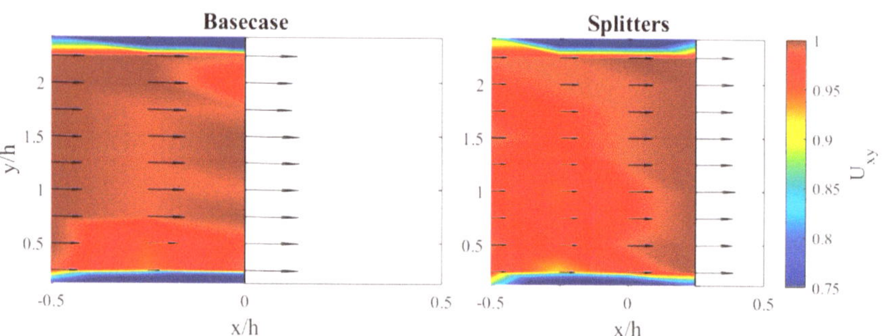

Figure 16. Non-dimensional time-averaged velocity magnitude U_{xy} (–) presented as color scheme for horizontal planes at water surface for the D-Model.

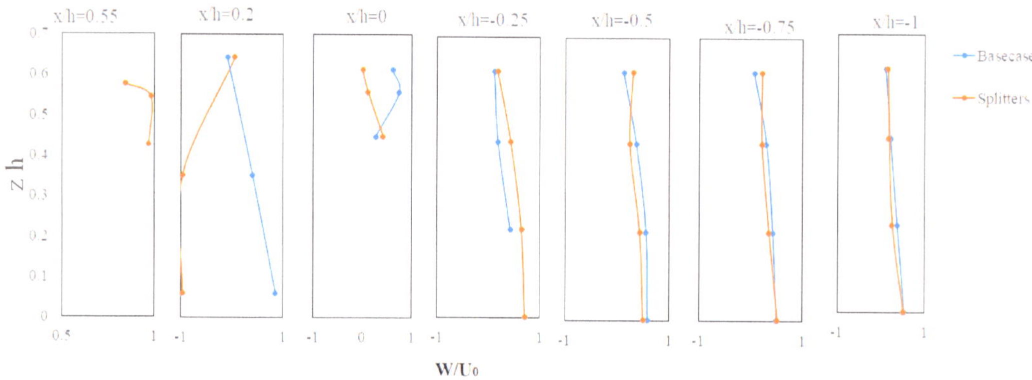

Figure 17. Comparison of normalized vertical w/u_0 one-dimensional profiles for stream-wise locations $x/h = 0.55, 0.2, 0, -0.25, -0.5, -0.75,$ and -1.0 in base case and splitters cases for the D-Model.

3.5. Energy Dissipation

The energy head in the stepped spillway flow region was examined based on total load measurements. H_0 represents the vertical distance from the point of inception. The energy dissipation rate ΔH was calculated as follows:

$$\Delta H = \frac{1}{d} \times \int_0^d (H_t - z_0) \times dy \tag{12}$$

where d is the flow depth, H_t is the total head, and z_0 is the step edge elevation above the datum. Experimental data on relative energy dissipation in the C-Model and D-Model, estimated using the hydraulic jump method, are presented in Figures 18a and 18b, respectively.

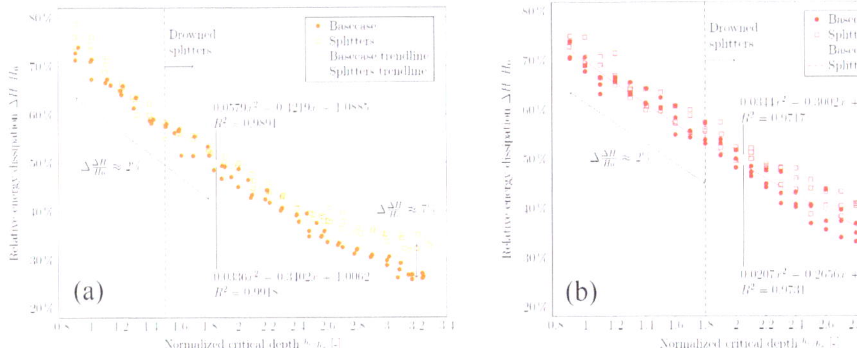

Figure 18. Experimental data of relative energy dissipation compared to normalized critical depth in the (**a**) C-Model and (**b**) D-Model.

It was observed that energy dissipation occurred in the boundary layer flow through turbulence and viscosity in the flow region that developed over both stepped spillway models. In the experiments conducted with the D-Model, when $h_c/h_s < 1.8$, the energy dissipation difference between the base case and the splitters in the stepped spillways was 2%, increasing to 7% when $h_c/h_s > 1.8$. The trend is similar in both the base case and splitters flow profiles, as demonstrated by polynomial curve fits with $R^2 > 0.97$:

$$0.02x^2 - 0.27x + 0.94 \ \text{ for Base case}$$

$$0.03x^2 - 0.3x + 0.98 \ \text{ for Splitters}$$

A similar situation was observed in the experiments with the C-Model, but the splitters became submerged at a reduced discharge. More stable measurements with less scattering were observed in the C-Model, with a normalized correlation coefficient of $R^2 = 0.99$ compared to the D-Model with its polynomial curve:

$$0.03x^2 - 0.34x + 1 \ \text{ for Base case}$$

$$0.06x^2 - 0.42x + 1 \ \text{ for Splitters}$$

4. Discussion

4.1. Flow Regimes and Velocity Distribution in the Step Edges

The velocity results from the BIV analysis indicated that the water column consists of a developing boundary layer with an ideal flow region above it. The flow accelerates downstream due to gravitational forces. Regarding the new deflecting nappe flow regime observed at intermediate flow rates (Figure 7a–c), the splitter configuration presents two drawbacks: intense spray and unstable transitions. These issues could potentially be resolved by optimizing the splitter geometry, which warrants further investigation. Another observation in the deflecting nappe regime is that the increase in relative energy dissipation is less substantial than the increase observed in the skimming flow region following the implementation of splitters.

For the profiles over step edges presented in Figures 12 and 13, the best data fit yielded a trendline value of $n = 3.6$ with a normalized correlation coefficient of $R^2 = 0.99$. This value of n is close to that suggested by other studies for similar chute slopes (Kramer [5] and Amador et al. [9]) for the stepped model. It should be noted, however, that these power-law coefficients are calculated from the non-aerated blackwater region. The analytic solution using Equation (9) gives $n = 4.7$ which shifts the graph to the right and generally results in larger dimensionless velocities for all values of $\frac{z}{\delta}$. The velocity profiles developed with BIV appear to correspond well with established mixing layer theory in the mixing zone. The velocity profiles for the base case from the inner corner of the steps are almost identical to those reported by Sánchez-Juny et al. [4]. There are also similarities between the results of Amador et al. [9] and the mixing layer theory equation proposed by Kramer [5], although minor differences exist, particularly below the pseudo-bottom. At the bottom of the graph, negative streamwise velocity values are observed, with a minimum occurring at approximately $z/\delta \approx -1$. From this point, the velocities evolve towards positive values, passing through the zero point, which is presumably at the center of the cavity vortex. The velocities seem to decrease after the end of the boundary layer, most likely due to the increasing air concentration in this zone, which reduces the number of bubbles available for tracking by the BIV algorithm.

Above the boundary layer δ, the velocity appears to decrease. This phenomenon was also reported by Sánchez-Juny et al. [4], who suggested that it might be due to insufficient lighting rather than actual flow velocities. However, observations from the present study indicate that the reduction in surface velocities is more likely a consequence of the difficulty in accurately defining the free surface. The probable cause is that PIVlab tracks sporadic droplets in the air above the surface rather than air bubbles within the air–water volume; as a result, the velocity decreases when the upper region of the frames, which includes these sporadic droplets and stationary air, is time-averaged.

The streamwise velocity distribution resulting from the deflecting nappe regime with the splitters configuration exhibits somewhat different characteristics. The greatest negative values are observed at the lowest part of the graphs, and the zero point is positioned further down, indicating that, compared to the base case, the cavity vortex is located closer to the inner corner of the step. This observation aligns closely with the visual observations shown in Figure 8. Due to the free-falling nappe deflecting off the splitters (Figure 7), the graphs are extended upwards.

Surface velocities captured from the top view (Figure 14) exhibit a tendency to oscillate until reaching the inception point, but downstream of this point, the corresponding raw data points show greater stability. This was possibly caused by light glares in the glossy blackwater region. Sánchez-Juny et al. [4] reported much larger differences in velocity magnitude before and after the inception point, attributing these differences to poor lighting; however, no such considerable differences were observed in this study.

4.2. Length of Inception and Cavitation Potential

The comparison of the point of inception showed relatively good agreement between the C-Model base case and Boes & Hager [16], as well as between D-Model base case and Matos [26] (Figure 11). The prototype data from the Hinze Dam also closely correspond, particularly with the primary D-Model. When the splitters configuration was added to the models, the inception point distance decreased, resulting in a shift of approximately the dimensionless distance $L_i/k_s = 10$. This indicates that adding the splitters accelerates the growth of the turbulent boundary layer. The base case hydraulic models correlate well with the established empirical relations for the growth of the turbulent boundary layer toward the water surface and the subsequent free-surface aeration.

Air concentration near the pseudo-bottom is a critical parameter in reducing cavitation. The air concentration at the pseudo-bottom can be calculated using the equation provided by Boes & Hager [16]:

$$C_b(x_i) = 0.015x_i^{\sqrt{tan\theta}/2} \qquad \text{for } 26^\circ \leq \theta \leq 55^\circ \tag{13}$$

where $x_i = (x - L_i)/z_{m,i}$ is the non-dimensional distance from the inception point, x is the longitudinal streamwise distance from the crest of the spillway, and z_{mi} is the air–water mixture depth at the point of inception. At the point of inception, the air concentration at the pseudo-bottom $C_{bi} = 0.01$ and the depth-averaged air concentration $\overline{C}_i = 0.226$ with a slope of $\theta = 51°$ result from the increasing non-linear relationship between water depth and air concentration (as derived from Equations (4) and (13)).

Frizell et al. [30] made the following recommendation regarding cavitation in their research on stepped spillways within a closed system with reduced ambient pressure:

$$\sigma_c > \sigma_{cr} = 4f \tag{14}$$

where σ_c is the cavitation index and σ_{cr} is the critical cavitation index. Boes & Hager [16] provide the friction factor f, including sidewall correction, as follows:

$$\frac{1}{\sqrt{f}} = \frac{1}{\sqrt{0.5 - 0.42sin(2\theta)}}\left[1 - 0.25log\left(\frac{h_s cos(\theta)}{D_{h,w,u}}\right)\right] \tag{15}$$

where $D_{h,w,u}$ corresponds to the equivalent clear water hydraulic diameter in the quasi-uniform zone of the flow and h_s is the height of step. The cavitation index is defined as follows:

$$\sigma_c = \frac{P_0 - P_v}{\frac{1}{2}\rho_w u_0^2} \tag{16}$$

where P_0 is the pressure at flow surface, P_v, is the vapor pressure, ρ_w is the water density and u_0 is the mean velocity. According to the cavitation control for the experiments, cavitation was determined at a value of 3.3, corresponding to a unit discharge of $q_w = 34.5 \text{ m}^2/\text{s}$ and a normalized critical depth of h_c/h_s. The control was conducted for a unit width at the position of $L_x/x_s = 11$, meaning that the hydraulic radius was $R_h = z_w = z$ due to the non-aerated flow. The velocity at this position was not measured directly for this discharge but was assumed to be $\sqrt{\eta}u_1 = 20 \text{ m/s}$ based on the energy loss factor reported by André [31], where $\eta = 1.15$. u_1 is the mean velocity in the upstream cross-section of the hydraulic jump. The critical cavitation index was calculated as $\sigma_{cr} = 0.244$ for this position. $D_{h,w,u}$ was assumed to be $4R_h$ despite the uniform flow not being fully developed. This critical cavitation index is less than the cavitation index $\sigma_c = 0.487$, indicating that cavitation is not expected. The calculations were performed with a vapor pressure $P_v = 2.4 \text{ kP}_a$, a water density $\rho_w = 1000 \text{ kg/m}^3$, and a reference pressure P_0 as the atmospheric pressure. Further calculations show that the critical velocity for inducing cavitation in the prototype is approximately 28 m/s, significantly higher than the value proposed by Boes & Hager [16].

4.3. Effect of Crest Splitters on Velocity Patterns above the Ogee Crest and Energy Dissipation

With reference to Figure 15, the normalized horizontal velocity components (u, v) and the vertical velocity component (w) obtained from ADV measurements show that as the flow approaches the ogee crest, the average flow velocity increases due to the decreasing cross-sectional area, reaching a maximum at the crest. After passing the peak, the water velocity decreases on the downstream side. In general, it has been observed that the velocity vectors (u and w) play a dominant role in both the base case and splitter experiments. The splitters slightly increased the water depth on the upstream side, which appears to cause a further decrease in water velocity and a corresponding reduction in energy. Additionally, it was observed that the splitters somewhat reduced the velocity, and therefore the energy, in the upstream part of the ogee crest and towards the downstream flow. In the splitter experiments, u decreased by 65% when the element was slightly above the ogee crest. Normalized horizontal velocities indicate that the zone of positive average vertical velocity v/u_0 increases toward the top of the ogee crest, while negative velocities occur toward the bottom. The splitters caused the velocities to shift upstream to $x/h = 0.5$ from the upstream side of the ogee crest, leading to a decrease in velocities both upstream and downstream of the ogee crest.

In Figure 17, changes in the vertical velocity vector direction w/u_0 can be seen after the flow passes over the ogee crest in the splitter configuration. The splitters were observed to increase water depth upstream of the crest, which slightly reduces velocity and, consequently, energy both upstream of the crest and toward the rear of the flow. In the upstream part of the flow, the velocity vectors in the base case and with splitters move in opposite directions. However, beyond the streamwise location $(x/h = -0.25)$, the profiles become almost identical.

Another advantage of the stepped spillway is the continuous energy dissipation along the spillway chute, with splitters contributing to a 7% increase in relative energy dissipation. This results in a large area over which energy is dissipated, reducing the stresses on the concrete and bedrock. In the C-Model, less turbulence, higher relative surface tension, and proportionally larger air bubbles could all contribute to reduced aeration and, to some extent, lower energy dissipation. As shown in Figure 18, a semi-linear decrease in normalized specific energy is observed, consistent with previous studies Zhang & Chanson [3], Hunt et al. [24] and Meireles et al. [28].

4.4. Scale Effects and Uncertainty

Scale effects are inevitable when downscaling highly turbulent two-phase flows. As anticipated, reduced energy dissipation and lower aeration efficiency were observed in the C-Model. In terms of flow transitions, the deflecting water jet over the splitters was observed across a wider range of normalized critical depths in the D-flume. The influence of the Weber number may explain why surface tension forces are more pronounced in the smaller C-Model, causing greater resistance for the nappe to deflect from the main flow.

The air bubbles in the two different scale models were observed to be of similar size, leading to proportionally larger and fewer bubbles in the C-Model relative to the prototype it represents. One practical issue arising from this situation is that the BIV algorithm had fewer bubbles to track, making the C-Model less suitable for BIV analysis.

The relative energy dissipation in the different scale models diverged more significantly at larger dimensionless discharges, with substantial deviations observed at design discharges. At these discharges, the flow in the stepped spillway is not aerated, as seen for $h_c/h_s = 3.3$ in Figure 7, indicating that the issue is not related to the two-phase characteristics in the spillway. However, the hydraulic jump at the dam toe is highly turbulent and aerated, making it subject to drastic scale effects, which could explain the deviation in energy dissipation across models at higher flow rates.

While the C-Model appears unsuitable for studying two-phase flow, the two different scale models exhibited very similar changes in relative energy dissipation from the base case to the splitter configuration, supporting the findings from the primary D-Model.

When working with video recordings and BIV analysis, distortion can occur due to the deformation of angles and distances in images caused by the camera lens and perspective. Another issue is the suitability of air bubbles as tracers, given the upward vertical velocity component of the bubbles. Sánchez-Juny et al. [4], who conducted similar experiments, concluded that air bubbles can be used as tracers but did not seem to address this aspect in detail. The findings of lower velocities in dark regions are consistent with those reported by Sánchez-Juny et al. [4]. A possible approach to quantifying the effects of rising velocity, lighting conditions, wall effects, and varying distances to the traced bubbles is to conduct parallel experiments using PIV with seeding particles and a laser-illuminated sheet, as demonstrated by Amador et al. [9]. Using video recordings with 1000 FPS was found to be sufficient for the use of PIVlab. Sánchez-Juny et al. [4] successfully used PIVlab with 400 FPS, suggesting the possibility of reducing the frame rate in this study, thereby saving processing time.

5. Conclusions

This study focused on practical, cost-effective, and feasible measures to improve energy dissipation in stepped spillways using the BIV technique. To investigate scale effects, laboratory tests were conducted with two different scale models, 1:50 and 1:17, while prototype data and a literature review were used to guide the analysis. The following conclusions can be drawn from this research:

- In the experiments, it was observed that the flow region developed in shear flow consists of a turbulent boundary layer, with ideal fluid flow above it. A rapidly changing flow movement was observed along the steps. Bernoulli's principle was used to calculate the energy distribution, assuming that the flow moved in the downward direction.
- The splitters were found to increase relative energy dissipation by 7%. In addition to enhancing energy dissipation, the splitters reduced the length of inception to $L_i/k_s = 10$, thereby lowering the potential for subsequent cavitation. This resulted in an increase in the maximum allowable unit discharge. Consequently, crest splitters can be considered a practical, viable, and cost-effective measure to improve energy dissipation in stepped spillways, applicable to both existing dams and new projects.
- In the experiment with the splitter, the horizontal velocity vector u around the ogee crest showed a significant decrease. Additionally, the vertical velocity vector v contributed to turbulence by changing direction, resulting in decreased average velocities in front of and behind the ogee crest, thereby reducing energy on the downstream side of the spillway.
- BIV was found to be a straightforward and highly effective technique for accurately measuring and determining flow characteristics in aerated flow conditions, likely the best practice for such applications. BIV enables the collection of large quantities of high-quality data with standard camera equipment, allowing for the extraction of detailed information and the production of dense plots without the need for commercial software or extensive processing capabilities.
- The scale effects observed in this study align well with established theory. Severe scale effects were noted in the air–water flow in the smaller-scaled model, underscoring the importance of using a larger scale to accurately study highly turbulent two-phase flow conditions. The findings closely correspond to established theory regarding the valid length scales for studying highly turbulent two-phase flow under Froude similitude.
- While the smaller C-Model appears unfit for the study of two-phase flow, the change in relative energy dissipation from the base case to the splitter configuration is nearly identical for the two different scale models, supporting the findings in the primary D-Model.

Author Contributions: Conceptualization, A.B. and L.L.; Methodology, A.B. and L.L.; Software, L.M.M., K.H.T. and A.B.; Formal analysis, L.M.M., K.H.T. and A.B.; Investigation, L.M.M., K.H.T. and A.B.; Resources, A.B.; Writing—original draft, L.M.M., K.H.T. and A.B.; Writing—review & editing, A.B. and L.L.; Supervision, A.B. and L.L. All authors have read and agreed to the published version of the manuscript.

Funding: This research received no external funding.

Data Availability Statement: Data are contained within the article.

Acknowledgments: The researchers like to thank Filmon Tquabo Gebremariam for his great help during experiments. The authors also would like to thank Simon Edward Mumford for his help with language editing and proofreading.

Conflicts of Interest: Author Lars Marius Mikalsen was employed by the company Dr.techn. Olav Olsen AS. Author Kasper Haugaard Thorsen was employed by the company Multiconsult Norge AS. The remaining authors declare that the research was conducted in the absence of any commercial or financial relationships that could be construed as a potential conflict of interest.

Notations

B_f	Flume width [L]
C_b	Air concentration at the pseudo-bottom [-]
C_{bi}	Air concentration at the pseudo-bottom at the point of inception [-]
$\overline{C}_i$	Mean air concentration at point of inception [-]
D	Flow depth [-]
$D_{h,w,u}$	Equivalent clear water hydraulic diameter in the quasi-uniform zone [-]
f	Darcy–Weisbach friction factor [-]
F_r	Froude number, $F_r = u/\sqrt{gz}$ [-]
F_*	Roughness Froude number [-]
g	Gravitational acceleration [L/T^2]
H	Ogee height [L]
H	Local energy head [L]
H_0	Vertical distance from the starting point of the inception [L]
H_f	Height of flume [L]
H_t	Total head [L]
h_c	Critical depth, $h_c = \sqrt[3]{q_w^2/g}$ [L]
h_s	Height of the step [L]
k_s	Normal height of the step [L]
L_e	Characteristic mixing layer length scale [L]
L_i	Longitudinal distance from the top of the crest to the inception point [L]
l_s	Length of the step [L]
L_x	Longitudinal distance from the top of the first step [L]
n	Power-law exponent [-]
N_s	Number of steps along the chute [-]
P_0	Pressure at flow surface [N/m^2]
P_v	Vapor pressure [M/LT^2]
Q	Water discharge [L^3/T]
Q_{max}	Maximum water discharge [L^3/T]
q_w	Unit water discharge [L^2/T]
R_1	Upstream radius 1 for ogee weir [L]
R_2	Upstream radius 2 for ogee weir [L]
R_e	Reynolds number [-]
R_h	Hydraulic radius [L]
u	Streamwise water velocity [L/T]
$\overline{u}$	Time-averaged streamwise velocity [L/T]
u_1	Incoming velocity to the hydraulic jump [L/T]
u_{90}	Air–water mixture velocity at the flow depth z_{90} [L/T]
u_{FS}	Free-stream velocity [L/T]
$\overline{u}_{if}$	Velocity at the infection point [L/T]
u_m	Air–water mixture velocity [L/T]
$\overline{u}_m$	Time-averaged mixture velocity [L/T]
$\overline{u}_{min}$	Minimum velocity in the mixing layer [L/T]
$\overline{u}_{ML}$	Mixing layer velocity [L/T]
W	Weber number [-]
X, Y, Z	Distances in x, y, z directions, respectively [L]
x_i	Dimensionless distance from the inception point [-]
x_s	Longitudinal distance between two step edges [L]
z_0	Step edge elevation above the datum [L]
z_{90}	Flow depth with 90% air concentration [L]
z_{if}	Inflection point elevation above pseudo-bottom [L]
z_{mi}	Mixture depth at point of inception [L]
u, v, w	Instantaneous velocity components [L/T]

$\overline{u}$, $\overline{v}$, $\overline{w}$	Mean velocity components [L/T]
u_0	Average approach flow velocity [L/T]
U_{xy}	Non-dimensional mean resultant velocity [-]
ΔH	Energy dissipation rate [-]
δ	Boundary layer thickness [L]
η	Correction factor for the singular loss when the flow changes direction from the stepped spillway to the horizontal bed [-]
θ	Angle of the slope of the chute [°]
λ_f	Length scale factor in Froude similitude [-]
ρ_w	Density of water [M/L^3]
σ_c	Cavitation index [-]
σ_{cr}	Critical cavitation index [-]

References

1. Chanson, H. *Hydraulics of Stepped Chutes and Spillways*; Balkema Publ: Lisse, The Netherlands, 2001; ISBN 9058093522.
2. Kokpinar, M.A. Flow over a stepped chute with and without macro-roughness elements. *Can. J. Civ. Eng.* **2004**, *31*, 880–891. [CrossRef]
3. Zhang, G.; Chanson, H. Application of local optical flow methods to high-velocity free-surface flows: Validation and application to stepped chutes. *Exp. Therm. Fluid Sci.* **2018**, *90*, 186–199; ISSN 0894-1777. [CrossRef]
4. Sánchez-Juny, M.; Estrella, S.; Matos, J.; Bladé, E.; Martínez-Gomariz, E.; Bonet Gil, E. Velocity Measurements in Highly Aerated Flow on a Stepped. *Water* **2022**, *14*, 2587. [CrossRef]
5. Kramer, M. Velocities and Turbulent Stresses of Free-Surface Skimming Flows over Triangular Cavities. *J. Hydraul. Eng.* **2023**, *149*, 04023012. [CrossRef]
6. Rajaratnam, N. An experimental study of air entrainment characteristics of the hydraulic jump. *J. Inst. Eng.* **1962**, *42*, 247–273.
7. Murillo, R.E. Experimental Study of the Development Flow Region On Stepped Chutes. Ph.D. Thesis, University of Manitoba, Winnipeg, MB, Canada, 2006.
8. Ohtsu, I.; Yasuda, Y. Characteristics of flow conditions on stepped channels. In Proceedings of the 27th IAHR Congress, San Francisco, VA, USA, 10–15 August 1997; pp. 583–588.
9. Amador, A.; Sánchez-Juny, M.; Dolz, J. Characterization of the Nonaerated Flow Region in a Stepped Spillway by PIV. *ASME J. Fluids Eng.* **2006**, *128*, 1266–1273. [CrossRef]
10. Bung, D.; Valero, D. Optical flow estimation in aerated flows. *J. Hydraul. Res.* **2016**, *54*, 575–580. [CrossRef]
11. Kramer, M.; Chanson, H. Optical flow estimations in aerated spillway flows: Filtering and discussion on sampling parameters. *Exp. Therm. Fluid Sci.* **2019**, *103*, 318–328. [CrossRef]
12. Lopes, P.; Leandro, J.; Carvalho, R.; Daniel, D. Alternating skimming flow over a stepped spillway. *Environ. Fluid Mech.* **2017**, *17*, 303–322. [CrossRef]
13. Leandro, J.; Bung, D.; Carvalho, R. Measuring void fraction and velocity fields of a stepped spillway for skimming flow using non-intrusive methods. *Exp. Fluids* **2014**, *55*, 1732. [CrossRef]
14. Wright, H.C.; Cameron-Ellis, D.G. Energy dissipation provisions for stepped spillways: Are additional measures necessary? A southern african perspective. In Proceedings of the SANCOLD Annual Conference 2018, Pretoria, South Africa, 3–7 November 2018.
15. Chanson, H. Energy dissipation on stepped spillways and hydraulic challenges—Prototype and laboratory experiences. *J. Hydrodyn.* **2022**, *34*, 52–62. [CrossRef]
16. Boes, R.M.; Hager, W.H. Two-Phase Flow Characteristics of Stepped Spillways. *J. Hydraul. Eng.* **2003**, *129*, 661–670. [CrossRef]
17. USBR Design of Small Dams. Water Resources Technical Publication, third edition. 1987. Available online: https://www.osti.gov/biblio/29108 (accessed on 19 August 2024).
18. Mikalsen, L.; Thorsen, K. Improved Energy Dissipation in Stepped Spillways—Hydraulic Model Study Applying Bubble Image Velocimetry. Master's Thesis, NTNU, Trondheim, Norway, 2023.
19. Robert, D. The dissipating of the energy of a flood passing over a high dam. *Civ. Eng. = Siviele Ingenieurswese* **1943**, *1943*, 48–92.
20. William, T.; Stamhuis, E. PIVlab—Towards User-friendly. Affordable and Accurate Digital Particle Image Velocimetry in MATLAB. *J. Open Res. Softw.* **2014**, *2*, 30. [CrossRef]
21. Bor, A. Experimental investigation of 90° intake flow patterns with and without submerged vanes under sediment feeding conditions. *Can. J. Civ. Eng.* **2022**, *49*, 452–463. [CrossRef]
22. Goring, D.G.; Nikora, V.I. Despiking acoustic doppler velocimeter data. *J. Hydraul. Eng.* **2002**, *128*, 117–126. [CrossRef]
23. Hunt, S.; Kadavy, K. Inception Point for Embankment Dam Stepped Spillways. *J. Hydraul. Eng.* **2012**, *139*, 60–64. [CrossRef]
24. Hunt, S.L.; Kadavy, K.C.; Hanson, G.J. Simplistic Design Methods for Moderate-Sloped Stepped Chutes. *J. Hydraul. Eng.* **2014**, *140*, 04014062. [CrossRef]
25. Wood, I. Uniform Region of Self—Aerated Flow. *J. Hydraul. Eng.* **1983**, *109*, 447–461. [CrossRef]
26. Matos, J. Hydraulic design of stepped spillways over RCC dams. In *Hydraulics of Stepped Spillways*; Minor, E.H., Hager, W., Eds.; CRC Press: London, UK, 2000; ISBN 9781003078609.

27. Amador, A.; Sánchez-Juny, M.; Dolz, J. Developing Flow Region and Pressure Fluctuations on Steeply Sloping Stepped Spillways. *J. Hydraul. Eng.* **2009**, *135*, 1092–1100. [CrossRef]
28. Meireles, I.; Renna, F.; Matos, J.; Bombardelli, F. Skimming, Nonaerated Flow on Stepped Spillways over Roller Compacted Concrete Dams. *J. Hydraul. Eng.* **2012**, *138*, 10. [CrossRef]
29. Takahashi, M.; Ohtsu, I. Aerated flow characteristics of skimming flow over stepped chutes. *J. Hydraul. Res.* **2012**, *50*, 427–434. [CrossRef]
30. Frizell, K.; Renna, F.; Matos, J. Cavitation Potential of Flow on Stepped Spillways. *J. Hydraul. Eng.* **2015**, *141*, 630–636. [CrossRef]
31. André, S. High Velocity Aerated Flow on Stepped Chutes with Macro-Roughness Elements. Ph.D. Thesis, EPFL, Lausanne, Switzerland, 2004. [CrossRef]

Article

The Effect of Pipeline Arrangement on Velocity Field and Scouring Process

Fereshteh Kolahdouzan [1], Hossein Afzalimehr [1], Seyed Mostafa Siadatmousavi [1], Asal Jourabloo [1] and Sajjad Ahmad [2],*

[1] School of Civil Engineering, Iran University of Science and Technology, Tehran 16846-13114, Iran
[2] Department of Civil and Environmental Engineering and Construction, University of Nevada, Las Vegas, NV 89154, USA
* Correspondence: sajjad.ahmad@unlv.edu

Abstract: This experimental study investigates the effect of changes in the arrangement of horizontal pipelines on changes in the velocity pattern in three dimensions and the scouring process around these submarine pipelines. Experiments have been carried out in four cases: single pipe, two pipes with a distance of 0.5 D, two pipes with a distance of D, and three pipes with a distance of 0.5 D (D is the diameter of the pipes). The velocity upstream, downstream, and on the pipes have been measured by the Acoustic Doppler Velocimeter (ADV). The results show that a single pipe's scouring depth in the first case is more significant than in the other cases. In the second case, the presence of the second pipe at a distance of 0.5 D from the first pipe significantly reduced the scour depth (28.6%) compared to the single pipe condition by changing the velocity pattern around the pipelines. By increasing the number of pipes to 3 with a distance of 0.5 D, this reduction in scouring depth has reached 47.6% compared to the single pipe condition. However, in the case of two pipes with a distance of D, the reduction of scouring depth was 21.4% compared to the case of a single pipe, and compared to the case of two pipes with a distance of 0.5 D, it increased by 10%.

Keywords: local scour; horizontal pipelines; scour depth; velocity field; ADV

Citation: Kolahdouzan, F.; Afzalimehr, H.; Siadatmousavi, S.M.; Jourabloo, A.; Ahmad, S. The Effect of Pipeline Arrangement on Velocity Field and Scouring Process. *Water* 2023, 15, 1321. https://doi.org/10.3390/w15071321

Academic Editors: Charles R. Ortloff and Jianguo Zhou

Received: 8 February 2023
Revised: 19 March 2023
Accepted: 24 March 2023
Published: 28 March 2023

1. Introduction

Submarine pipelines are important infrastructures for transporting water, natural gas, oil, and petroleum products. Due to the increasing extraction of oil and gas resources, the use of submarine pipelines is rapidly increasing. When a pipe is placed on an erodible seabed, scouring occurs around the pipeline due to the interaction between the pipeline and the erodible seabed, under a current or a wave, or a combination of both. As a result, parts of the pipe become suspended and have no support. Over time, the length of free openings increases, and the pipe may rupture or structurally fail under severe oscillating loads due to vortices formed around the pipeline. Therefore, the mechanism of occurrence and expansion of the scour cavity, its depth, and the factors affecting it have received considerable attention from researchers and designers.

The flow pattern during scouring around the pipe is complex. One of the most important phenomena that occur in the pipe wake and are effective in the development of scour holes is vortex shedding. When the distance between the pipe and the bed is relatively small, a weak shear layer is created in the lower part of the pipe, and due to the weak interaction of the upper and lower shear layers of the pipe, the phenomenon of vortex shedding does not occur [1]. Jensen and Samer (1990) [2] investigated the flow and velocity changes around the pipe, without the initial distance of the pipe from the bed in different stages of scouring. Their study shows that, before scouring downstream of the pipeline, the horizontal velocity values are negative in the height range of the pipe's presence. Then, at levels higher than the upper surface of the pipe, the velocity profile is accompanied by a strong gradient, and the horizontal velocity components change direction, and their magnitude increases. Additionally, after the scour balance is reached, the outflow from under the pipe has caused S-shaped profiles downstream of the pipeline to form. In the case where e/D (e is the distance between the bottom surface of the pipe and the bed; D is the diameter of the pipe) is smaller than 0.3, the vortex shedding phenomenon stops [3]. Oner et al. (2008) [4], have also investigated the flow interaction with the pipe for different values of the distance between the pipe and the bed using the particle image velocimetry (PIV) technique. The results show that the effect of the bed on the flow field around the pipe is very small for e/D $\geq$ 1, but for small values of e/D, a vortex is created upstream of the pipe in the vicinity of the bed. In the condition of e/D = 0, the size of this vortex decreases with the increase in the Reynolds number. Additionally, the phenomenon of vortex shedding occurs for e/D $\geq$ 0.2. Lin et al. (2009) [5] stated that the phenomenon of vortex shedding occurs at a Reynolds number greater than 40, and a boundary layer adjacent to the pipe is separated due to the reverse pressure gradient. The separation of the boundary layer that occurs in the upper and lower parts of the pipe creates two shear layers in the upper and lower part of the pipe, and when the two shear layers interact with each other, vortex shedding occurs. Lin et al. (2009) used the PIV method to investigate the characteristics of the velocity field and velocity time series and observed that, within the range of 0 $\leq$ e/D $\leq$ 0.3, this vortex is formed upstream of the pipe and in the vicinity of the bed and the pipe. As e/D decreases, the size of the vortex and the vertical distance between the center of the vortex and the bed increases. In cases where the size of the vortex is large, this vortex acts as an obstacle against the flow, and by preventing the flow from entering under the pipe, it leads to the weakening of the flow under the pipe.

Abbaszadeh Tavassoli and Haji Kandi (2010) [6] simulated the flow around the pipe after creating a scour hole with Fluent software and stated that the vortices downstream of the pipeline after creating a scour hole causes an increase in the flow velocity downstream of the pipe. This increase near the bed causes the formation of scour cavity downstream of the pipe. This increase in velocity will continue until the scour hole reaches equilibrium, and after that, the effect of the vortices on the bed downstream of the pipeline will decrease. Yeganeh Bakhtiari et al. (2011) [7] modeled the flow around the pipeline in the case of e/D = 0.3. Observing the S-shaped horizontal velocity profiles downstream of the pipeline, they concluded that the maximum horizontal velocity in the jet flow (below the pipeline), is 1.3 times the velocity of the free surface of the incoming flow. In addition, an examination of the sections under the pipe and downstream of the pipe has shown that the horizontal velocity gradient is high near the bed. Therefore, in the conditions of an erodible bed, it is expected that the jet flow under the pipe plays a major role in the erosion of the bed under the pipe. This study also states that, with the increase in the distance between the pipe and the bed, the maximum velocity and the average velocity of the flow under the pipe gradually increase. This is due to the positive pressure gradient in the flow direction when the distance between the pipe and the bed is small, which causes a decrease in the flow velocity and also a decrease in the flow flux in the vicinity of the bed. Zhang and Shi (2016) [8] investigated the flow around the pipe with initial distances of 0.1 D, 0.3 D, and 0.5 D from the bed in sections x/D = 1 and x/D = 3.5 downstream of the pipe. They observed S-shaped horizontal velocity profiles and reported the occurrence of two vortices downstream of the pipe. This study also states that in the two cases of distance from the bed 0.1 D and 0.3 D, due to the small distance of the pipe from the bed, the interaction between the shear layer of the upper and lower surfaces of the pipe downstream is not complete. For this reason, the lower vortex is not fully developed and is smaller in size than the upper vortex. At the same time, in the case of the distance from the bed 0.5 D, the effect of the bed is reduced and both vortices are almost the same size. Many previous studies have mentioned the presence of S-shaped profiles downstream of the pipeline. They have explained the existence of these profiles along with the reverse flow due to the presence of two vortices in that area. By moving downstream from the pipe, the S-shape of the profiles gradually disappears at a relatively far distance from the pipe (where the effect of the pipe on the flow is negligible). The shape of the horizontal velocity profiles becomes similar to the profiles of the flow upstream of the pipe [2,9,10]. Penna et al. (2020) [11], investigated the flow around a single pipe in shallow flow conditions (H < 5 D). They observed two vortices downstream of the pipe and reported the maximum values of positive and negative Reynolds stress at the place of formation of the upper and lower vortices, respectively. Leo et al. (2021) [12] investigated the time development of vortices downstream of the pipe after the formation of scour hole. They measured the length of the created vortex in the largest size, twice the diameter of the pipe. Chen et al. (2022) [13], stated that in $0.3 \le e/D \le 1.5$ states, vortices shed from the upper and lower surfaces of the pipe with different intensities and in $e/D \ge 1.5$ states with the same intensity. The frequency of vortex shedding is higher in the first state.

So far, most studies that have reported on the flow structures around the pipelines are about single-pipe states. Less research has been reported on the pipe groups scour, while in many recent pipeline projects, pipe groups are used. In such circumstances, understanding the effect of pipeline layout on the flow condition is essential. Although some studies have been reported, knowledge gaps regarding the impact of pipeline layout on flow conditions remain. Accordingly, the novelty of this study is to make a comparison between four layouts of pipes, (a) single pipe, (b) two pipes with a distance of 0.5 D, (c) two pipes with a distance of D, and d) three pipes with a distance of 0.5 D. In this study, the scouring mechanism and the components of the flow velocity in 3 D around the pipelines are compared in these four arrangements. The results of this study may help to highlight the best layout for parallel pipes where the pipes are straight on the erodible bed without using scour control methods, such as impermeable horizontal or vertical plates.

2. Materials and Methods

In this study, experiments have been conducted in a laboratory flume for four cases. First: one pipe; second: two pipes with a distance of 0.5 D from each other; third: two pipes with a distance of D from each other; and fourth: three pipes with a distance of 0.5 D from each other (D is the diameter of pipes). The flume is 12.0 m long, 0.9 m wide, and 0.6 m deep with a rectangular cross-section area and glass walls. A pump was used to circulate the water with a discharge of 31.6 L per second. An electromagnetic flowmeter was installed in the supply conduit to measure the discharge passing through the flume continuously. PVC pipes with a circular cross-section and a diameter of 4 cm have been used in this study, located 6.0 m downstream of the flume entrance. The pipes are attached to the walls on both sides and do not move during the runs. The experimental section (sandbox) was 2.0 m long, 0.17 m deep, and 0.9 m wide. In this testing section, the flow was fully developed. The sediment used in this study was uniform sand. The median diameter of the sediment particles was $d_{50} = 0.83$ mm, and the geometric standard deviation of particle size distribution was $\sigma_g = 1.33$ (<1.4). The approaching flow depth H was maintained at 20 cm. The experiments ran with an average approaching flow velocity of U = 18.1 cm/s, which satisfied the clear-water condition of $u^*/u^*_c = 0.71$ where u^* is the shear velocity. The point gauge with an accuracy of ± 1 mm was used to measure flow depth. After the scour process reached the equilibrium state for each experimental run, a depth of 2200 to 3500 points was measured using a mobile point gauge with an accuracy of 0.5 mm, and the 3 D topography was drawn using Surfer software with a surface resolution of 1.0 cm.

The instantaneous three-dimensional velocity components were measured at different sections using a down-looking Acoustic Doppler Velocimeter (ADV) and Vectrino Plus made by Nortek with a duration of 120 s. The sampling frequency was set at 200 Hz [14]. The accuracy and quality of the collected data were controlled by two parameters, the correlation coefficient (COR) and the signal-to-noise ratio (SNR). WinADV software [15] has been used to filter the inappropriate recorded data with SNR and COR less than 15 dB and 70%, respectively. In this study, we use this filter to obtain the desired data. Moreover, the filter provided by Goring and Nikora [14] for the phase-space threshold despiking has been used to detect and eliminate spurious data.

Velocity profiles are collected at 11–15 sections in the longitudinal direction of the flume and 5 sections in the transverse direction of the flume in each run. Along each vertical axis, data have been collected at 20–30 measuring points. Observations have been collected from 3 mm above the bed to the point 5 cm below the water surface. In this study, due to the limitation of ADV for data collection under the pipes, there are no observations in these areas. Figure 1 shows how the pipe is placed in the flume. Additionally, the location of data collection in the longitudinal direction of the flume for the single-pipe state can be seen in this figure.

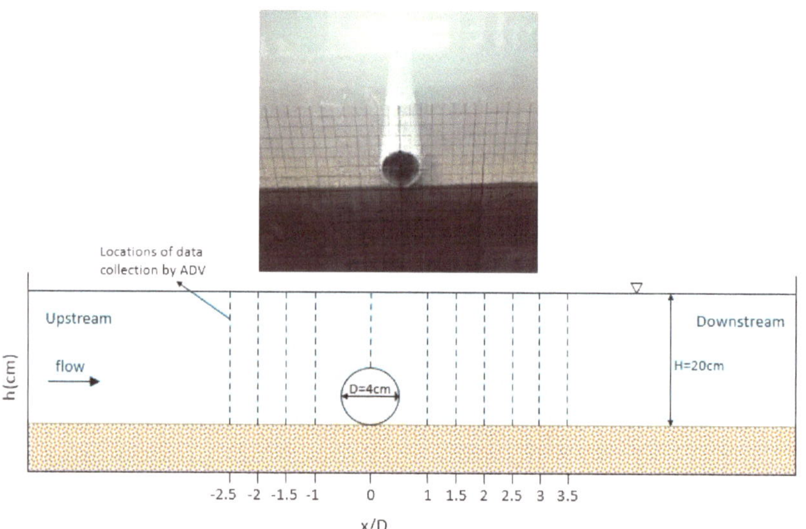

Figure 1. The location of the single pipe and measuring points in the longitudinal direction of the flume.

3. Results

3.1. Scour Process

At the start of the test, the exit of the sediment from under the pipe towards the downstream side is evident, which shows the start of the scouring process. At the very beginning, the rapid rotational movement of sand particles is observed in a small space upstream and downstream of the pipe. Additionally, at a short distance from the rapid rotational movement downstream of the pipe, the movement of sand grains can be seen with less intensity but in a relatively larger space. For the first time, Mao (1986) [16], stated that this phenomenon is because of the existence of three vortices A, B, and C around the pipe at the beginning of scouring, where vortex A is upstream and vortices B and C are downstream of the pipe. Zhang and Shi (2016) [8], also observed these vortices.

The sediments that were transported downstream of the pipes showed two behaviors: (a) Part of the sediments was piled up at a short distance from the downstream pipe and formed a sand dune. (b) Some of them entered the vortices formed in the distance between the sand dune formed downstream and the pipe, until they were somehow removed from that flow and formed the sand dune, such as in the first category or were transferred to the downstream. Over time, the scour hole became wider and the formed sand dunes were moved downstream and gained more distance from the pipe. Gradually, the speed of sediment transfer from the scour hole and the movement of the sand dune downstream decreased until the scour hole reached equilibrium.

In the tests conducted in the second and third cases, where the two pipes are placed at distances of 0.5 D and D, respectively, the presence of the upstream pipe caused a delay in the scouring process under the downstream pipe. Lee et al. (2020) [17] stated in their study that this delay to start scouring the downstream pipe is due to the presence of the upstream pipe. Like the process mentioned above for the two-pipe modes, in the three-pipe mode, scouring under the second pipe started with a delay compared to the first pipe, and scouring under the third pipe also started with a delay compared to the second pipe. Figure 2 shows the three-dimensional topography of the bed after the completion of the tests.

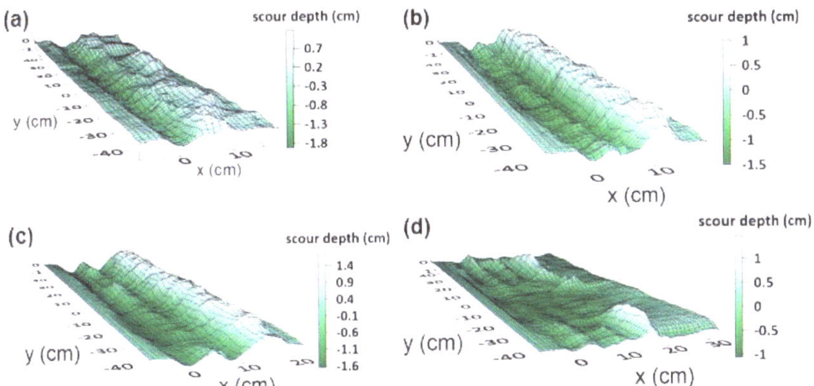

Figure 2. Topography of the final scouring hole around the pipelines: (**a**) single pipe, (**b**) two pipes G/D = 0.5, (**c**) two pipes G/D = 1, (**d**) three pipes G/D = 0.5.

In all four tested cases, scouring started from the sides of the walls and gradually progressed toward the center of the channel. Although scouring in the central area of the channel started with a delay compared to the side of the walls, it has progressed and the sand dunes downstream of the pipes in the central area of the channel have moved further downstream and lost their height. In other words, the scour hole is wider in the central area of the channel. The maximum scouring depth is also created in this area. The sand dunes created downstream of the pipes have a higher height near the walls and are closer to the pipe. This difference in the progress of scouring in the central area of the channel and walls, in the tests performed with one pipe and three pipes Figure 2a,d, is greater than in the two pipes (Figure 2b,c).

3.2. Scour Depth and Time

The results of scouring depth and equilibrium time are shown in Table 1. In the following study, the central area in the width of the channel is a range of 60 cm wide (−30 cm < y < 30 cm). The areas next to the walls are the areas that are, at most, 15 cm away from the wall.

Table 1. Scour depth and test time for experimental runs.

States	S (cm)	t_e (min)
Single pipe	2.1	645
Two pipes-0.5 D	1.5	525
Two pipes-D	1.65	555
Three pipes-0.5 D	1.1	480

To validate the tests, the results of the single-pipe test have been compared with some important studies conducted in the field of pipeline scouring. Westerhorstmann et al. (1992) [18], during an experiment on the scouring of a single pipe with a diameter of 3 cm, expressed the maximum scouring depth as 0.52 D. Zhao et al. (2016) [19], also recorded this depth as 0.53 D during an experiment on a single pipe with a 3.2 cm diameter. In Zhang and Shi (2016)'s study [8], the maximum scour depth for a pipe with a diameter of 10 cm was stated as 0.55 D. According to Table 1 in this study, the maximum scouring depth in the case of a single pipe with a diameter of 4 cm is obtained as 0.52 D, which is consistent with the results of the other studies. In these studies, the maximum scour depth formation location is recorded at a short distance from the center of the pipe downstream, which is also evident in the present study's results. Penna et al. (2020) [11], obtained the maximum scour depth for a single pipe with a diameter of 3 cm as 1.8 D and explained the reason for this increase by performing experiments in shallow flow conditions. (H < 5 D)

According to the results of Table 1, the highest scouring depth occurred in the case of a single pipe. In the case of two pipes with distance D, the scouring depth has decreased by 21.4%, and test time by 13.95%, compared to the case of a single pipe. Further, by reducing the distance between two pipes to 0.5 D, the scour depth has decreased by 9.1% compared to the case of two pipes with a distance of D and by 28.6% compared to the case of a single pipe. This result is consistent with those of similar studies. Maddah et al. (2021) [20], during an experiment they conducted on two pipes with a diameter of 4 cm in live bed conditions, stated that, by reducing the distance between the two pipes from D to 0.5 D, the maximum scour depth decreased by 19.7 percent. They repeated their experiments using pipes with diameters of 3.2 and 2 cm and saw a decrease of 13.9% and 11.1%, respectively, in the amount of the maximum scouring depth. Westerhorstman et al. (1992) [18] also recorded a 25% decrease in the maximum scour depth by reducing the distance between two pipes with a diameter of 3 cm from D to 0.5 D. The test time was also reduced by 5.4% in the case of two pipes with a distance of 0.5 D, compared to the case of two pipes with a distance of D, and by 18.6% compared to the case of a single pipe.

For three pipes with a distance of 0.5 D from one another, the maximum scour depth was reduced by 26.7% and 47.6%, respectively, compared to the two-pipe test with a distance of 0.5 D and the single-pipe test. The test time also shows a reduction of 8.57% and 25.58%, respectively. The bed profile in the equilibrium state in the central axis of the channel for each state is shown in Figure 3.

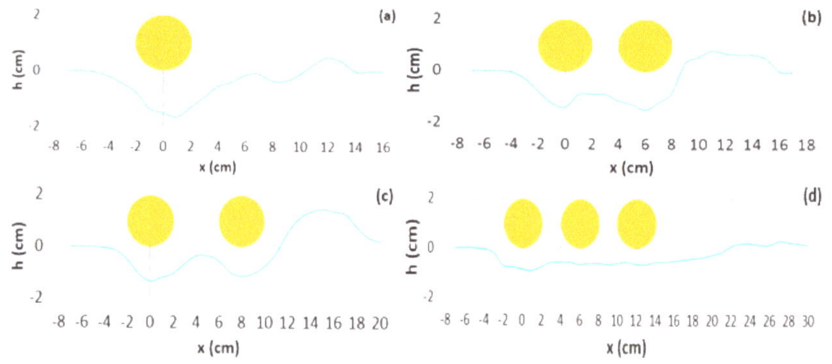

Figure 3. Centerline longitudinal bed profile at equilibrium: (**a**) single pipe, (**b**) two pipes G/D = 0.5, (**c**) two pipes G/D = 1, (**d**) three pipes G/D = 0.5.

3.3. Velocity Field around the Pipelines

3.3.1. Velocity Field in X-Y Plane

Figure 4 shows the contours of u velocity at 0.6 cm from the bed (z = 0.6 cm). Figure 4a, for the single-pipe test, shows that the horizontal component of the flow velocity (u) upstream decreases when approaching the pipe. As the flow passes through the scour hole created below the pipe, the horizontal flow velocity increases greatly, which indicates the outflow jet from under the pipe. By increasing the distance from the pipe in the area between the pipe and the sand dunes formed downstream, the horizontal velocity of the flow decreases. Additionally, after passing through the area of sand dunes, the decrease of horizontal velocity near the bed continues. For example, upstream of the pipe at the coordinates x = −10 cm, y = −23 cm, the horizontal velocity of the flow is u = 10.6666 cm/s, which near the pipe at the coordinates x = −4 cm, y = −23 cm to the value of u = 7.9887 cm/s has decreased. Downstream, the outflow jet from under the pipe, at coordinates x = 4 cm, y = −23 cm increases speed up to u = 14.1914 cm/s. Then, during a decreasing process in coordinates x = 14 cm, y = −23 cm, the value of u = 5.1011 cm/s has been reached. In this figure, reverse flow is not observed upstream and downstream of the single pipeline. By approaching the walls, the horizontal velocity values decrease, which indicates the effect of the walls. For example, in the upstream of the pipe from coordinates x = −10 cm, y − 0 to coordinates x = −10 cm, y = −42 cm, the velocity has decreased from u = 12.2335 cm/s to u = 10.1111 cm/s. Downstream of the pipe from coordinates x = 4 cm, y = 0 to coordinates x = 4 cm, y = −42 cm, the velocity decreases from u = 13.4359 cm/s to u = 9.4478 cm/s. In addition, the maximum value of the horizontal velocity at z = 0.6 cm, in this case, is u = 14.6849 cm/s, which is in the central cross-section of the channel at coordinates x = 4 cm and y = 23 cm.

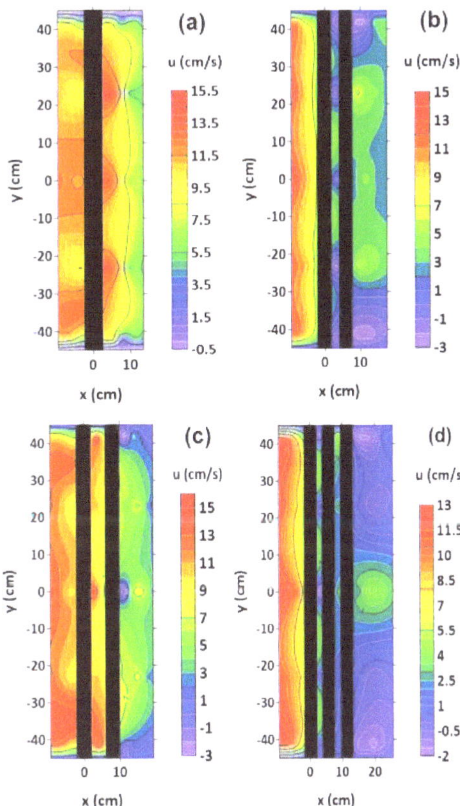

Figure 4. Velocity field around the pipelines in the X-Y plane at height of 0.6 cm from the scour bed (u, cm/s): (**a**) single pipe, (**b**) two pipes G/D = 0.5, (**c**) two pipes G/D = 1, (**d**) three pipes G/D = 0.5. Note that the range of the color scale is not equalized for better illustration.

In Figure 4b (state 2—two pipes with a distance of 0.5 D), the horizontal component of the velocity near the bed upstream has decreased by approaching the pipes. For example, from the coordinate x = −10 cm, y = 0 to the coordinate x = −4 cm, y = 0, it has decreased from the value of u = 13.1805 cm/s to the value of u = 8.5098 cm/s. The horizontal velocity of this flow has been significantly reduced by entering the scour hole formed in the distance between the two pipes. For example, the velocity value at coordinates x = 3 cm, y = 0 is equal to u = 0.6363 cm/s. Additionally, negative values of horizontal velocity can be seen in this area, which indicates reverse flow. For example, in the coordinates, x = 3 cm, y = −23 cm and x = 3 cm, y = −42 cm, the horizontal velocity values are u = −2.7319 cm/s and u = −2.0174 cm/s, respectively. As the flow passes through the scour hole under the second pipe to the downstream side, an increase in the velocity values is observed. Still, these values are smaller compared to the upstream flow of pipelines. For example, in the coordinates x = 10 cm, y = 0, u = 3.2508 cm/s, which show an increase compared to u = 0.6363 cm/s in the coordinates x = 3 cm, y = 0, but compared to u = 8.5098 cm/s in the upstream of the pipe at coordinates x = −4 cm, y = 0 is much less. Next, by moving away from the downstream pipe, a decreasing trend occurs in the velocity values. For example, from coordinates x = 10 cm, y = 0 to x = 18 cm, y = 0, the velocity reaches from u = 3.2508 cm/s to u = 2.3106 cm/s. According to the mentioned materials, in this tested case, due to the short distance between the two pipes, the flow has been blocked, which causes the velocity values to be small, and the reverse flow is created in this area. Comparing Figure 4a with Figure 4b shows the effect of the presence of the second pipe at a short distance from the first pipe. In other words, the second pipe's presence has significantly reduced the horizontal velocity of the flow passing through the scour hole near the bed downstream of the pipe. The examination of the horizontal velocity values in the central axis of the channel (y = 0) show that the velocity of the outflow from under the pipe downstream in the second state at a distance of 0.5 D from the second pipe is u = 3.2508 cm/s, which compared to the first state, at the same distance from the pipe in the downstream, with the velocity u = 13.4359 cm/s, it has decreased by 75.8%. This can be a justification for less scouring depth in the second case compared to the first. The point of commonality between the two states is that the horizontal velocity values near the bed are higher in the central area of the channel, compared to the area near the walls. For example, in the upstream part, from coordinates x = −10 cm, y = 0 to coordinates x = −10 cm, y = −42 cm, the horizontal velocity value has decreased from u = 13.1805 cm/s to u = 11.0879 cm/s.

Figure 4c, which is related to the third case with two pipes with distance D, shows that the values of horizontal velocity near the bed decreased as it approached the upstream pipe. A decreasing trend is also observed downstream of the pipelines, but in the distance between two pipes, some increase in the velocity values can be seen. Downstream of the pipelines, in the area of the central axis, as well as near the walls, negative values of the horizontal velocity can be seen. For example, in the central axis of the channel (y = 0), the horizontal velocity at x = −10 cm is equal to u = 13.8888 cm/s. Then, it decreases at x = −4 cm to u = 9.1231 cm/s. At the distance between two pipes, it has values of u = 14.3719 cm/s and u = 9.1892 cm/s at x = 3.5 cm and x = 4.5 cm, respectively. Immediately downstream of the second pipe at x = 12 cm, it had a negative value of u = −4.0123 cm/s, then again from x = 16 cm to x = 20 cm, the horizontal velocity decreased from u = 8.4674 cm/s to u = 1.3925 cm/s. The maximum values of positive horizontal velocity have also been observed in the central transverse area of the channel (velocity u = 14.3719 cm/s in coordinates x = 3.5 cm, y = 0). The comparison of Figure 4a,c shows that the presence of the second pipe downstream of the first pipe has significantly reduced the horizontal velocity of the flow near the bed in the distance between the two pipes and downstream of the pipes. For example, in the central axis of the channel, at a distance of 0.5 D from the downstream pipe, u has decreased by 77.5% compared to the same distance from the downstream pipe in the first case. The same issue justifies the reduction in scouring depths in the third case compared to the single pipe case. The comparison of Figure 3b,c shows that increasing the distance between the two pipes from 0.5 D to D has caused the effect of the presence of the second pipe on reducing the horizontal flow velocity in the space between the two pipes, as well as the downstream of them, it should be weaker so that in the distance between two pipes and at the central axis of the channel, the magnitude of the horizontal velocity in the third state is 22.6 times compared to the second state. This issue is in harmony with the greater scouring depth in the third state compared to the second.

In Figure 4d (case 4: three pipes with a distance of 0.5 D) in the upstream part of the pipelines similar to previous three cases, as the flow approaches the first pipe (upstream pipe), the horizontal velocity of the flow has decreased. For example, in the central axis, it has decreased from u = 12.1336 cm/s at x = −10 cm to u = 10.6877 cm/s at x = −4 cm. When the flow enters the scour hole formed in the distance between the first and second pipe, the horizontal velocity values become negative, which indicates reverse flow. Nevertheless, in the distance between the second and third pipes, there is no effect of reverse flow, and the horizontal velocity is positive, but with a small magnitude. Downstream of the pipes, in the central axis of the channel, there has been some increase in the horizontal flow velocity (up to u = 4.4238 cm/s at x = 18 cm). In other parts, however, the horizontal velocity values are positive or negative and close to zero. By comparing this state with the second state, the values of the horizontal velocity downstream of the pipes and also in the area between the first and second pipes are almost similar to each other and a significant decrease in the horizontal velocity of the flow has been observed along with the presence of reverse flow. The difference is that the velocity values are closer to zero in the case of three pipes. Additionally, in both mentioned cases, in the central area of the channel downstream of the pipes, the velocity has increased a little. This velocity increase occurred in the case of three pipes in the central axis of the channel and other parts, there is a positive velocity close to zero or negative, but in the second case, two pipes with a distance of 0.5 D, in a larger part of the central area of the channel and the negative horizontal velocity values are limited to the parts close to the walls.

In the following, the lateral component of the velocity (v) at 0.6 cm from the bed will be investigated. In Figure 5a–d, this component is shown for the four tested cases. In all four arrangements of pipes, in the negative-y areas, v is positive (flow deviation in the counterclockwise direction) and in positive-y areas, v is negative (flow deviation in the clockwise direction). This shows the deviation of the current towards the transverse central axis of the channel (y = 0). By approaching the transverse central axis of the channel, the lateral velocity values (both positive and negative) decrease and tend to zero.

In the first test case, near the walls, at a short distance from the upstream and downstream of the pipe, some deviation of the flow towards the walls is observed. For example, at the coordinates x = −4 cm, y = −42 cm, v = −0.4678 cm/s, and at the coordinates x = 4 cm, y = −42 cm, v = −0.0118 cm/s. In the second case of the test, this deviation continues to a greater distance downstream of the pipes and the maximum value of this deviation (lateral velocity deviation towards the wall) occurred at a distance of 1.5 D from the pipe to the downstream side (value v = −1.0465 cm/s in coordinates x = 14 cm, y = −42 cm and the value of v = 1.0555 cm/s in coordinates x = 14 cm, y = 42 cm). In the third case of the test, the deviation of the flow towards the wall is observed in the upstream parts, the distance between the two pipes, and the downstream. The maximum value of deviation in the distance of 1.5 D downstream of the second pipe at the coordinates x = 16 cm, y = −42 cm is equal to v = −0.8228 cm/s, and between the pipes at the coordinates x = 4.5 cm, y = 42 cm to the value of v = 1.6554 cm/s. In the fourth case of the test, the deviation of the lateral component of the flow velocity towards the wall was created only in the upstream and downstream of the pipelines. (The maximum value of deviation at x = −4 cm, y = −42 cm is equal to v = −2.0067 cm/s and at x = −4 cm, y = 42 cm is equal to v = 2.3256 cm/s).

In Figure 6a–d, the contours are presented for the vertical component of the velocity at 0.6 cm from the bed. Results show that an upward flow is formed upstream of the pipes, which can be caused by the separation of the flow. By approaching the pipeline, negative values of the vertical velocity are observed, which can be due to the flow entering the scour hole. Downstream of the pipelines, an upward flow has occurred due to the exit from the scour hole. In the following, this flow faces the sand dunes formed downstream of the pipe(s), and the values of the vertical component of the velocity remain positive. Gradually, passing through the positive slope of the sand dunes and approaching their peak, the values of the vertical velocity tend to zero, and then, upon reaching the negative slope of the sand dunes, the values of the vertical component of the velocity become negative.

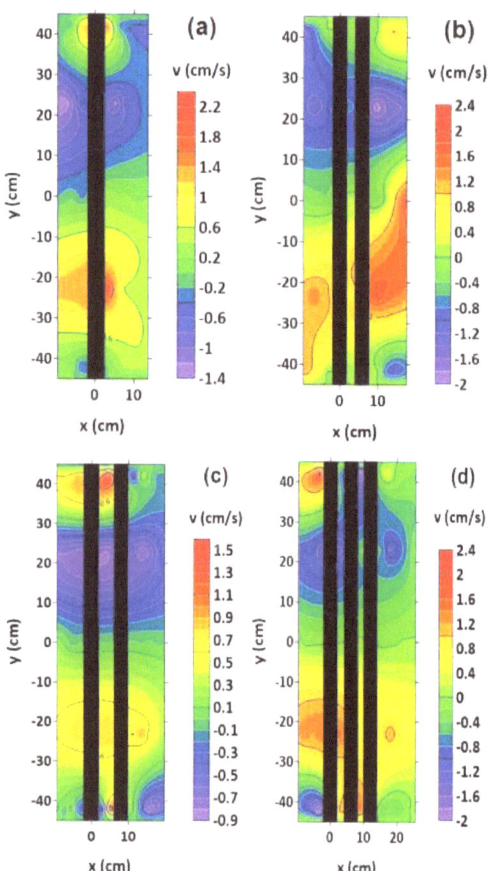

Figure 5. Velocity field around the pipelines in the X-Y plane at height of 0.6 cm from the scour bed (v, cm/s): (**a**) single pipe, (**b**) two pipes G/D = 0.5, (**c**) two pipes G/D = 1, (**d**) three pipes G/D = 0.5. Note that the range of the color scale is not equalized for better illustration.

The highest values of the vertical component of the velocity downstream of the pipe are observed in the first state (single pipe) (the maximum value of w = 9.8877 cm/s in coordinates x = 4 cm, y = −23 cm). This problem can be justified considering that the depth of the scour hole in the first case is greater than the other three cases, and the flow downstream of the single pipe must exit from a deeper hole. The maximum positive values of the vertical component of the velocity among the other three states, as expected, are related to the third state (two pipes with section D), which has the highest scour depth after the first state. These large positive values of vertical velocity, in addition to the downstream of the pipelines (the maximum value of w = 3.6357 cm/s at the coordinates x = 14 cm, y = 0 at the distance D from the downstream pipe), are also observed in the area between the two pipes. (The maximum value of w = 6.8709 cm/s in coordinates x = 3.5 cm, y = −42 cm). This can be caused by the outflow of the flow from the scour hole created under the first pipe. Additionally, the vortices created in the distance between two pipes can be the cause of these vertical velocity values. In the second case of the test, the maximum positive values of the vertical component of the flow velocity formed downstream of the pipes are lower than the first and third cases (the maximum value of w = 2.4222 cm/s in coordinates x = 10 cm and y = −23 cm in the distance 0.5 D from the downstream pipe). Between the two pipes, unlike the third case, the velocity in the vertical direction has positive values but with a small magnitude. In the fourth case, the maximum positive values of the vertical component of the velocity created downstream of the pipes are lower than in the second case (the maximum value of w = 1.9882 cm/s in coordinates x = 18 cm, y = −23 cm at the distance D from the downstream pipe). This issue indicates the gentle slope of the scour hole in this area. In the distance between the pipes in the fourth state, like the second state, the vertical velocity has positive values but with a small magnitude. The mentioned discussion and results are in agreement with the results obtained from the scour hole in all cases.

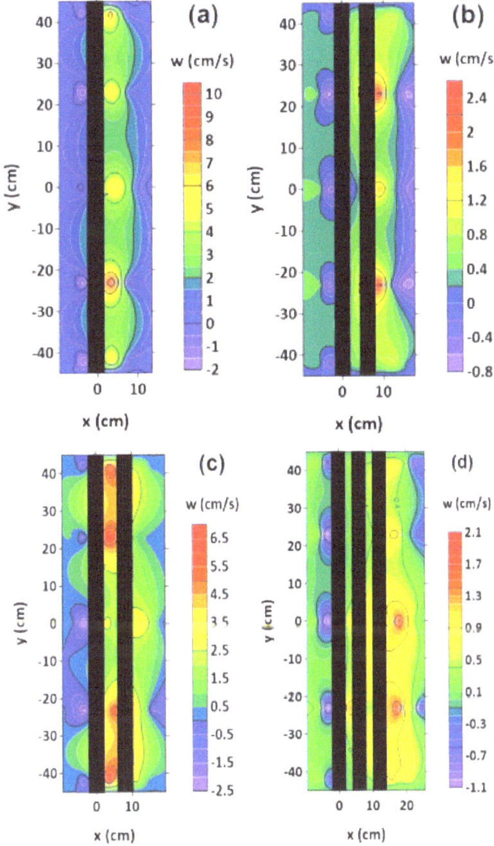

Figure 6. Velocity field around the pipelines in the X-Y plane at height of 0.6 cm from the scour bed (w, cm/s): (**a**) single pipe, (**b**) two pipes G/D = 0.5, (**c**) two pipes G/D = 1, (**d**) three pipes G/D = 0.5. Note that the range of the color scale is not equalized for better illustration.

3.3.2. Velocity Field in X-Z Plane

Figure 7a shows that upstream of the pipe (x = −10 cm to x/D = −2.5), the effect of the presence of the pipe on the flow path is less, the horizontal component of the flow velocity near the bed becomes larger as z increases, and by moving away from the bed, the horizontal velocity gradient decreases. When approaching the pipe, the effect of the pipe on blocking the flow is quite evident. The gradient of the horizontal component of velocity u is intense on the pipe and the value of the velocity has increased from the number close to zero to the velocity u = 20.7743 cm/s at 0.5 cm from the upper surface of the pipe. Furthermore, with the increase of z, not much change occurs in the horizontal velocity values. The shear layer of the flow passing over the pipe, by separating from its surface downstream, has formed two vortices during the interaction with the shear layer exiting from under the pipe. Additionally, in Figure 7a, the negative values of the horizontal velocity in this area indicate the existence of reverse flow. According to Zhang et al. (2016) [9], the formation of reverse flow indicates the presence of the two mentioned vortices in the flow area of the pipeline wake. Gradually, with the distance from the pipe downstream, the effect of the pipe's presence on the flow is reduced, and the horizontal velocity changes at different depths, becoming like the flow at a large distance upstream of the pipe. Downstream of the pipe and near the bed, the u values increase. Abbaszadeh Tavassoli and Haji Kandi. (2010) [6] point out the increase in velocity in this area and considered it to be the cause of the scour hole downstream of the pipeline. Yeganeh Bakhtiari et al. (2011) [7] also state that vortex shedding occurs downstream. Brors (1999) [21] has investigated the velocity contours around the single pipe after the scour cavity has reached equilibrium which are very similar to the contour lines of Figure 7a.

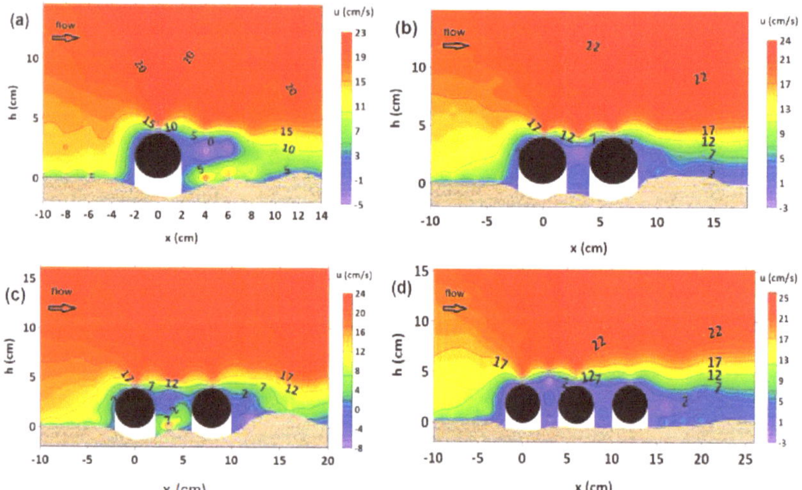

Figure 7. Velocity field around the pipelines in the X-Z plane (u, cm/s): (**a**) single pipe, (**b**) two pipes G/D = 0.5, (**c**) two pipes G/D = 1, (**d**) three pipes G/D = 0.5. Note that the range of the color scale is not equalized for better illustration.

In Figure 7b, contours for the horizontal component of the velocity in the second stage of the test are shown. In this case, like the first case, the horizontal component of the velocity has decreased by approaching the pipelines. In the areas on the pipes, a strong horizontal velocity gradient is also observed, and this gradient is higher on the first pipe than on the second pipe. At 0.5 cm from the upper surface of the first pipe, u = 20.4537 cm/s, and at the same distance from the upper surface of the second pipe, u = 15.6927 cm/s. With a sharp increase in velocity in these areas, there is no significant velocity gradient further when increasing z. Downstream of the pipes, the reverse flow was not observed, but in the distance between the pipes, with increasing distance from the bed, the positive values of the horizontal velocity are initially small. At the distances of 3.3 and 3.8 cm from the bed, the reverse flow is observed with the values of u = −1.7599 cm/s and u = −1.9538 cm/s. In the following, the values of the horizontal velocity are again positive but with a small magnitude. In this area, approaching the height level of the upper surface of the pipes (h = 4 cm), a strong gradient has occurred in the velocity values, so that the velocity u = 3.7643 cm/s at z = 4.4 cm is reached u = 19.0666 cm/s at z = 5 cm. After reaching the maximum velocity, there is no significant change in the velocity values. Hu et al. (2019) [22], during an experiment on the scouring of two parallel pipes in a shallow flow, pointed out the formation of the maximum velocity at a short distance from the upper surface of the pipes.

In Figure 7c, third test case, by approaching pipelines from the upstream side, the horizontal velocity values decrease. Additionally, a strong gradient of u is observed in the area on the pipes. In this case, like the second case, the velocity gradient on the upper surface of the first pipe is stronger than the second pipe, so that, up to a distance of 0.5 cm from the upper surface of the first pipe, the horizontal velocity is u = 19.1342 cm/s and up to the same distance from the upper surface in the second pipe, the velocity has reached u = 15.6726 cm/s. After this sharp increase in velocity in these areas, the u changes are small as z increases. Downstream of the pipelines, in the areas near the bed up to z = 2.2 cm, there was a reverse flow with the maximum value of u = −7.8705 cm/s at x = 12 cm, z = 0.3 cm. Then, with the increase of z, there is no effect of the reverse flow, and the horizontal component of the velocity is faced with a high gradient in the positive direction. Additionally, in the area between the second pipe and the sand dune, a reverse flow is formed, but with the increase of x, there is no effect of the reverse flow in the areas on the dune. Moving downstream, negative values of horizontal velocity near the bed are also observed downstream of the sand dune (with a maximum value of u = −0.1539 cm/s at x = 20 cm, z = 0.3 cm). In the area between the two pipelines, a sharp increase in the positive u values near the bed has occurred. A little above this area, in the range of z = 2 cm to z = 4 cm, negative u values have been created. This issue shows the interactions caused by the collision of shear layers separated from the upper and lower surfaces of the first pipeline. This process, like the process occurring downstream of the single pipe (case 1), has caused the creation of two vortices in this area. By comparing the third and second states, it seems that the increase in the distance between the two pipes has caused the shear layer to separate from the upper and lower surfaces of the first pipe to have enough space to mix in between the two pipes and form two vortices. Near the bed, there has been enough space to create a strong horizontal velocity gradient by the outflow jet from under the first pipe. These vortices can be considered as the Carman vortices that Ishigai et al. (1972) [23] mentioned in their study.

Figure 7d shows the u contours for the fourth tested case. In the upstream of the pipelines, like the previous cases, the horizontal component of velocity u has decreased as it approaches the upstream pipe. On the upper surface of the pipes, a strong horizontal velocity gradient is observed, which decreases on the second and third pipes, respectively. After this extreme gradient, with the increase of z, no noticeable change in u values is observed. In the distance between the first and second pipe, like the second case, u values are small and do not fluctuate significantly. At z = 0.9, 4.3, 4.8 cm, the reverse flow was created with the values of u = −1.8244, −2.9317, −2.9228 cm/s, respectively, and in the rest of the elevated levels of this area, the velocity was positive, but its magnitude was small. The reason for this can be seen in the previous cases, because of the flow's stagnation due to the small distance between the pipes. This flow stagnation is much greater in the distance between the second and third pipes. In this area, there is no sign of reverse flow and the horizontal velocity values are all positive, and at the same time, very close to zero. In the downstream of the pipelines, the values of the horizontal component of the velocity are positive with a small magnitude, and only in the cross section of x = 16 and in the height range of 1.5 cm ≤ z ≤ 2.3 cm, the presence of reverse flow is observed (−2.0353 cm/s ≤ u ≤ −0.6308 cm/s). Near the bed, the horizontal velocity values also increase slightly.

Figure 8 shows the dimensionless profiles of the horizontal component of the velocity in the direction of the flow u, in the upstream, downstream, and on the single pipe, after the bed balance. Examining the velocity profiles in sections x/D = −1.5, −2, and −2.5 shows that the effect of the pipe's presence on the horizontal velocity profile far upstream is insignificant. At section x/D = 0, a high velocity gradient can be seen on the pipe. In this area, the velocity at a distance close to the pipe almost reaches the maximum value

($u = 0.97\,u_{max}$), and further, with the increase of z, no significant changes in the velocity values are observed. In sections $x/D = 1.5$ and $x/D = 1$, S-shaped profiles are formed, which indicate two vortices in this area of the flow due to the presence of negative horizontal velocity values in them. In section $x/D = 2$, the S-shaped profile is still observed, but there is no reverse flow. In the next sections ($x/D = 2.5, 3, 3.5$), gradually, there is no more S-shaped profile, and the effect of the pipe's presence on the velocity profiles becomes less and less. A similar general trend in the changes of the velocity profiles formed around the single pipe can be seen in the studies of researchers such as Jensen et al. (1990) [2], Zhang et al. (2016) [9], and Chen et al. (2020) [10], so the results of the single pipe case are similar to previous studies in this field.

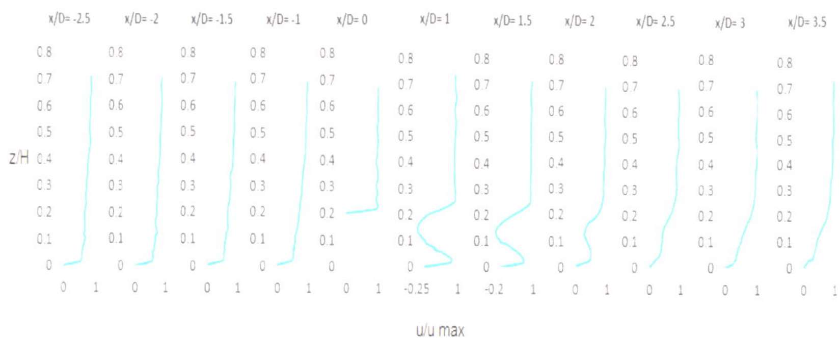

Figure 8. Profiles of normalized velocity u in different sections for the case of single pipe.

By comparing the velocity profile at $x/D = 3.5$ and $x/D = -2.5$, the velocity gradient near the bed at $x/D = 3.5$ is much lower compared to $x/D = -2.5$. At $x/D = 3.5$, the velocity has approached the maximum value at a greater distance from the bed (at $z/H = 0.26$, u value has reached $u = 0.82\,u_{max}$). The reason for this can be the continuation of the effect of the shear layer separated from the upper surface of the pipe to a significant distance downstream of the pipe, which Figure 7a also illustrates. Moving downstream, the thickness of the boundary layer will be similar to its thickness at a long distance upstream of the pipe, and the effect of the pipe's presence on the flow will disappear to a great extent. This matter is seen in the study of Chen et al. (2022) [13]. Zhang and Shi (2016) [8] have observed the S-shaped horizontal velocity profile and the occurrence of two vortices, during the investigation of the flow around a single pipe with an initial distance of 0.1 D, 0.3 D and 0.5 D from the bed, at the sections of $x/D = 1$ and $x/D = 3.5$ (downstream of the pipe).

Figures 9 and 10 show the dimensionless profiles of the horizontal component of the velocity around the pipelines for the study's second and third cases. Examining these profiles in sections of $x/D = -2.5, -2$ and -1.5 shows that the effect of the presence of pipelines on the horizontal component of velocity in this range is negligible in both tested cases. In section $x/D = -1$, due to the presence of the pipe in the flow path, a decrease in the horizontal velocity values is observed in the range of $z/H \leq 0.2$. By comparing the velocity profiles of this section in Figures 9 and 10, near the bed, the velocity values are higher in Figure 10 (state 3). For example, in $z/H = 0.015$ in the third state, $u = 0.4275\,u_{max}$ and at the same z/H in the second case, the horizontal velocity is $u = 0.3606\,u_{max}$, which can be due to the absence of the flow stagnation between the pipes in the third case. As a result, the flow passes more freely through the scour cavity formed below the upstream pipe. The profiles of sections $x/D = 0, 1.5$ in Figure 9 and sections $x/D = 0$ and 2 in Figure 10 show the extreme velocity gradient on the pipes. The profile of section $x/D = 0.75$ in Figure 9 (between two pipes) shows irregularities in the values of u near the bed. In this area, these values are positive and have a small magnitude. With the increase in z, first, the horizontal velocity moved towards negative values, reaching $u = -0.08\,u_{max}$, and then in the range close to the level of the upper surface of the pipelines, a sharp increase was found in the positive direction so that, at z/H = 0.25, the value of u = 0.83 u$_{max}$. After that, with the increase of z, there was no significant change in u values. At the same time, in sections x/D = 0.875 and 1.125 in Figure 10, S-shaped profiles were formed, which have negative velocity values and indicate the presence of Carman vortices. In the section x/D = 2.5, in Figure 9, which is the area immediately downstream of the downstream pipe (corresponding to the second state of the test), an S-shaped profile is observed, but there is no negative value in the velocity values. While in section x/D = 3 in Figure 10, which shows the area immediately downstream of the pipelines (related to the second test case), the S-shaped profile is not formed, and at the same time, near the bed, the velocity values are negative. In the height range of the presence of the pipe (z/H ≤ 0.2), irregularity is observed in the horizontal velocity profile. In the next sections and moving downstream (Figures 9 and 10), the influence of the pipelines on the horizontal velocity profiles gradually decreases. Of course, as mentioned in the description of the velocity profiles related to the first case (single pipe), until a relatively large distance downstream of the pipe(s), the velocity profiles are still unlike the upstream of the pipes, and the thickness of the shear layer is greater. This thickness gradually decreases, until the velocity profiles become like the velocity profiles upstream away from the pipelines.

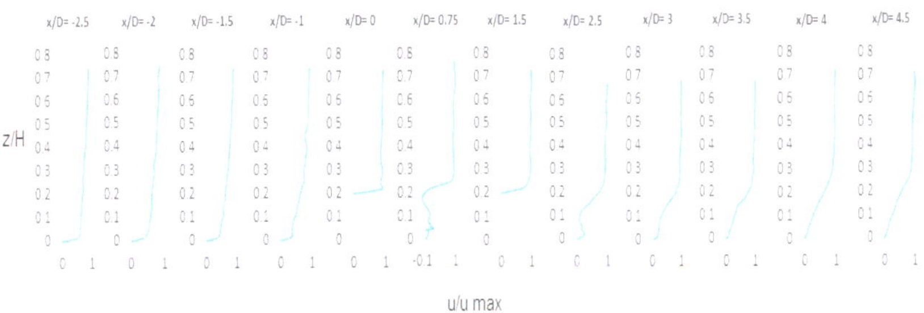

Figure 9. Profiles of normalized velocity u in different sections for the case of two pipes G/D = 0.5.

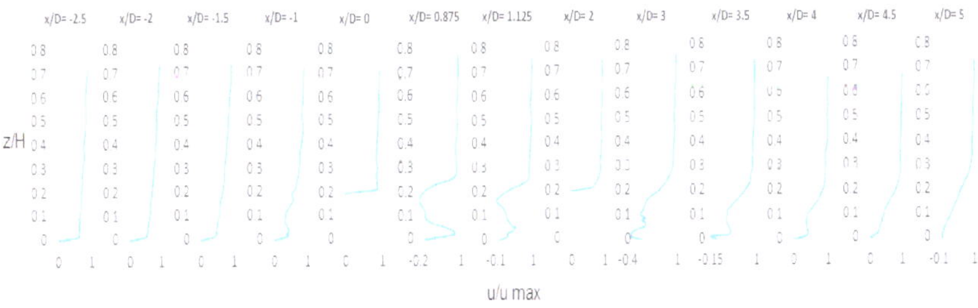

Figure 10. Profiles of normalized velocity u in different sections for the case of two pipes G/D = 1.

Figure 11 shows the dimensionless profiles of the horizontal velocity, corresponding to the fourth case. In this case, in the section x/D = −1, in the height range of the presence of the pipe, a decrease in the horizontal velocity values is observed. The comparison of the velocity profiles in the sections at x/D = 0, x/D = 1.5, and x/D = 3 shows that the intensity of the vertical gradient of the horizontal velocity $\left(\frac{du}{dz}\right)$ on the upper surfaces of the first to third pipes (respectively) decreases relatively. In the cross section x/D = 0.75, which corresponds to the area between the first and second pipes, as z increases from near the surface of the bed, there have been fluctuations in the horizontal velocity values that do not follow a specific order. At the same time, except for the two areas where the graph entered negative values, in the rest of the range between the two pipes (z < 4), the graph is very close to the vertical axis and only fluctuates in positive values and close to zero. This can be attributed to the stagnation effect of the flow in this area due to the small distance of the pipes. This effect is more observable in the cross section of $x/D = 2.25$ (between the second and third pipes), so that in part $z < 4$, the graph is close to the vertical axis and the horizontal velocity values have very small fluctuations. The maximum value of the horizontal velocity in this part is equal to $u = 0.084\,u_{max}$ at $z/H = 0.045$. The reason for this can be considered the presence of the third pipe at 0.5 D from the second pipe, which has aggravated the flow stagnation in the area between the pipes. Downstream of the pipes, S-shaped profiles are also observed, but only in the section of $x/D = 4$ reverse flow with the maximum value of $u = -0.084\,u_{max}$ is created at $z/H = 0.105$. By moving downstream, the S-shaped profiles gradually disappear in the next sections, and no trace of the S-shaped profile can be observed in sections $x/D = 6$ and 6.5.

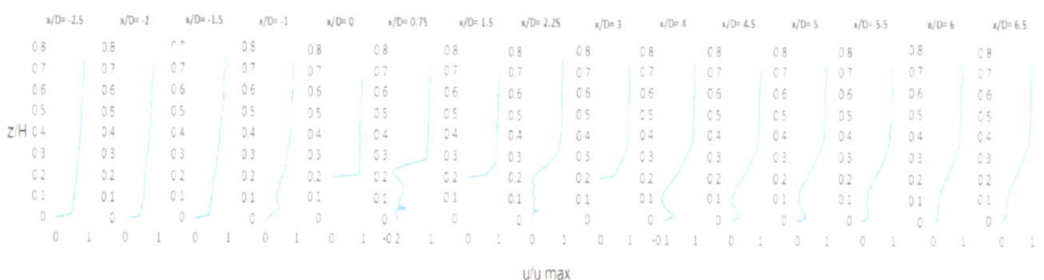

Figure 11. Profiles of normalized velocity u in different sections for the case of three pipes $G/D = 0.5$.

Figure 12 shows the contours of the transverse component of velocity (v) for four test conditions. As expected, due to the symmetry of the elements in the transverse direction of the channel in the experiments of this study (horizontal pipelines), the flow examination in the x-z plane in the transverse central axis of the channel shows that the values of v are very small and there is only a slight flow deviation to the left or right. In Figure 12a, which is related to the first case (single pipe), an almost two-dimensional flow is formed upstream of the pipeline. Of course, in the area behind the pipe, the transverse deviation of the flow increases slightly. Downstream, the v values increase slightly, which can be attributed to the outflow jet from under the pipe, the presence of Carman vortices in the pipe wake, and the presence of sand dunes. In Figure 12b (the second case), in the gap between the two pipes, some deviation of the flow in the transverse direction is observed. Additionally, in the downstream areas of the sand dunes and near the pipe and the bed upstream of the first pipe, an increase in the lateral deviation of the flow is observed. In the rest of the parts, the values of the transverse component of the velocity are very low or zero. In Figure 12c, the third case, the transverse velocity tends to zero in most parts. More deviation in the transverse flow is observed in parts such as the upstream flow near the bed (behind the first pipe), the distance between two pipes (with the presence of Carman vortices), and downstream of the pipelines. In Figure 12d, the fourth case, in the downstream parts of the pipelines, upstream points near the first pipe, and the space between the pipes, the lateral deviation of the flow is slightly increased.

Figure 13 shows the contours of the vertical component of the velocity (w) for four test conditions. In the upstream part and close to the pipe, the separation of the flow into two parts is clearly defined. The first part is the downward flow near the bed ($w < 0$), which can indicate the flow entering the scour cavity, and the second part is the upward flow ($w > 0$) that passes over the upper surface of the pipe. Downstream, the flow passing over the upper surface of the pipe is first horizontal and then downwards. The upward flow of the outlet from under the pipe is also observed near the bed. The values of the vertical component of the velocity in the distances between the pipes (in the second to fourth states) near the bed have become positive, and with a slight distance from it (increasing z), the vertical velocity has become negative. In the second and fourth case, these values are small. In the third case, when the two pipes are placed at a greater distance from each other, the shear layers separate from the upper and lower surfaces of the first pipe have more room to move in the space between the two pipes. In this case, an upward jet flow is observed near the bed, and immediately above it, a downward flow is observed, which has led to the formation of Carman vortices in this area. The maximum downward velocity in this part is w = −2.1905 cm/s in coordinates x = 3.5 cm, z = 1.9 cm, and the maximum upward velocity in this part is w = 4.5353 cm/s in coordinates x = 3.5 cm, z = 0.6 cm.

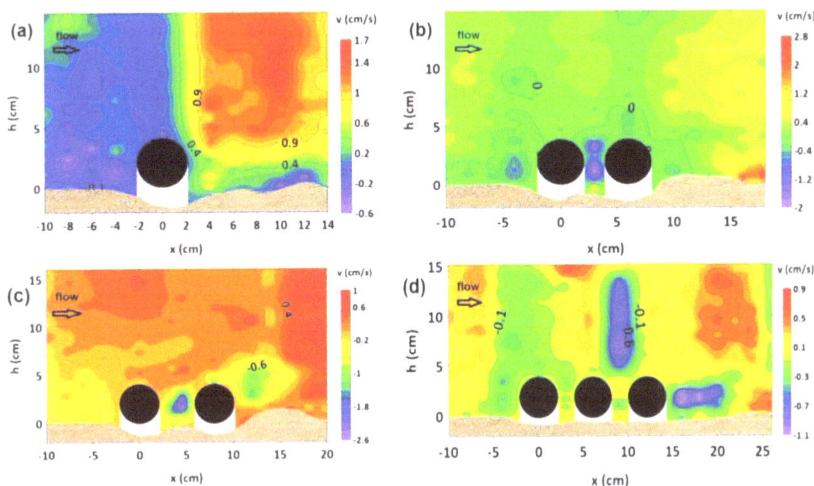

Figure 12. Velocity field around the pipelines in the X-Z plane (v, cm/s): (**a**) single pipe, (**b**) two pipes G/D = 0.5, (**c**) two pipes G/D = 1, (**d**) three pipes G/D = 0.5. Note that the range of the color scale is not equalized for better illustration.

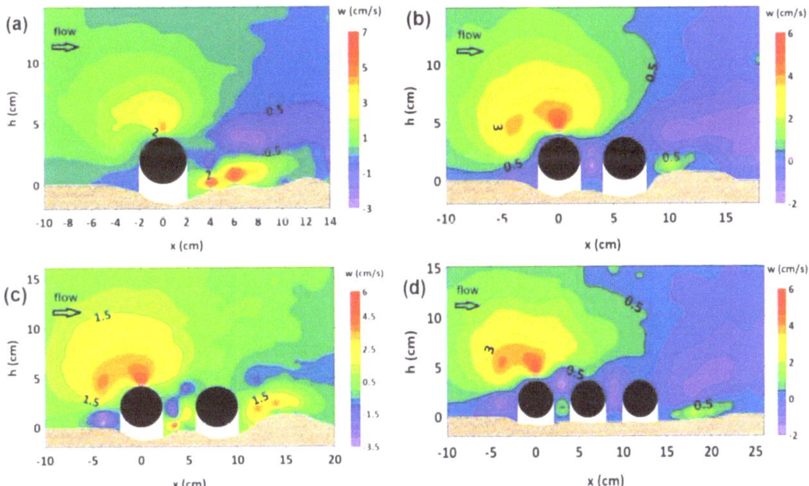

Figure 13. Velocity field around the pipelines in the X-Z plane (w, cm/s): (**a**) single pipe, (**b**) two pipes G/D = 0.5, (**c**) two pipes G/D = 1, (**d**) three pipes G/D = 0.5. Note that the range of the color scale is not equalized for better illustration.

4. Discussion

In this study, scouring under the pipelines in four arrangements (cases) was investigated: (1) single pipe, (2) two pipes with a distance of 0.5 D, (3) two pipes with a distance of D, (4) three pipes with a distance of 0.5 D, in a direct channel with a uniform flow and H/D = 5

(H is the depth of flow). Additionally, horizontal, vertical, and lateral components of flow velocity were measured using ADV around the four arrangements. The results show:

1- The magnitude of the horizontal component of the velocity near the bed for four tested cases matches with the results of their maximum scour depth.

2- In the first case, downstream of the pipeline, the existence of S-shaped profiles along with the negative values of horizontal velocity in sections $x/D = 1$ and $x/D = 1.5$ show that vortices are formed in this area.

3- In the second case, the reverse flow was not observed downstream of the pipelines, but in the distance between the two pipes, at distances of 3.3 and 3.8 cm from the bed. In general, in the area between the two pipes, the magnitude of the velocity is close to zero, which is caused by the small distance between the pipes and, as a result, the flow stagnated to a large extent in this area.

4- In the third case, in the distance between two pipes, large positive horizontal velocity values were observed near the bed, and reverse flow in the range of $z = 2$ cm to $z = 4$ cm. S-shaped horizontal velocity profiles can also be seen in this area, which are formed due to the presence of two vortices. This process is similar to the process that happens downstream of the single pipe and is caused by the sufficient distance between the pipes for the interaction of the shear layers. Downstream of the pipelines, in this case, near the bed, the reverse flow was observed up to $z = 2.2$ cm, and after that, positive horizontal velocity was observed. S-shaped profiles are not formed in this area.

5- In the fourth case, in a distance between the pipes, the effect of flow stagnation is observed. Downstream of the pipelines, there are S-shaped profiles of horizontal velocity, but negative horizontal velocity was observed only in the section of $x/D = 4$, 1.5 cm $\leq z \leq 2.3$ cm.

5. Conclusions

This study represents an advancement in the current understanding of the flow-structure interaction at scoured horizontal pipelines (especially in two-pipe and three-pipe arrangements) subjected to currents. It was demonstrated that the maximum and minimum scour depth among the four pipe arrangements examined in this study is formed in single-pipe and three-pipe arrangements, respectively. As a result, the use of pipe groups at a distance of 0.5 D has effectively reduced the maximum scour depth. This study indicates that the scouring depth under the pipelines is in harmony with the velocity components in the horizontal and vertical directions (u and w), and changing the arrangement of pipelines significantly changes the rate of these velocity components. Additionally, by comparing the arrangements of two pipes with a distance of 0.5 D and three pipes with a distance of 0.5 D, it seems that increasing the number of pipes will lead to lower scouring. Considering that in most practical projects, the pipe groups are used under the sea, where the bed material is fine sand. The results of this study will therefore be helpful for designers.

Over the last few decades, various methods have been proposed and investigated to protect pipelines from scouring. Some designers prefer to bury the pipelines under the seabed in order to protect them. One of the advantages of this method is to protect the pipeline against lateral movements caused by currents and sea waves. However, with the burial of the pipeline, the possibility of physical inspection of the pipeline is eliminated and some problems are created in its maintenance stages. A number of researchers have used other protection methods, such as using horizontal or vertical impermeable plates, to control erosion under pipelines, particularly for the one and two pipeline cases. The results of their studies show that the use of these plates is effective in reducing scouring.

In this study, the focus is on the best arrangement of pipelines that are directly placed on the erodible bed with sand materials with a size of $d_{50} = 0.83$ mm. Additionally, the results show that the way the pipelines are placed on the bed is very important. However, in cases where the bed materials are finer and more exposed to scouring, it is recommended to use the best arrangement of pipelines along with an efficient protection method, which is expected to bring better results.

Author Contributions: F.K. laboratory works, methodology, software, writing—original draft, preparation; H.A. and S.M.S. supervision, writing—review, methodology, validation and editing; S.A. writing and editing, and A.J. laboratory works, editing. All authors have read and agreed to the published version of the manuscript.

Funding: This research received no external funding.

Data Availability Statement: The data presented in this study are available on request from the first author (Kolahdouzan F).

Conflicts of Interest: The authors declare no conflict of interest.

Nomenclature

The following symbols are used in this paper:

D	diameter of the pipe [L]
e	distance between the lower surface of the pipe(s) and the bed [L]
d_{50}	median diameter of the sediment particles [L]
σ_g	geometric standard deviation of particle size distribution [dimensionless]
H	approaching flow depth [L]
U	average approaching flow velocity [LT^{-1}]
u^*	shear velocity [LT^{-1}]
u^*_c	critical shear velocity [LT^{-1}]
h	vertical axis
S	maximum scour depth [L]
t_e	duration of the test [T]
G	distance between pipelines [L]
z	vertical distance from the scoured bed [L]
u	horizontal component of velocity [LT^{-1}]
v	lateral component of velocity [LT^{-1}]
w	vertical component of velocity [LT^{-1}]

References

1. Bearman, P.W.; Zdravkovich, M.M. Flow around a circular cylinder near a plane boundary. *J. Fluid Mech.* **1978**, *89*, 33–47. [CrossRef]
2. Jensen, B.L.; Sumer, B.M.; Jensen, H.R.; Fredsoe, J. Flow around and forces on a pipeline near a scoured bed in steady current. *J. Offshore Mech. Arct. Eng.* **1990**, *112*, 206–213. [CrossRef]
3. Lei, C.; Cheng, L.; Kavanagh, K. Numerical Flow Visualization of Vortex Shedding Flow Over a Circular Cylinder Near a Plane Boundary. In Proceedings of the Ninth (1999) Interna onal Offshore and Polar Engineering Conference, Brest, France, 30 May–4 June 1999.
4. Oner, A.A.; Kirkgoz, M.S.; Akoz, M.S. Interaction of a current with a circular cylinder near a rigid bed. *Ocean. Eng.* **2008**, *35*, 1492–1504. [CrossRef]
5. Lin, W.J.; Lin, C.; Hsieh, S.C.; Dey, S. Flow characteristics around a circular cylinder placed horizontally above a plane boundary. *J. Eng. Mech.* **2009**, *135*, 697–716. [CrossRef]
6. Abbaszadeh Tavassoli, A.; Haji Kandi, H. Investigating the flow after creating scour holes under oil pipelines using Fluent software. In Proceedings of the 9th Hydraulic Conference of Iran, Tarbiat Modares University, Tehran, Iran, 8 November 2010.
7. Yeganeh Bakhtiari, A.; Kazemi Nejad, M.; Hosseini Ghahi, H. Investigation of scouring and hydrodynamic forces on transmission pipelines under the effect of flow. In Proceedings of the National Conference of New Finding in Civil Engineering, Nagaf Abad, Iran, 23 February 2010.
8. Zhang, Z.; Shi, B. Numerical simulation of local scour around underwater pipelines based on fluent software. *J. Appl. Fluid Mech.* **2016**, *9*, 711–718. [CrossRef]
9. Zhang, Z.; Shi, B.; Guo, Y.; Chen, D. Improving the prediction of scour around submarine pipelines. *Marit. Eng.* **2016**, *169*, 163–173. [CrossRef]
10. Chen, L.; Wang, Y.; Sun, S.; Wang, S. The effect of boundary shear flow on hydrodynamic forces of a pipeline over a fully scoured seabed. *Ocean Eng.* **2020**, *206*, 107326. [CrossRef]
11. Penna, N.; Coscarella, F.; Gaudio, R. Turbulent flow field around horizontal cylinders with scour hole. *Water* **2020**, *12*, 143. [CrossRef]

12. Liu, M.M.; Jin, X.; Wang, L.; Yang, F.; Tang, J. Numerical investigation of local scour around a vibrating pipelines under steady currents. *Ocean Eng.* **2021**, *221*, 108546. [CrossRef]
13. Chen, W.; Ji, C.; Alam, M.M.; Xu, D.; Zhang, Z. Three-dimensional flow past a circular cylinder in proximity to a stationary wall. *Ocean Eng.* **2022**, *247*, 110783. [CrossRef]
14. Goring, D.G.; Nikora, V.I. Despiking acoustic Doppler velocimeter data. *J. Hydraul. Eng.* **2002**, *128*, 117–126. [CrossRef]
15. Wahl, T. Analyzing ADV data using WinADV. In Proceedings of the Joint Conference on Water Resources Engineering and Water Resources Planning and Management, Minneapolis, MN, USA, 30 July–2 August 2000; pp. 1–10.
16. Mao, Y. *The Interaction between a Pipeline and an Erodible Bed*; Series Paper Technical University of Denmark; Technical University of Denmark: Lyngby, Denmark, 1986.
17. Li, Y.; Ong, M.C.; Fuhrman, D.R.; Larsen, B.E. Numerical investigation of wave-plus-current induced scour beneath two submarine pipelines in tandem. *Coast. Eng.* **2020**, *156*, 103619. [CrossRef]
18. Westerhorstmann, J.H.; Machemehl, J.L.; Jo, C.H. Effect of pipe spacing on marine pipeline scour. Presented at the Second International Offshore and Polar Engineering Conference, San Francisco, CA, USA, 14 June 1992; Volume II, pp. 101–109.
19. Gao, F.P.; Yang, B.; Wu, Y.X.; Yan, S.M. Steady current induced seabed scour around a vibrating pipeline. *Appl. Ocean Res.* **2006**, *28*, 291–298. [CrossRef]
20. Maddah, S.; Kolahdouzan, F.; Eftekhari, A.; Singh, V.P.; Afzalimehr, H. Experimental Investigation of Scouring in Groups of Parallel Pipelines. *Int. J. Hydraul. Eng.* **2021**, *10*, 27–34.
21. Brors, B. Numerical modeling of flow and scour at pipelines. *J. Hydraul.* **1999**, *125*, 511–523. [CrossRef]
22. Hu, D.; Tang, W.; Sun, L.; Li, F.; Ji, X.; Duan, Z. Numerical simulation of local scour around two pipelines in tandem using CFD–DEM method. *Appl. Ocean Res.* **2019**, *93*, 101968. [CrossRef]
23. Ishigai, S.; Nishikawa, E.; Nishimura, K.; Cho, K. *Experimental Study of Structure of Gas Flow in Tube Bank Swith Tube Axes Normal to Flow (Part I, Karman Vortex Flow from Two Tubes at Various Spacings)*; Bulletin of the JSME 15; The Japan Society of Mechanical Engineers: Tokyo, Japan, 1972; pp. 949–956.

 **water**

Article

Structure Integrity Analysis Using Fluid–Structure Interaction at Hydropower Bottom Outlet Discharge

Mohd Rashid Mohd Radzi [1,2], Mohd Hafiz Zawawi [1,*], Mohamad Aizat Abas [3], Ahmad Zhafran Ahmad Mazlan [3], Mohd Remy Rozainy Mohd Arif Zainol [4], Nurul Husna Hassan [1], Wan Norsyuhada Che Wan Zanial [1], Hayana Dullah [1] and Mohamad Anuar Kamaruddin [5]

1   Department of Civil Engineering, College of Engineering, Universiti Tenaga Nasional,
    Kajang 43000, Selangor, Malaysia
2   Hydro Life Extension Program (HELP), Business Development (Asset) Unit, TNB Power Generation Division,
    Petaling Jaya 46050, Selangor, Malaysia
3   School of Mechanical Engineering, Engineering Campus, Universiti Sains Malaysia,
    Nibong Tebal 14300, Pulau Pinang, Malaysia
4   School of Civil Engineering, Engineering Campus, Universiti Sains Malaysia,
    Nibong Tebal 14300, Pulau Pinang, Malaysia
5   School of Industrial Technology, Universiti Sains Malaysia, Nibong Tebal 14300, Pulau Pinang, Malaysia
*   Correspondence: mhafiz@uniten.edu.my

Citation: Mohd Radzi, M.R.; Zawawi, M.H.; Abas, M.A.; Ahmad Mazlan, A.Z.; Mohd Arif Zainol, M.R.R.; Hassan, N.H.; Che Wan Zanial, W.N.; Dullah, H.; Kamaruddin, M.A. Structure Integrity Analysis Using Fluid–Structure Interaction at Hydropower Bottom Outlet Discharge. *Water* **2023**, *15*, 1039. https://doi.org/10.3390/w15061039

Academic Editor: Charles R. Ortloff

Received: 13 January 2023
Revised: 14 February 2023
Accepted: 21 February 2023
Published: 9 March 2023

Abstract: Dam reliability analysis is performed to determine the structural integrity of dams and, hence, to prevent dam failure. The Chenderoh Dam structure is divided into five parts: the left bank, right bank, spillway, intake section, and bottom outlet, with each element performing standalone functions to maintain the overall Dam's continuous operation. This study presents a numerical reliability analysis of water dam reservoir banks using fluid–structure interaction (FSI) simulation of the bottom outlet structures operated at different discharge conditions. Three-dimensional computer-aided drawings were used to view the overall Chenderoh Dam. Next, a two-way fluid–structure interaction (FSI) model was developed to explore the influence of fluid flow and structural deformation on dam systems. The FSI modeling consists of Ansys Fluent and Ansys Structural modules to consider the boundary conditions separately. The reliability and performance of the reservoir bottom outlet structure was effectively simulated and recognised using FSI. The maximum stress on the bottom outlet section is 18.4 MPa, which is lower than the yield stress of mild steel of 370 MPa. Therefore, there will be no structural failure being observed on the bottom outlet section when the butterfly valve is fully closed. With a few exceptions, the FSI models projected that bottom outlet structures would be able to run under specified conditions without structural collapse or requiring interventions due to having lower stress than the material's yield strength.

Keywords: fluid-structural interaction; computational fluid dynamics (CFD); fluid flow dynamic; bottom outlet; Ansys

1. Introduction

Dams are hydraulic structures that are used to store water in reservoirs, pool water for agriculture, provide bed control, or divert flow away from crumbling banks or into diversion channels for flood control [1]. As such, a dam is designed to withstand the forces exerted by both static and dynamic water loadings [2]. Furthermore, the dam must withstand deterioration, ageing, and stresses caused by weather extremes and vibrations for longer than its design life span [3]. Dams have long been a major part of society's infrastructure, contributing to socioeconomic development and wealth through hydroelectric generating and residential water supply, for example [4]. A dam project typically includes a water-retaining structure (dam), a water-releasing structure (spillway), a water-conveying structure (conduits), and other components (such as turbines, power plants, etc.) [5].

The focus of this study is the Chenderoh Dam, a hydraulic structure situated in Tasik Chenderoh, Kuala Kangsar District, Perak, Malaysia. The dam construction began in 1928 and was completed in 1930, which was later officiated on 28 June 1930. The purpose of the Chenderoh Dam is for the hydroelectric scheme for the lower Perak region. The Chenderoh Dam continues its operation to date, making it the oldest hydroelectric dam and power station in Malaysia.

The Chenderoh Dam was the first dam to be built in the river (between 1927 and 1930), some 52 km downstream from the Kenering Dam. This dam is an Ambursen-type concrete hollow buttress dam, with gated and ungated overflow spillway sections over the crest. The gated section is equipped with a lower sector gate and an upper radial gate, and the ungated spill section was provided in 1972, with flashboards up to an elevation of 60.45 MASL, and, subsequently, a Japanese crest was anchored to the spillway crest downstream of the flashboards. The maximum height of the dam is 23 m and the crest length is about 390 m, with an angled axis. At full supply level (FSL), the reservoir volume is 95 million m^3, and the surface area is 20.5 km^2. The main powerhouse, with three Francis turbines of 10 MW each, together with the switchyard, is located on the right abutment. In 1981, a fourth turbine (Boving propeller-type) was installed at the dam toe, adjacent to the gated section, using one of the previous bottom outlets as the intake. The dam and appurtenant structures have been refurbished and rehabilitated several times.

Generally, the Chenderoh Dam structure can be divided into five parts, namely, the left bank, right bank, spillway, intake section, and bottom outlet, with each part performing standalone functions to ensure the continuous operation of the overall Chenderoh Dam.

Both the left bank and right bank resist and withstand the upstream water in Chenderoh Lake. The spillway structure regulates the water level of the reservoir from the upstream to the downstream parts through the control of sector gates. The primary power generation is located in the intake section, which consists of turbines and penstocks. Moreover, the bottom outlet serves as the secondary power generator and releases water from the upstream to downstream.

The operating condition of the left bank and right bank are dependent on the upstream water level. For the spillway's water regulating system, there are four operating conditions in which the sector gate could be open at a height of 4 feet, 10 feet, 12 feet, and 16 feet, respectively, yielding different discharge rates.

As for the intake section, its operation is based on the opening of the head gate and wicket gate. Under normal circumstances, both the head gate and wicket gate are either fully closed or fully opened. Nonetheless, under special conditions when required, the head gate could be fully opened or half-opened, while the wicket gate remains fully closed. Lastly, the operation of the bottom outlet is based on the opening condition of the butterfly gates, either fully closed or fully opened.

Computational fluid dynamics (CFD) focuses on computational transport phenomena, such as computational fluid dynamics, mass transfer, and heat transfer, as well as any other process that involves transportation phenomena [6,7]. CFD is useful in a wide variety of applications, for example, meteorological events, environmental risks, and the interaction of numerous objects with the air or water environment, to name a few. Numerical experiments can be carried out in a virtual flow laboratory [8]. Nowadays, the majority of CFD software programs have advanced to the point where they can simulate some types of fluid–structure interaction (FSI) applications [9]. CFD is research that uses computer technology to combine all of the equations in fluid flow (Richter, 2012), and it is a sophisticated numerical approach that is used in conjunction with physical modeling to model hydraulic processes [10]. Due to the general rapid growth of computer technologies, CFD has received more attention in recent years [11].

Nowadays, with the use of high-performance computers and more efficient algorithms, CFD can be a viable choice for making any analysis or experimentation easier, particularly when multiple prototypes are required throughout the design and testing process, as well as saving time and money [12,13]. The use of commercial CFD software to estimate time-averaged media velocity and pressure distributions along workpiece surfaces is now possible, thanks to the development of a rigorous approach for generating continuum media flow equations [14]. In addition, CFD solutions can describe the dynamic interaction between fluid flow, wind turbines, and floating platforms, allowing for full-scale simulations [12].

These equations show how a flowing fluid's velocity, pressure, temperature, and density are connected. Humans will be able to grasp and solve the Navier–Stokes equation more easily because it is analytical. Both compressible and incompressible fluids can be treated with these equations [13]. Understanding the physical events that occur in the flow of fluids around and within the chosen item or structure is the ultimate goal of CFD [15,16].

The Navier–Stokes equation is a partial differential equation (PDE) in fluid mechanics that describes the flow of incompressible fluids [16]. It is a term that defines the motion of viscous fluids [15]. Differential equations represent the link between the flow variables and their evolution in space and time in fluid flow equations [17]. The dynamic equilibrium of a fluid element can be used to derive the Navier–Stokes equations. To answer this problem using a computer, it must be converted to a discretized form [18]. Numerical discretization methods, such as the finite difference method (FDM), finite element method (FEM), and finite volume method (FVM), are used as translators (FVM) [19,20]. As a result, because discretization is dependent on them, the entire domain problem must be broken into several little portions [13]. The governing equations of CFD are the Navier–Stokes equations. The continuity equation, momentum equation, and energy equation can be obtained using mass, momentum, and energy conservation [20].

The exchange of energy between moving fluid and solid structures causes fluid–structure interaction (FSI) [15]. A fluid–structure interaction (FSI) is a multidimensional physics interaction between the rules of fluid dynamics and structural mechanics [21]. Fluid–structure coupling can occur in a variety of engineering domains, and it is considered critical in the design of many engineering systems [22]. In the case of FSI, a fluid and structural problem can be solved in conjunction with boundary conditions, described as a connected part of the boundary [21]. In order to compute the numerical solution, the strongly coupled equations of both problems must be solved simultaneously. Usually, FSI problems have a strong dependency between fluid and structure [23]. An FSI problem can be approached in one of two ways: monolithic or partitioned [22]. The monolithic technique involves solving the flow equations and structural equations at the same time, allowing for consideration of their mutual influence throughout the solution process [24,25].

The flow equations and the structure equations are solved independently in a partitioned FSI simulation, which means that the flow does not change while the structural equations are solved, and vice versa [25]. As a result, the partitioned approach necessitates the use of a coupling algorithm to incorporate the fluid–solid interaction into the system [26]. The partitioned approach, on the other hand, keeps software modularity and diversity [22]. For flow equations and structural equations, more efficient solution approaches are likely to be applied [27].

CFD models and other numerical models are increasingly being employed in engineering investigations [28]. Benchmark testing is commonly used to determine the validity of these models [29]. This is done to quantify the agreement between the model's predictions and the real world, which is represented by observations in experiments [30]. This approach implies that all real-world variables important to the investigation are adequately measured in the experiments and in the model's predictions [31].

Structural analysis is important for a hydraulic structure. A hydraulic structure needs to be analyzed to ensure its stability and integrity. Structural analysis is a method of analyzing a structural system in order to predict its behavior and consequences using mathematical equations and physical laws [19]. All structures that must withstand varying loads are subject to this type of study. Structural analysis computes a structure's deformations, internal forces, stresses, support responses, accelerations, and stability using applied materials science, mechanics, and applied mathematics [32]. The analysis findings are utilized to confirm the structure's strength and usability [33].

Stress is a physical quantity that expresses the internal forces that contiguous particles of a continuous material exert on each other in continuum mechanics, whereas strain is the measure of the material's deformation [33]. The force per unit area applied to a material is referred to as stress [19].

Failure due to structural weakness is referred to as structural uncertainty [19]. Hydraulic structure failures can be caused by soil saturation and instability, erosion, hydraulic soil failures, wave action, hydraulic overloading, structural collapse, material failure, and so on [34–36]. Hydraulic erosion, high pore-water pressure, seismic stresses, and other variables all contribute to embankment failures [34].

Every construction should be designed with environmental, ecological, and public safety in mind [36]. Hydraulic structures have diverse characteristics, such as shape and size, depending on the project. This is dependent on the discharge and the function to be carried out properly [33]. Hydraulic physical modeling or CFD modeling may be useful for the design of unique structures that do not meet the guidelines offered [9]. In the case of FSI problems, the structure's equations should be written in such a way that substantial deformations of the structure are unlikely [37].

The main objective of this paper is to study the fluid–structure interactions of the bottom outlet structures of Chenderoh Dam operated at different discharge conditions.

2. Methodology

2.1. Three-Dimensional Computer-Aided Drawings of Chenderoh Dam

The overall Chenderoh Dam consists of five sections, namely, the intake, right bank, sector gate, bottom outlet, and left bank. A top view of the overall Chenderoh dam is shown in Figure 1, based on the provided drawing plans, which can be segregated into the structures of intake, right bank, sector gate, bottom outlet, and left bank. Three-dimensional drawings of the overall dam were generated using SolidWorks 2017 software, based on the build drawing plans and dimensions provided by the TNB Chenderoh Hydropower Station. Figures 2 and 3 exhibit the entire three-dimensional sketch of the Chenderoh Dam from downstream and upstream perspectives, respectively.

Figure 4 depicts the detailed cross-sectional views of the bottom outlet: (a) downstream/front view of the bottom outlet including the turbine house, penstock, etc.; (b) top view of the bottom outlet including the turbine house, penstock, etc.; (c) backside of the bottom outlet (foundation, trash rack, etc.); (d) tented gate connection to the penstock, butterfly gate no. 1 and turbine, and (e) penstock and butterfly gate no. 2. The bottom outlet is another important feature of the dam structure since it is the second location for the source of power generation for the station and the location for the release of water from the upstream to the downstream. For the bottom outlet, the propeller-type turbine is used for the power generation located inside the turbine house of the dam. The release of water at the bottom outlet location is controlled by two butterfly gates, which are separately shown in Figure 4d,e. These gates are essential since there will be another case study of water surging inside the penstock, and the effects of flow-induced vibration to the whole bottom outlet part.

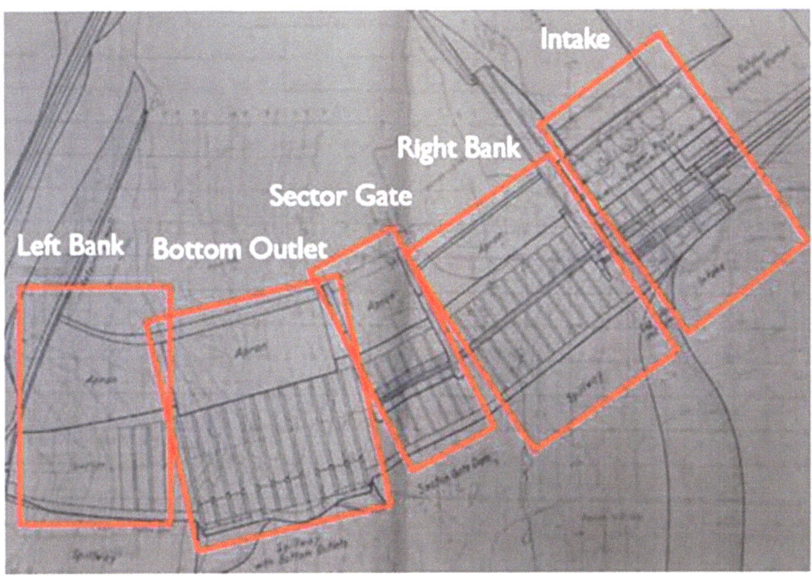

Figure 1. Top view of the overall Chenderoh Dam based on the provided drawing plans.

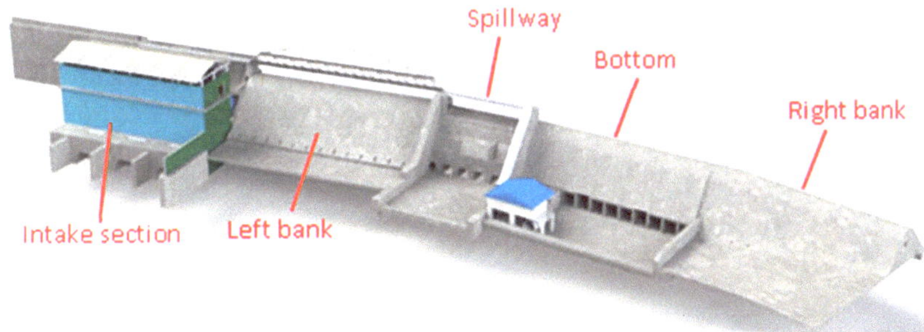

Figure 2. Downstream view of the overall Chenderoh Dam.

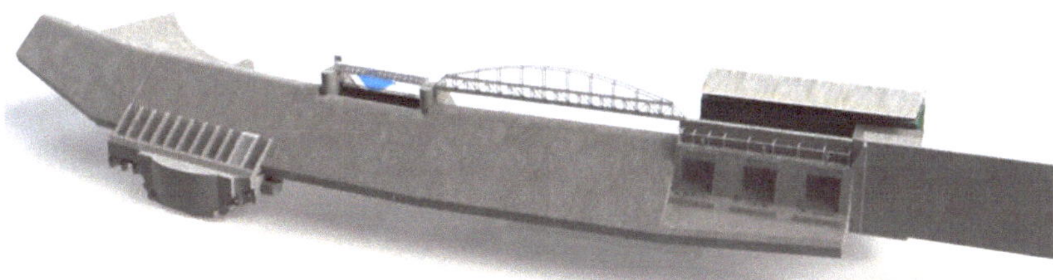

Figure 3. Upstream view of the overall Chenderoh Dam.

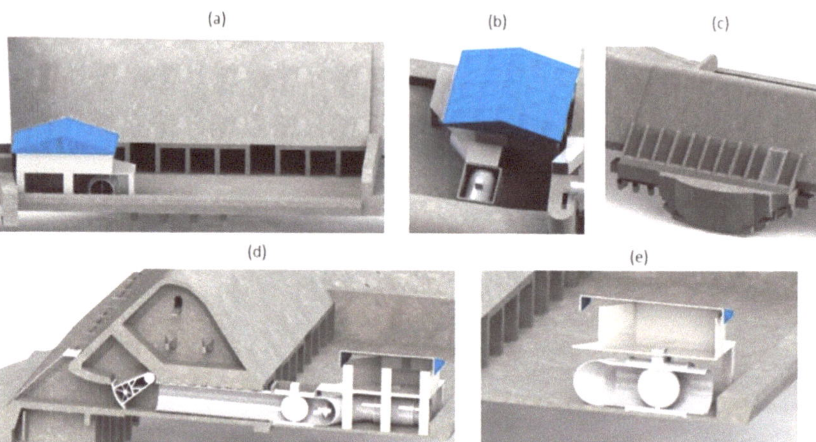

Figure 4. Detailed cross-sectional views of bottom outlet. (**a**) Downstream/front view of bottom outlet including the turbine house, penstock, etc.; (**b**) top view of bottom outlet including the turbine house, penstock, etc.; (**c**) backside of bottom outlet (foundation, trash rack, etc.); (**d**) tented gate connection to the penstock, butterfly gate no. 1, and turbine, and (**e**) penstock and butterfly gate no. 2.

2.2. Fluid–Structure Interaction Numerical Simulation

The two-way fluid–structure interaction (FSI) was developed to investigate the coupling impact of fluid flow and structural deformation in dam systems. The flow chart in Figure 5 summarises the general sequences involved in the current FSI modelling, from the domain and mesh generation to the boundary condition assignment. Ansys Fluent and Ansys Structural modules were used to investigate the fluid and structural domains separately.

Both the fluid domain and the structural domain were investigated independently using the Ansys Fluent and Ansys Structural modules, respectively. The fluid flow in the present simulation is a rectangular enclosure surrounding the bottom outlet structure, which is the location where the fluids are predicted to occur, as shown in Figure 6. The Navier–Stokes equations, which include a momentum equation and a continuity equation, regulate the fluid phase. The linear elastic equation guides the structural phase.

The numerical fluid and structural domains are then discretized via mesh generation in the next stage. Using the improved mesh size, structural hexagonal meshes were created on both the mesh and fluid domains. The dynamic mesh on the fluid flow is enabled in the present FSI simulation to account for structural domain movement and deformation.

A multiphase volume of fluid model with an implicit scheme regulated by the transport equation was used to follow the flow front of the water. Furthermore, the k-model simulates water flow turbulence. First order upwind, least square cell method, and SIMPLE pressure-velocity coupling are the solutions employed in this paper. Following that, the Ansys Fluent and Ansys Mechanical settings were connected using the system coupling module to provide two-way data transmission.

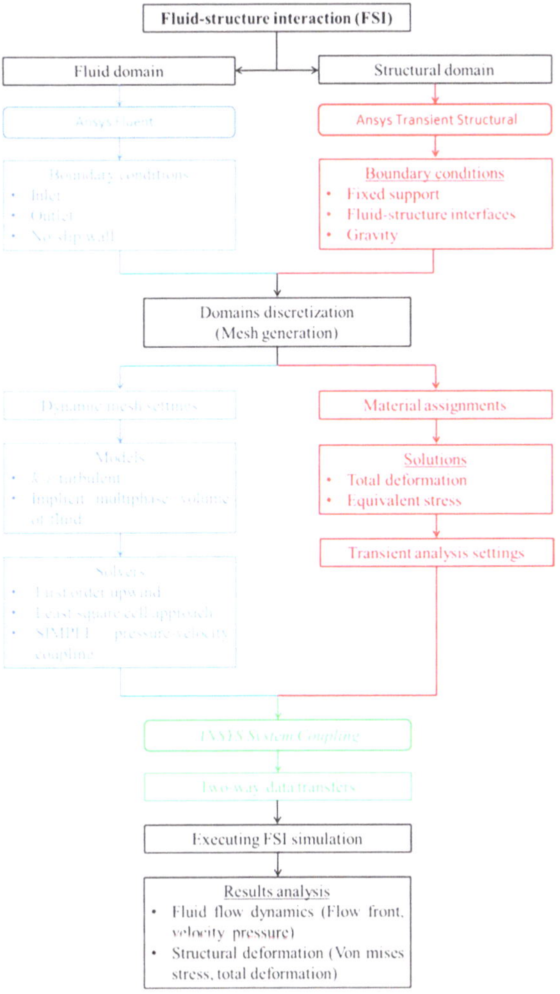

Figure 5. Flow sequences of the fluid–structure interaction numerical simulation.

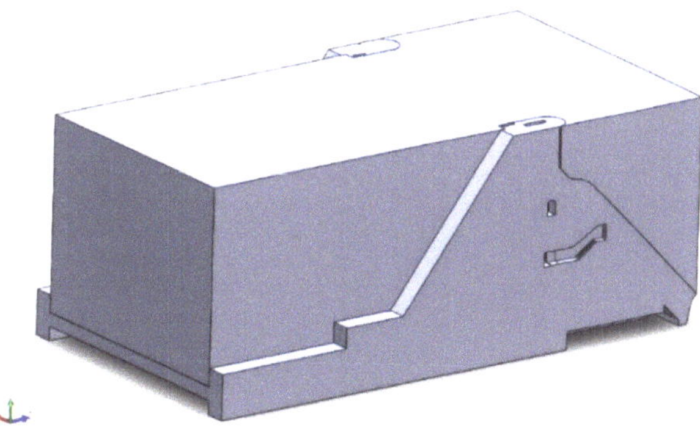

Figure 6. Schematics of fluid and structure domain.

2.2.1. Governing Equations

In the Ansys Fluent module, the current flow simulation was governed by the incompressible and isothermal Navier–Stokes equations, with the respective governing continuity and momentum equations as follows:

$$\frac{\partial \rho}{\partial t} + \nabla \cdot \left(\rho \vec{v} \right) = 0 \tag{1}$$

$$\frac{\partial \rho}{\partial t} \left(\rho \vec{v} \right) + \nabla \cdot \left(\rho \vec{v} \vec{v} \right) = -\nabla p + \nabla \cdot \overline{\overline{\tau}} + \rho \vec{g} + \vec{F} \tag{2}$$

where p is the static pressure; ρ is the density; $\vec{g}$ denotes the gravitational acceleration; $\vec{F}$ is the external body force; and $\overline{\overline{\tau}}$ is the stress tensor.

The standard k-ε turbulence model was employed in the current simulation, which solves the following transport equations:

$$\frac{\partial}{\partial t} (\rho k) + \frac{\partial}{\partial x_i} (\rho k u_i) = \frac{\partial}{\partial x_j} \left[\left(\mu + \frac{\mu_t}{\sigma_k} \right) \frac{\partial k}{\partial x_j} \right] + G_k + G_b - \rho \varepsilon - Y_M + S_k \tag{3}$$

$$\frac{\partial}{\partial t} (\rho \varepsilon) + \frac{\partial}{\partial x_i} (\rho \varepsilon u_i) = \frac{\partial}{\partial x_j} \left[\left(\mu + \frac{\mu_t}{\sigma_\varepsilon} \right) \frac{\partial \varepsilon}{\partial x_j} \right] + C_{1\varepsilon} \frac{\varepsilon}{k} (G_k + G_{3\varepsilon} G_b) - C_{2\varepsilon} \frac{\varepsilon^2}{k} + S_\varepsilon \tag{4}$$

where G_k is the generation of turbulence kinetic energy due to the mean velocity gradients; G_b is the generation of turbulence kinetic energy due to buoyancy; Y_M is the contribution of the fluctuating dilatation in compressible turbulence to the overall dissipation rate; σ denotes the turbulent Prandtl number; S is the user-defined source term; and the constant terms are $C_{1\varepsilon}$ and $C_{2\varepsilon}$.

To track the water–air interface, the multiphase volume of fluid (VOF) model was employed, which is governed by the transport equation:

$$\frac{\partial \alpha}{\partial t} + \vec{V} \bullet \nabla \alpha = 0 \tag{5}$$

Numerical software packages solve problems using a series of discrete points. Each point, or node, adds a degree of freedom (DOF) to the system. Therefore, the more DOFs in the model the better it will capture the structural behaviour. Each DOF adds complexity and increases solving time. The simulation needs to balance the complexity of the model with the solving time. Too few DOFs, and the response could be incorrect, and too many DOFs, and the model could take days to run. The Ansys Mechanical module was used to run the simulation on the structural domains. The findings of the static structural analysis for the dam gates were fed as input, after the pressure loads were obtained from the fluid simulations. The following equation was used by the Ansys Mechanical APDL for the static linear analyses.

$$[K]\{u\} = \{F^a\} + \{F^r\} \tag{6}$$

where $\{u\}$ is the nodal degree of freedom (DOF) vector, and $[K]$ is the total stiffness or conductivity matrix, which defined as

$$[K] = \sum_{m=1}^{N} [Ke] \tag{7}$$

with the number of elements, N, and the element stiffness or conductivity matrix, $[Ke]$. On the right-hand side of equation (7), $\{F^r\}$ is the nodal reaction load vector, and $\{F^a\}$ is the total applied load vector, which is the sum of the applied nodal load vector, $\{F^{nd}\}$, and the total of all the element load vector effects (pressure, acceleration, thermal, gravity), and $\{F^e\}$:

$$\{F^a\} = \{F^{nd}\} + \{F^e\} \tag{8}$$

Ansys Mechanical uses finite element (FE) techniques on all structural models to construct a system of simultaneous linear equations, as detailed in the previous section. After that, either a direct elimination technique or an iterative method is used to solve the equations. The preconditioned conjugate gradient (PCG) solver was used in this simulation as well. General CG methods cast the solution in the form of a series of vectors, $\{p_i\}$, to solve standard systems of equations in the form of previous equations recursively.

$$\{u\} = \alpha_1\{p_1\} + \alpha_2\{p_2\} + \alpha_3\{p_3\} + \ldots + \alpha_m\{p_m\} \tag{9}$$

2.2.2. Boundary Conditions

The setup of the numerical simulation was initiated with the assignation of suitable boundary conditions on the fluid domain and structural domain. Generally, the boundary conditions imposed on the fluid domain are no-slip wall, inlet and outlet, while, for the structural domain they are fixed support, gravity, and fluid structure interface. The velocity inlet was assigned to the reservoir inlet at one end of the fluid domain, while the pressure outlet was assigned to the opposite end. The water would constantly flow in the fluid domain at the inlet surface with a constant velocity of 1 m/s at the reservoir intake, with a water level height of 44.8 m.

Furthermore, the entire fluid domain was adjusted to a 0 Pa atmospheric pressure state (gauge). The no-slip boundary condition was enforced together with the fluid-structure interface condition on all the contacting surfaces between the fluid and structure domains. Furthermore, the dam spillway's foundation was designated as a fixed support. There is also gravity of a magnitude of 9.81 m/s^2 acting downward enabled on both domains.

2.2.3. Mesh

The grid-generation procedure was then used to discretize both the fluid and structural domains into tiny elements. Meshes are divided into two categories: structured and unstructured grids. In structured grids, the cells are ordered and numbered according to indices, such as I, j, and k. Planar cells with four edges (2-D) or volumetric cells with six faces make up a structured grid (3-D). Although the cells are ordered according to indices, they might be geometrically deformed. Unstructured grid cells, on the other hand, cannot be uniquely identified by indices, and the relationship between adjacent cells must be accounted for by other means, which usually involves another set of memory storage. To overcome this deficiency, the grid density was changed so that an extremely tiny mesh was allotted at the necklace vortices, detached shear layers, and near-wake zone areas [38]. Furthermore, the cells come in a variety of shapes, although the most common are triangles or quadrilaterals (2-D), and tetrahedrons or hexahedrons (3-D).

Unstructured grids are often used for complex geometries because they are easier to design by the user with fewer tedious grid generation processes, are more robust, and can fit steep angles in the geometry without compromising grid skewness. Structured grids, on the other hand, are preferable in other ways. For example, with a structured grid, fewer cells are normally generated than with an unstructured grid.

Furthermore, for the same number of cells, structured grids allow better control over local grid refinement and sharper resolution in border layers than unstructured grids. Calomino et al. [39] divided the domain into pieces depending on direction using the basic geometric decomposition technique [39]. In this method, the domain is divided into segments by direction [40]. Due to the generally huge space for the water dam geometry and the need to replicate complicated geometry, the meshes were generated using an unstructured grid. This is mostly owing to the FSI interactions, which occur frequently in complex geometries. Combining tetrahedrons and hexahedrons mesh types is also possible, thanks to the unstructured mesh allowances. As a result, during the FSI simulation, better findings will be captured.

To determine the optimum mesh types for numerical modal analysis, a mesh independent analysis was conducted. Table 1 shows the number of elements, with their corresponding curvature and proximity minimum size, for the fluid mesh of the bottom outlets. The mesh with the curvature and proximity minimum size of 0.55 m produced the highest maximum velocity of 124.1 m/s. It can be observed in Figure 7 that the maximum velocity value starts to saturate at a 0.45 m mesh size. This indicates that this size is the optimum mesh size, as using the smaller mesh size will not affect the results greatly. The generated mesh models of the dam structures to be used in the numerical modal analysis are shown in Figure 8.

Table 1. Comparison between mesh size, number of elements, and maximum velocity for the fluid domain of the intake section.

Mesh Size (m)	Element Number	Maximum Velocity (m/s)
0.55	386,358	124.1
0.5	452,317	122.467
0.45	534,276	122.367
0.35	797,630	122.332
0.3	1,027,179	122.3

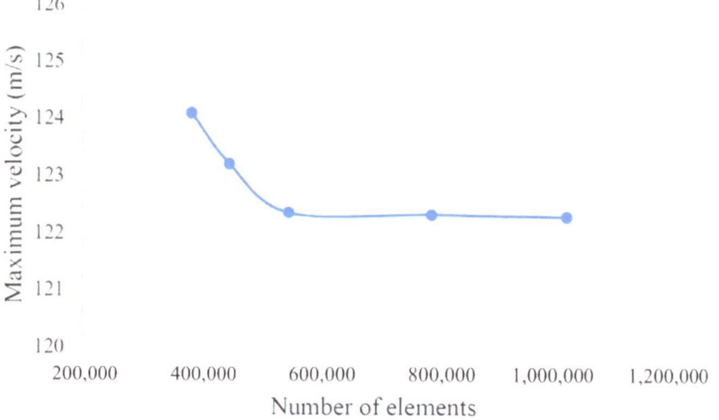

Figure 7. Mesh sensitivity analysis of five different mesh sizes for the fluid domain.

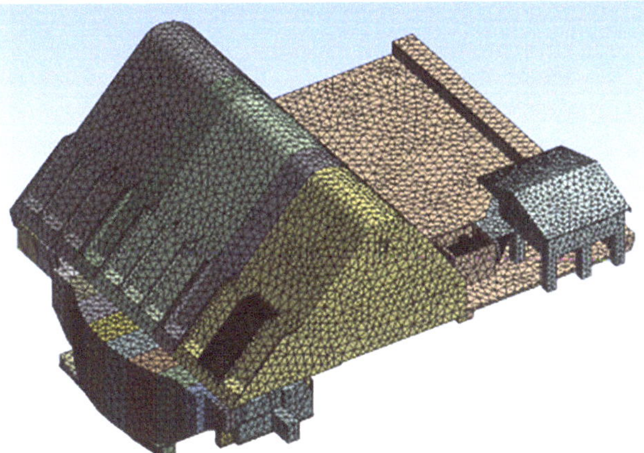

Figure 8. Meshing picture of bottom outlet 2.2.4 Ansys fluid flow (Fluent).

Referring to the second objective of this study, which is to analyze the fluid dynamics from the reservoir to the sector gate spillway of Chenderoh Dam at different sector gate openings, the calculations of the model are performed using the robust CFD-solver Fluent. Fluent provides various multiphase models that are based in the Eulerian–Eulerian approach [41]. Although a Eulerian model that has been selected for the numerical approach may require more computational effort, it can handle a wider range of particulate loading values and is more accurate than the other available multiphase models in Fluent [42].

In this multiphase model, the different phases are treated mathematically as interpenetrating continuous, and therefore the concept of phasic volume fraction is introduced, where the volume fraction of each phase is assumed to be a continuous function of space and time. The sum of the volume fractions of the various phases is equal to unity. An accordingly modified set of momentum and continuity equations for each phase is solved. Pressure and inter-phase exchange coefficients are used in order to achieve coupling for these equations [42].

In Ansys Fluent, pressure and inter-phase exchange coefficients are used to achieve coupling between the different equations used to model fluid flow and inter-phase interactions. The pressure coefficient is used to couple the Navier–Stokes equations, which describe the fluid flow, with the pressure equation, which enforces mass conservation. The pressure coefficient is a numerical parameter that relates the change in pressure to the change in the velocity of the fluid. It is used to ensure that the velocity and pressure fields are consistent with each other. The inter-phase exchange coefficient is used to couple the different phases in a multiphase flow simulation, such as gas and liquid. It represents the rate at which mass, momentum, and energy are exchanged between the phases, and it is used to ensure that the different phases are in thermal and mechanical equilibrium. The inter-phase exchange coefficient is dependent on the physical properties of the phases, as well as the geometry of the system being simulated.

The type of solver used in this simulation is the SIMPLE (Semi-Implicit Method for Pressure-Linked Equations) family of algorithms used for introducing pressure into the continuity equation. The SIMPLE algorithm uses a relationship between velocity and pressure corrections to enforce mass conservation and to obtain the pressure field. If the momentum equation is solved with a guessed pressure field, p^*, the resulting face flux, J_f^*, is computed from Equation (10).

$$\sum_{f}^{N_{faces}} J_f A_f = 0 \tag{10}$$

$$J_f = \hat{J}_f + d_f(p_{c0} - p_{c1}) \tag{11}$$

where J_f = the face flux; $\hat{J}_f$ = contains the influence of velocities in these cells; d_f = the function of $\tilde{a}_P$; the average of the momentum equation = a_P; the coefficient for the cells on either side of face f; and p_{c0} and p_{c1} = the pressure between two cells on either side of the face.

$$J_f^* = \hat{J}_f^* + d_f\left(p_{c0}^* - p_{c1}^*\right) \tag{12}$$

does not satisfy the continuity equation. Consequently, a correction, J_f', is added to the face flux, J_f^*, so that the corrected face flux

$$J_f = J_f^* + J_f' \tag{13}$$

satisfies the continuity equation. The SIMPLE algorithm postulates that J_f' be written as

$$a_P p' = \sum_{nb} a_{nb} p_{nb}' + b \tag{14}$$

where the source term, b, is the net flow rate into the cell:

$$b = \sum_{f}^{N_{faces}} J_f^* A_f \tag{15}$$

The pressure-correction equation (Equation (6)) may be solved using the algebraic multigrid (AMG) method. Once a solution is obtained, the cell pressure and the face flux are corrected using:

$$p = p^* + \alpha_p p' \tag{16}$$

$$J_f = J_f^* + d_f\left(p_{c0}' - p_{c1}'\right) \tag{17}$$

Here, α_p is the under-relaxation factor for pressure. The corrected face flux, J_f, satisfies the discrete continuity equation identically during each iteration.

2.2.4. Ansys Transient Structural

This type of analysis is used to determine the dynamic response of a structure under the action of any general time-dependent loads. It could be used to determine the time-varying displacements, strains, stresses, and forces in a structure as it responds to any transient loads [43]. The time scale of the loading is such that the inertia or damping effects are considered to be important. It also could be used to examine deflections, deformations, stresses, and strains on assemblies, part-by-part or at the feature level [15]. The input parameters for a few properties, such as the density of the structural concrete or Young's modulus for structural steel, were set in the engineering data before the simulation was run.

2.2.5. Ansys System Coupling

After the transient structural and fluid flow have been analyzed, system coupling was then run to analyze the FSI. System coupling is a process where the interpretations are made from the solutions given by the numerical models. Then, the relationship between the patterns of flow and input parameters or structure may be concluded.

3. Results and Discussion

3.1. Fluid–Structure Interaction Numerical Simulation on Bottom Outlet Structures

The performance and reliability of Chenderoh Dam's structures were assessed quantitatively based on the findings attained from the numerical FSI simulations, in terms of the flow dynamics and structural associated parameters. This is to ensure the bottom outlet structures can withstand the enormous pressures from the rapid and continuous water flow from the reservoir dam, without losing their structural integrity, ultimately, to eliminate the risk of dam failures. The bottom outlet is based on the open condition of a butterfly valve opening.

The investigated operating conditions for the bottom outlet section are based on the opening and closing the butterfly valve. The sudden closing of the butterfly valve is expected to cause minor surging to occur, which might lead to slight vibrations and a high region of stresses. It should be noted that the drawings for the butterfly valve and penstock pipes are not provided by the hydropower station and TNB. Therefore, the current numerical findings presented are based on the assumed geometry and dimensions of the penstock pipe and butterfly valve, and the material properties of concrete with Young's modulus of 30 GPa and mild steel of 210 GPa.

Figure 9 shows the condition of the butterfly valve. When the butterfly valve is fully opened, the face of the butterfly valve is parallel to the water flow. When the butterfly valve is fully closed, the face of butterfly valve is perpendicular to the water flow.

(**a**) Fully opened

(**b**) Fully closed

Figure 9. Condition of butterfly valve when fully opened and fully closed.

Figure 10 presents the numerical contours of the bottom outlet section when the butterfly valve is fully opened. A high-velocity region is located on the penstock pipe, with values in the ranges of 14.1–9.4 m/s, due to the water flow in the confined, narrow penstock pipe. Such a rapid flow will induce a large force on the inlet section and internal structures of the penstock pipes. Beyond the penstock region, the water velocity subsided to a value of 4.7 m/s. A hydrostatic pressure contour was observed on the upstream. A high deformation was observed on the region of the concrete wall in the downstream section, with values of 11.9–17.9 mm. As the higher deformation was located on the outer fin of the concrete wall, care must be taken to monitor the relevant structure for potential failure and crack. There is a high-stress region of about 73.9 MPa located on the inner section of the penstock. Since the maximum stress of the penstock is less than the yield stress of mild steel of 370 MPa, no structural failure will be observed.

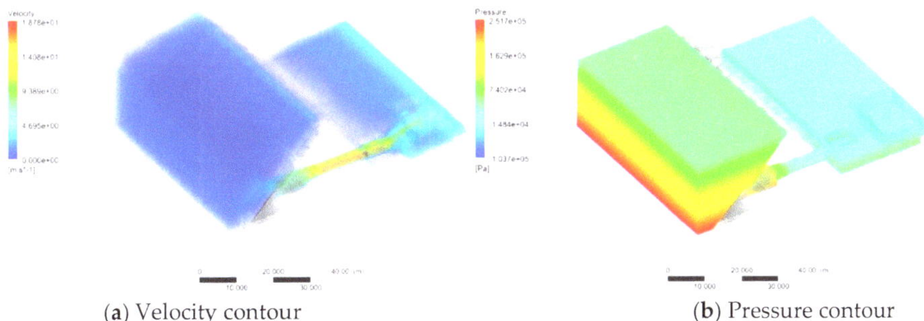

(**a**) Velocity contour

(**b**) Pressure contour

Figure 10. *Cont.*

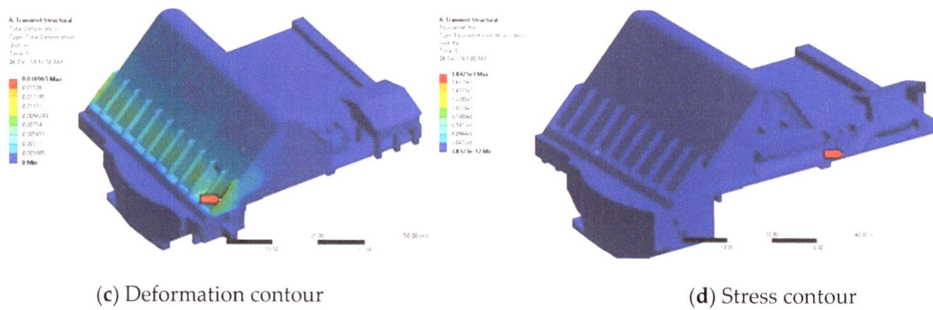

(c) Deformation contour   (d) Stress contour

Figure 10. Contours on the bottom outlet with fully opened butterfly valve.

Figure 11 presents the numerical contours of the bottom outlet section when the butterfly valve is fully closed. A high-velocity region is observed in the penstock pipes, with values in the range of 19.97–13.32 m/s, which later subsides to a value of 6.66 m/s beyond the penstock region. The high-deformation region is observed at the concrete wall of the downstream section, in the range of 13–17 mm. These deformations are caused by the nearly stagnant upstream water and are expected to have minimal impact to the structure. Furthermore, the maximum stress on the bottom outlet section is 18.4 MPa, which is lower than the yield stress of mild steel of 370 MPa. Therefore, there will be no structural failure observed on the bottom outlet section when the butterfly valve is fully closed.

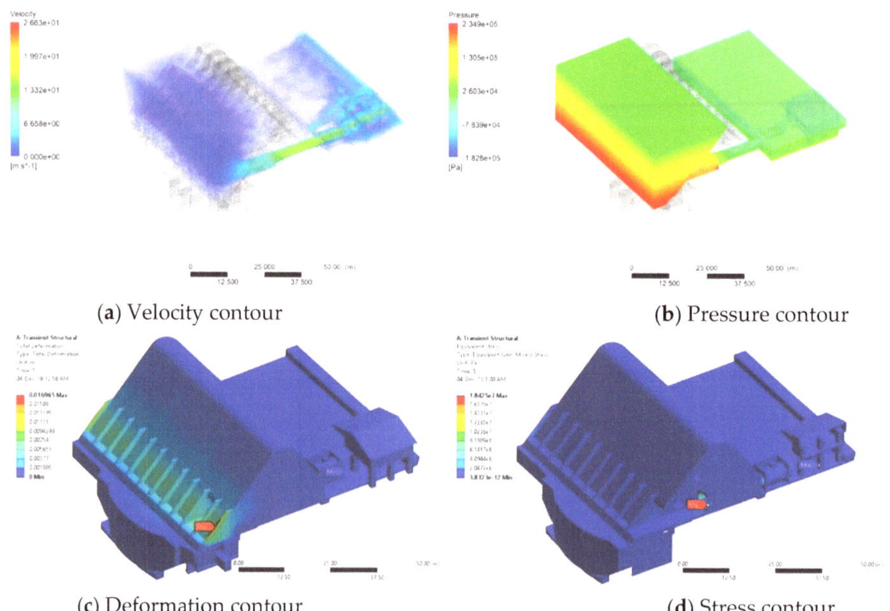

(a) Velocity contour   (b) Pressure contour

(c) Deformation contour   (d) Stress contour

Figure 11. Contours on the bottom outlet with fully closed butterfly valve.

3.2. Structural Analysis on Bottom Outlet Structures

In Figure 12, the bottom outlet operates at Case 1. Figures 12 and 13 show the stress contours on the bottom outlet when the butterfly valve is fully opened (Case 1) and fully closed (Case 2), respectively. The bottom outlet comprises the penstock adjoining two butterfly gate valves. The maximum stresses of 73.92 MPa occur at the gate valve during 100% opening of the gate valve. Meanwhile, a maximum stress value of 18.425 MPa, when the gate valve is closed, occurs at the penstock surface intake. Since the maximum stress of

221

the penstock is less than the yield stress of mild steel of 370 MPa, no structural failure will be observed.

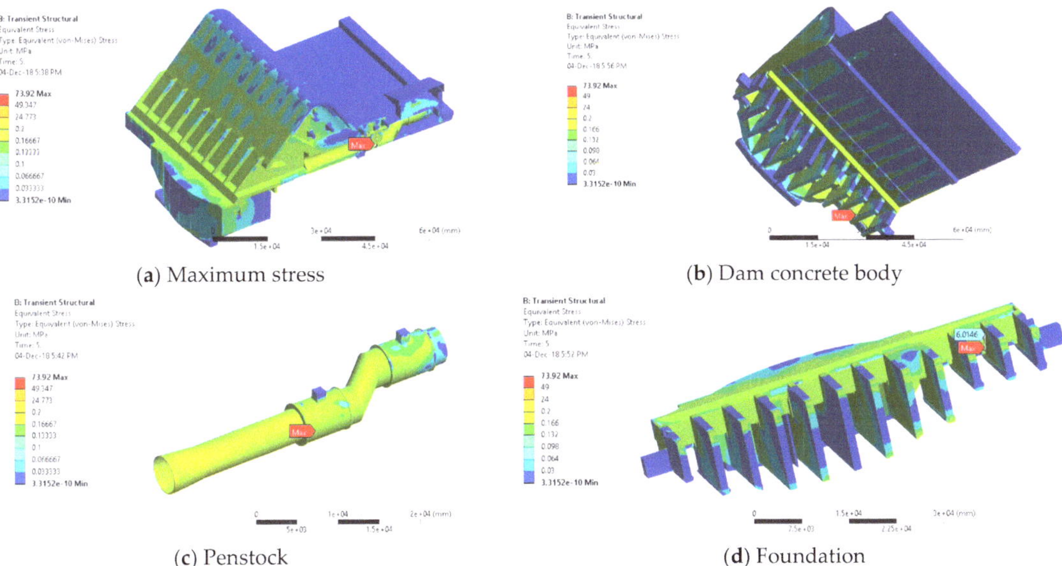

(**a**) Maximum stress

(**b**) Dam concrete body

(**c**) Penstock

(**d**) Foundation

Figure 12. Bottom outlet operates at Case 1.

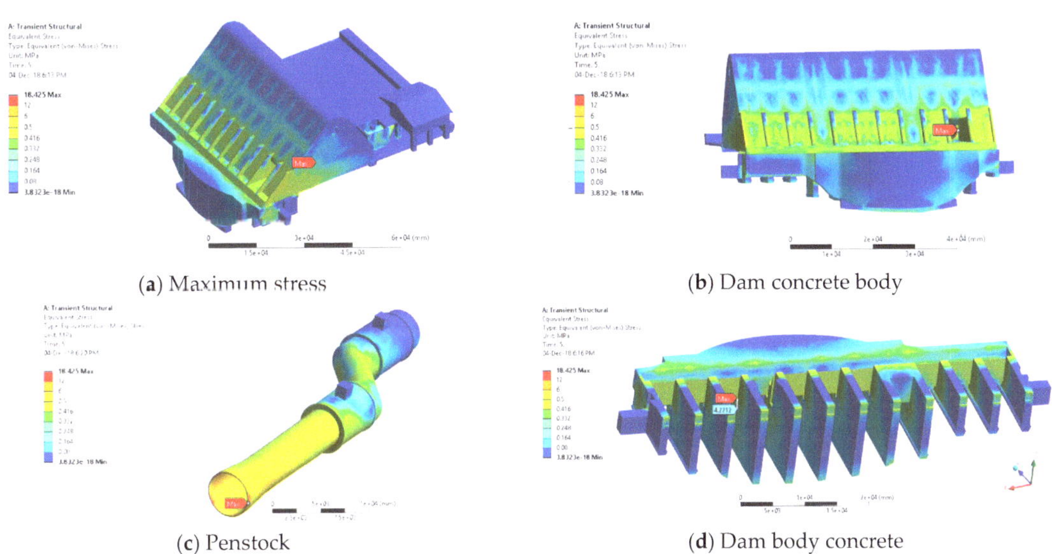

(**a**) Maximum stress

(**b**) Dam concrete body

(**c**) Penstock

(**d**) Dam body concrete

Figure 13. Bottom outlet operates at Case 2.

These stresses are related to the flow velocity at the bottom outlet, by which greater velocity will produce maximum stresses to the penstock surface intake. This is proven by Manafpour and Rovesht, 2017 in their study, where the flow velocity increased with a smaller gate opening, but it gradually increases with the gate closure, as shown in Table 2 [44]. Hence, the greater the velocity, the greater the stress created on the penstock surface. The extraction of the data is conducted; therefore, the location of maximum stress is found at the intake. The values at the mechanical equipment and stilling basin are also high.

Table 2. Comparison of study conducted with that of Manafpour and Rovesht, 2017.

Case	Chenderoh Dam–Bottom	Case	Seymareh Dam [44]
Butterfly valve is fully closed	14.1	Gate opening–10%	2.0
Butterfly valve is fully opened	19.97	Gate opening–30%	4.0
-	-	Gate opening–70%	8.0
-	-	Gate opening–100%	18.0

4. Conclusions

By using fluid-structure interaction (FSI), which utilized both Ansys Fluent and Ansys Transient Structural, both the fluid domain and structural domain aspects of the reservoir bank of a dam have been successfully investigated. The impact of the upstream water flow on the dam structures, alongside their interaction phenomena, were numerically simulated by a fluid–structure interaction (FSI) numerical approach. The model's computations in Ansys fluid flow were carried out with the help of the robust CFD-solver Fluent, whereas Ansys Transient Structural analysis was used to determine the dynamic response of the structure under the influence of any time-dependent loads. After analyzing the transient structural and fluid flow, system coupling was performed to analyze the FSI. System coupling is a process in which interpretations are derived from numerical model results. The relationship between the flow patterns and the input parameters or structure may then be determined. To assess the reliability of the FSI simulations on the dam structures, a quantitative validation was conducted by comparing both the numerical and experimental volume flow rates at the penstocks at the intake sections.

From the FSI numerical simulation reports, it was assessed that the high-stress region located at the inner section of the penstock is about 73.9 MPa. The numerical contours of the bottom outlet section, when the butterfly valve was fully opened and fully closed, had a maximum stress on the penstock less than the yield stress of mild steel of 370 MPa, so no structural failure will be observed. Generally, the FSI simulations predicted that the bottom outlet structures will be able to operate under the prescribed conditions without structural failure or required interventions, due to having lower stress that the material's yield strength, with only a few exceptions. It is recommended for future studies to consider other parts of Chenderoh Dam to be analyzed, namely the left bank, right bank, spillway, and intake section to fully validate the Dam's reliability and performance.

Author Contributions: Writing—review and editing, M.R.M.R., M.H.Z., M.A.A., A.Z.A.M., M.R.R.M.A.Z., N.H.H., W.N.C.W.Z., H.D. and M.A.K. All authors have read and agreed to the published version of the manuscript.

Funding: This research was funded by UNITEN R&D Sdn. Bhd, U-TG-RD-21-19.

Data Availability Statement: Not applicable.

Acknowledgments: The authors would like to acknowledge Universiti Tenaga Nasional Berhad under UNITEN R&D Sdn. Bhd for providing the facilities and financial assistance of project code U-TG-RD-21-19.

Conflicts of Interest: The authors declare no conflict of interest.

References

1. Chin, D.A.; Mazumdar, A.; Roy, P.K. *Water-Resources Engineering*; Prentice Hall: Englewood Cliffs, NJ, USA, 2000; Volume 12.
2. Balmer, M.; Spreng, D. Hydroelectric Power. In *Future Energy*; Elsevier: Amsterdam, The Netherlands, 2008; pp. 193–209.
3. Chen, S.-H. Rock Slopes in Hydraulic Projects. In *Hydraulic Structures*; Springer: Berlin/Heidelberg, Germany, 2015; pp. 813–868.

4. Michael, C. *Environmental Management and Concept During Construction of Dam: A Study Case of Murum Dam, Belaga District, Kapit Division, Sarawak*; Universiti Malaysia Sarawak: Sarawak, Malaysia, 2013.
5. Cook, C.B.; Richmond, M.C.; Serkowski, J.A. *The Dalles Dam, Columbia River: Spillway Improvement CFD Study*; Pacific Northwest National Lab.(PNNL): Richland, WA, USA, 2006.
6. Fu, C.; Hafliðason, B. Progressive Failure Analyses of Concrete Buttress Dams: Influence of Crack Propagation on the Structural Dam Safety. In *Concrete Structures*; Stockholm University: Stockholm, Sweden, 2015.
7. Zawawi, M.H.; Hassan, N.H.; Ramli, M.Z.; Zahari, N.M.; Radzi, M.R.M.; Saleha, A.; Salwa, A.; Sidek, L.M.; Muda, Z.C.; Kamaruddin, M.A. Fluid-Structure Interactions Study on Hydraulic Structures: A Review. *AIP Conf. Proc.* **2018**, *2030*, 20244.
8. Manikandan, R.; Jayashiri, R.; Indumathi, R.; Archana, J. Fluid Structure Interaction of Arch Dam on Full Reservoir Level under Seismic Loading. In Proceedings of the 2016 COMSOL Conference in Bangalore, Bangalore, India, 20–21 October 2016; Volume 39.
9. Richter, T. Goal-Oriented Error Estimation for Fluid–Structure Interaction Problems. *Comput. Methods Appl. Mech. Eng.* **2012**, *223*, 28–42. [CrossRef]
10. Bø, A.T. Fluid Structure Interaction in a Pipe. Master's Thesis, Norwegian University of Science and Technology, Trondheim, Norway, 2018.
11. Soucek, O. A Brief Introduction to Fluid- Structure Interactions Fluid-Structure Interactions Teaching Slides. 2012. Available online: https://geo.mff.cuni.cz/jednooci_slepym/os-FSI-intro.pdf (accessed on 1 February 2023).
12. Hellgren, R. Influence of Fluid Structure Interaction on a Concrete Dam during Seismic Excitation:-Parametric Analyses of an Arch Dam-Reservoir-Foundation System. In Proceedings of the Seconde International Dam World Conference, Lisbon, Portugal, 21–24 April 2015.
13. Hartmann, S.; Meister, A.; Schäfer, M.; Turek, S. *International Workshop on Fluid-Structure Interaction. Theory, Numerics and Applications*; Kassel University Press GmbH: Kassel, Germany, 2009; ISBN 3899586670.
14. International Atomic Energy Agency. *Iaea Integration of Tracing with Computational Fluid Dynamics for Industrial Process Investigation*; International Atomic Energy Agency: Vienna, Austria, 2004; p. 221.
15. Frisch, J. Numerical Modelling—Introductory Approach Ppt. In Proceedings of the 9th SimLab Course on Parallel Numerical Simulation, Belgrade, Serbia, 4–8 October 2010.
16. Fu, S.; Biwole, P.H.; Mathis, C. Numerical and Experimental Comparison of 3D Particle Tracking Velocimetry (PTV) and Particle Image Velocimetry (PIV) Accuracy for Indoor Airflow Study. *Build. Environ.* **2016**, *100*, 40–49. [CrossRef]
17. Parameshwaran, R.; Dhulipalla, S.J.; Yendluri, D.R. Fluid-Structure Interactions and Flow Induced Vibrations: A Review. *Procedia Eng.* **2016**, *144*, 1286–1293. [CrossRef]
18. Arias, I.; Knap, J.; Chalivendra, V.B.; Hong, S.; Ortiz, M.; Rosakis, A.J. Numerical Modelling and Experimental Validation of Dynamic Fracture Events along Weak Planes. *Comput. Methods Appl. Mech. Eng.* **2007**, *196*, 3833–3840. [CrossRef]
19. Hou, G.; Wang, J.; Layton, A. Numerical Methods for Fluid-Structure Interaction—A Review. *Commun. Comput. Phys.* **2012**, *12*, 337–377. [CrossRef]
20. Zawawi, M.H.; Saleha, A.; Salwa, A.; Hassan, N.H.; Zahari, N.M.; Ramli, M.Z.; Muda, Z.C. A Review: Fundamentals of Computational Fluid Dynamics (CFD). *AIP Conf. Proc.* **2018**, *2030*, 20252.
21. Singarella, P.N.; Adams, E.E. *Physical and Numerical Modeling of the External Fluid Mechanics of OTEC Pilot Plants*; Massachusetts Institute of Technology, Energy Laboratory: Cambridge, MA, USA, 1982.
22. Ng, F.C.; Abas, A.; Abustan, I.; Rozainy, Z.M.R.; Abdullah, M.Z.; Kon, S.M. Fluid/Structure Interaction Study on the Variation of Radial Gate's Gap Height in Dam. *IOP Conf. Ser. Mater. Sci. Eng.* **2018**, *370*, 12063. [CrossRef]
23. Fluent, I. *Modeling Turbulent Flows Manual*; University of Southampton: Southampton, UK, 2006; Volume 6-2, pp. 6–49.
24. Delafosse, A.; Line, A.; Morchain, J.; Guiraud, P. LES and URANS Simulations of Hydrodynamics in Mixing Tank: Comparison to PIV Experiments. *Chem. Eng. Res. Des.* **2008**, *86*, 1322–1330. [CrossRef]
25. Rock, A.; Zhang, R.; Wilkinson, D. *Velocity Variations in Cross-Hole Sonic Logging Surveys Causes and Impacts in Drilled Shafts*; United States Department of Transportation, Federal Highway Administration: Washington, DC, USA, 2008.
26. Jamshed, S. *Using HPC for Computational Fluid Dynamics: A Guide to High Performance Computing for CFD Engineers*; Academic Press: Cambridge, MA, USA, 2015; ISBN 0128017511.
27. Clayton, M.J.; Johnson, R.E.; Song, Y.; Al-Qawasmi, J. *Information Content of As-Built Drawings*; Texas A&M University: College Station, TX, USA, 1998.
28. Mohd Nasir, S.R.; Zahari, Z.; Isa, M.; Che Ibrahim, C.K. Delay of as-built drawings submission for malaysian toll highway. *J. Teknol.* **2016**, *78*.
29. Clayton, M.; Johnson, R.; Song, Y.; Al-Qawasmi, J. A Study of Information Content of As-Built Drawings for USAA, the Caudill Rowlett and Scott (CRS) Center. *USAA Proj. Coll. Stn. TX* **1998**.
30. Azman, A.; Zawawi, M.H.; Hassan, N.H.; Abas, A.; Razak, N.A.; Mazlan, A.Z.A.; Rozainy, M.A.Z.M.R. *Effect of Step Height on The Aeration Efficiency of Cascade Aerator System Using Particle Image Velocimetry*; EDP Sciences; MATEC Web of Conferences: Les Ulis Cedex A, France, 2018; Volume 217, p. 4005.
31. Zawawi, M.H.; Aziz, N.A.; Radzi, M.R.M.; Hassan, N.H.; Ramli, M.Z.; Zahari, N.M.; Abbas, M.A.; Saleha, A.; Salwa, A.; Muda, Z.C. Computational Fluid Dynamic Analysis at Dam Spillway Due to Different Gate Openings. *AIP Conf. Proc.* **2018**, *2030*, 20245.
32. Kiricci, V.; Celik, A.O. Modeling Hydraulic Structures With Computational Fluid Dynamics. In Proceedings of the International Scientific Conference People, Buildings and Environment 2014 (PBE2014), Kroměříž, Czech Republic, 15–17 October 2014.

33. Haga, K.; Terada, A.; Kaminaga, M.; Hino, R. Water Flow Experiment Using the PIV Technique and the Thermal Hydraulic Analysis on the Cross-Flow Type Mercury Target Model. *ETDEWEB* **2001**, *2*, 1293–1303.

34. Thanh, N.C.; Ling-Ling, W. Physical and Numerical Model of Flow through the Spillways with a Breast Wall. *KSCE J. Civ. Eng.* **2015**, *19*, 2317–2324. [CrossRef]

35. Chanson, H. *Hydraulics of Open Channel Flow*; Elsevier: Amsterdam, The Netherlands, 2004; ISBN 0080472974.

36. Duró, G.; de Dios, M.; López, A.; Liscia, S.O. *Physical Modeling and CFD Comparison: Case Study of a Hydro-Combined Power Station in Spillway Mode*; Utah State University: Logan, UT, USA, 2012.

37. El-Zayat, A. Physical Modeling For Complex Hydraulic Structures. 2016. Available online: https://www.semanticscholar.org/paper/PHYSICAL-MODELING-FOR-COMPLEX-HYDRAULIC-STRUCTURES-el-Zayat/fb9a73a267077740ade24cce36f18aad116c815b (accessed on 1 February 2023).

38. Tafarojnoruz, A.; Lauria, A. Large Eddy Simulation of the Turbulent Flow Field around a Submerged Pile within a Scour Hole under Current Condition. *Coast. Eng. J.* **2020**, *62*, 489–503. [CrossRef]

39. Calomino, F.; Alfonsi, G.; Gaudio, R.; D'Ippolito, A.; Lauria, A.; Tafarojnoruz, A.; Artese, S. Experimental and numerical study of free-surface flows in a corrugated pipe. *Water* **2018**, *10*, 638. [CrossRef]

40. Lauria, A.; Alfonsi, G.; Tafarojnoruz, A. Flow pressure behavior downstream of ski jumps. *Fluids* **2020**, *5*, 168. [CrossRef]

41. Ryan, E.M.; DeCroix, D.; Breault, R.; Xu, W.; Huckaby, E.D.; Saha, K.; Dartevelle, S.; Sun, X. Multi-Phase CFD Modeling of Solid Sorbent Carbon Capture System. *Powder Technol.* **2013**, *242*, 117–134. [CrossRef]

42. Georgoulas, A.; Angelidis, P.; Kopasakis, K.; Kotsovinos, N. 3D Multiphase Numerical Modelling for Turbidity Current Flows. In *Numerical Modelling*; IntechOpen: London, UK, 2012.

43. Björkmon, M. Evaluation of Finite Element Tools for Transient Structural Dynamic Simulations of Firing Systems. Master's Thesis, Chalmers University Of Technolog, Göteborg, Sweden, 2010.

44. Manafpour, M.; Rovesht, T.J. Numerical Investigation of the Flow Characteristics in the Vicinity of Pressure Conduit's Gates. In Proceedings of the Long-Term Behaviour and Environmentally Friendly Rehabilitation Technologies of Dams, Iran, Tehran, 17–19 October 2017.

Article

CFD Simulation of a Submersible Passive Rotor at a Pipe Outlet under Time-Varying Water Jet Flux

Mohamed Farouk [1,2,*], **Karim Kriaa** [1,3] **and Mohamed Elgamal** [1,4]

[1] College of Engineering, Imam Mohammad Ibn Saud Islamic University, IMSIU, Riyadh 11432, Saudi Arabia
[2] Irrigation and Hydraulics Department, Faculty of Engineering, Ain Shams University, Cairo 11517, Egypt
[3] Department of Chemical Engineering, National School of Engineers of Gabes, University of Gabes, Gabes 6029, Tunisia
[4] Irrigation and Hydraulics Department, Faculty of Engineering, Cairo University, Giza 12613, Egypt
* Correspondence: miradi@imamu.edu.sa

Abstract: During the past two decades, passive rotors have been proposed and introduced to be used in a number of different water sector applications. One of these applications is the use of a passive rotor at the outlets of pipe outfalls to enhance mixing. The main objective of this study is to develop a CFD computational workflow to numerically examine the feasibility of using a passive rotor downstream of the outlet of pipe outfalls to improve the mixing properties of the near flow field. The numerical simulation for a pipe outlet with a passive rotor is a numerical challenge because of the nonlinear water-structure interactions between the water flow and the rotor. This study utilizes a computational workflow based on the ANSYS FLUENT to simulate that water-structure interaction to estimate the variation in time of the angular speed (ω) of a passive rotor initially at rest and then subjected to time-varying water velocity (v). Two computational techniques were investigated: the six-degrees-of-freedom (6DOF) and the sliding mesh (SM). The 6DOF method was applied first to obtain a mathematical relation of ω as a function of the water velocity (v). The SM technique was used next (based on the deduced ω-v relation by the 6DOF) to minimize the calculation time considerably. The study has shown that the 6DOF technique accurately determines both maximum and temporal angular speeds, with discrepancies within 3% of the measured values. A number of numerical runs were conducted to investigate the effect of the gap distance between the passive rotor and the pipe outlet and to examine the effect of using the passive rotor on the near flow field downstream of the rotor. The model results showed that as the gap distance of the pipe outlet to the passive rotor increases, the rotor's maximum angular speed decreases following a decline power-law trend. The numerical model results also revealed that the passive rotor creates a spiral motion that extends downstream to about 15 times the pipe outlet diameter. The passive rotor significantly increases the turbulence intensity by more than 500% in the near field zone of the pipe outlet; however, this effect rapidly vanishes after four times the pipe diameter.

Keywords: turbulence closure; k-ε model; varying bed topography; flow over bedforms; turbulence intensity

Citation: Farouk, M.; Kriaa, K.; Elgamal, M. CFD Simulation of a Submersible Passive Rotor at a Pipe Outlet under Time-Varying Water Jet Flux. *Water* **2022**, *14*, 2822. https://doi.org/10.3390/w14182822

Academic Editor: Charles R. Ortloff

Received: 14 July 2022
Accepted: 7 September 2022
Published: 10 September 2022

Publisher's Note: MDPI stays neutral with regard to jurisdictional claims in published maps and institutional affiliations.

1. Introduction

Nowadays, over 150 countries utilize desalination in some form or another to meet their individual water needs and provide water to more than 300 million people. However, the effluent from those desalination facilities generally negatively impacts the environment and requires thermal or brine management [1,2].

The concept of utilizing passive rotors is found in the aeronautical engineering literature, where passive rotors are added to the original wide chord blade rotors to enhance the system's performance at large [3].

In the water engineering technology literature, a "passive rotor" generally refers to a rotor that naturally revolves due to the induced forces from the water–rotor interactions (without any external power sources to operate) [4]. Passive rotors differ from water turbines in that they contain only the energy capture mechanism (the rotor blades system), whereas water turbines have both the energy capture and conversion mechanisms. Therefore, passive rotors are not subjected to external loads (from generators). Consequently, it will rotate at a velocity (called the "runaway velocity") primarily determined by the rotating torque created by the fluid on the blades and by the resistance generated by the liquid on the revolving blades.

During the last decade, passive rotors have been proposed/used in different water and environmental engineering applications. Figure 1 explores the samples of these applications, which include:

- Use of a passive rotor at the water tank pipe outlet to increase the effluent drainage rate from the effluent tanks in water treatment facilities [5];
- Use of a passive rotor at the outlet of pipe outfall to improve mixing;
- Using a passive rotor upstream of a water level control gate to adjust the upstream water afflux by controlling the rotation of the passive rotor without the need to change the gate opening [3].
- Using passive rotors downstream of a sluice gate for energy dissipation

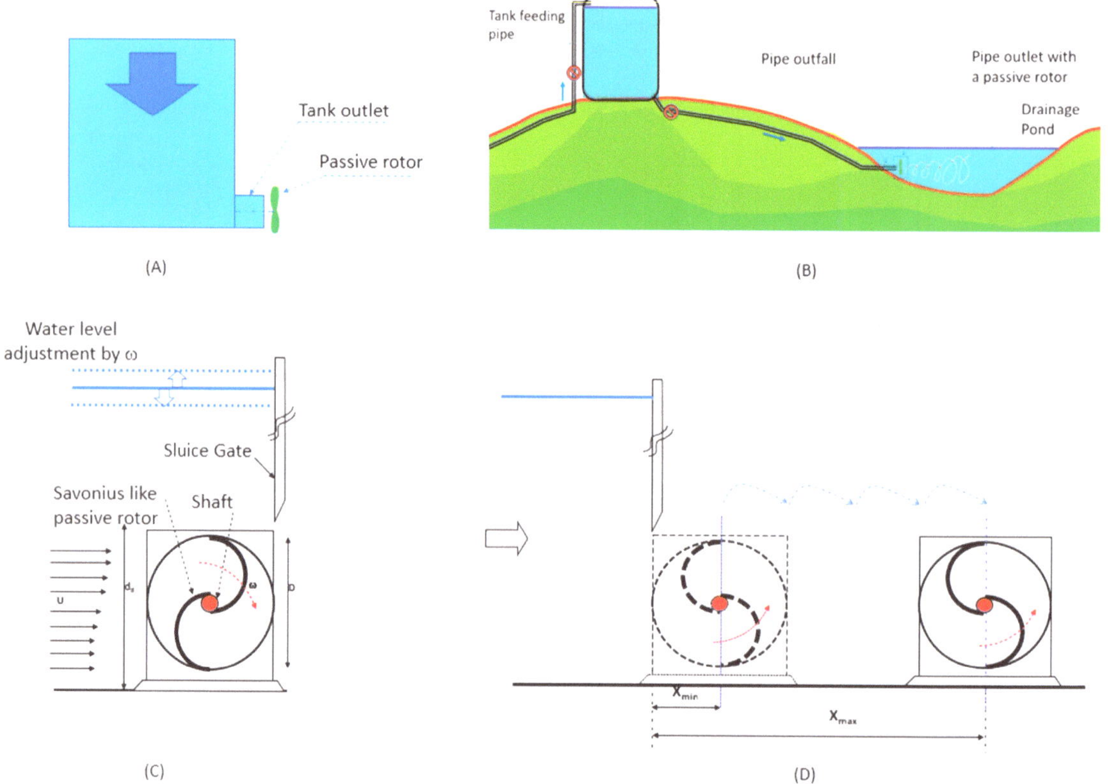

Figure 1. Samples of proposed applications where passive rotors could be utilized. (**A**) at the outlet of a drainage tank, (**B**) at the outlet of pipe outfall, (**C**) at the upstream of a sluice gate, (**D**) at the downstream of a sluice gate.

In 2017, a research work [3] studied a water sluice gate's hydraulic performance with a passive Savonius-like rotor. The study showed that adding a passive rotor affects the upstream water afflux, and the rotation of the rotor could control this effect. The study also proposed replacing the energy dissipater baffle blocks (frequently used in stilling basins to reduce the energy downstream of hydraulic structures) with some loaded rotors to generate a "useful part of energy while dissipating the harmful part of it". In 2021, the effect of using a passive rotor (at the pipe outlet of a draining tank) on the water drainage rate from the tank was experimentally investigated. The study has shown that adding a symmetric, four-blade, passive rotor increased the average water drainage rate from the tank by up to 9.0% due to the formation of a low-pressure region at the pipe outlet caused by the swirl flow induced by the water–rotor reciprocal interactions [5].

Currently, the authors are involved in a research project investigating the feasibility of using passive rotors at the outfalls of water facility plants (application presented in Figure 1b). The project includes some research objectives. One of these objectives is to experimentally and numerically study the mixing characteristics resulting from using the passive rotors in the near field flow zone. The authors experimentally studied the optimum number of blades and the effect of the rotor similarity in their first paper [5].

This study is conducted to answer the following four questions: (1) Based on the available CFD packages, what is the relevant and practical numerical workflow that could be followed to simulate the temporal variation of the angular speed of the passive rotor given an outlet subjected to time-varying effluent? (2) What is the suitable turbulence model(s) that could be used to obtain the most accurate results for such an application? (3) What is the effect of the pipe outlet–rotor gap distance on the rotor's induced angular speed? (4) From a numerical perspective, and based on the developed CFD workflow (in item 1 above) and the relevant turbulent model (in item 2), will the use of a passive rotor at the pipe outlet to improve mixing characteristics be technically feasible?

Before addressing the first question, it should be emphasized that the numerical manipulations for applications involving a passive rotor are considered challenging since the angular speed of the rotor is not known a priori, and it is affected by the mutual interactions between the fluid (the water) and the structure (the rotor) interactions.

The rest of the paper is organized as follows. The research methodology is presented in Section 2, where the problem statement, the experimental study (used for model validation), and the assumptions are described. In Section 3, the 3D CFD numerical model is established using FLUENT, with workflow development, mesh generation, grid discretization, and boundary conditions. In Section 4, the model validation is conducted, where the selected model discretization is examined, the sensitivity of the used turbulence models is assessed, and the required computational resources are discussed. Section 5 discusses the passive rotor's results on the near flow field of the outfall and its corresponding expected effects on the mixing characteristics. Section 6 concludes the findings of the present paper and states the recommendations.

2. Method Statement

2.1. Study Objectives

The first aim of the current study is to propose a numerical workflow that first helps simulate the variation in time of the angular speed of a passive rotor subjected to time-varied effluent flux. Then, the developed numerical workflow will be used to assess the technical feasibility of adopting the passive rotor element at the pipe outlets to enhance mixing in the near field downstream of the outfalls.

Therefore, the study objectives are to respond to the following questions:

What is the appropriate numerical method to deal with the moving rotor, and what is the practical numerical workflow that could be adopted to simulate the variation in time of the angular speed of the passive rotor given an outlet subjected to time-varying effluent?

What suitable turbulence model(s) could be used to obtain the most accurate results for such an application?

Is it technically feasible to use the passive rotors at the pipe outlets to enhance the mixing characteristics of the near flow field?

2.2. Physical Experiment

This subsection briefly presents the physical experiment used to verify the numerical model. More details about the conducted experiments are found in [5]. The experimental setup used to investigate the hydraulic performance of a submerged 18 mm PVC pipe outlet equipped with a passive rotor is presented in Figure 1. The pipe outlet is a drainage facility for a falling head vertical plexiglass tank. The tank's height and diameter are 68 cm and 14 cm, respectively. The tank is placed about 30 cm above the water surface of the mixing tank (tank 2, Figure 1b). The mixing tank has a constant level with horizontal dimensions and a height of 69 cm × 30.5 cm and 28 cm, respectively. This tank is composed of two similar chambers separated by a plexiglass baffle with a height of 24.2 cm (a side weir). The pipe outlet is submerged under the fixed headwater surface in tank No. 2. Water passes from the first to the second chamber of tank No. 2 via the internal side weir. A side circular outlet in the second chamber is used to maintain the water level in the tank.

Figure 1c,d present the side and front views of the pipe outlet with a four-bladed passive rotor. The diameter of the rotor and the distance between the pipe outlet face and the rotor centroid are 31 mm and 13.5 mm, respectively.

Tank 1 was filled with water for the experiment's preparation by opening the submerged pump, valves 1 and 2, while keeping valve 3 closed (Figure 1b).

2.3. Measurements via Video Tracking

The lab experiment requires the measurements of two parameters over the course of time. These parameters are the tank water level and the angular speed of the passive rotor. The angular speed can be measured using tachometers and stroboscopes. Recently, the video tracking (VT) approach has been successfully used in many applications [4]. The VT technique has many advantages. For instance, the VT system is relatively inexpensive, as the tracking can even be conducted by the available smart-phone cameras, it can be used to capture the temporal variations of the rotor angular speed, and it can also be used to measure the angular speed for submerged revolving objects (similar to the study on hand). Nevertheless, the VT system has its own limitations, which will be discussed later on in the following sections.

Two digital cameras (cameras 1 and 2) were used to capture the variations of the rotor's angular speed and the water elevation in the falling head tank (tank1), respectively. The first is a bridge-type Fuji Finepix digital camera (Tokyo, Japan), and the second is a Microsoft LifeCam Studio webcam (Redmond, WA, USA).

The freeware video analysis Tracker package (version 6.0.4, from Open Source Physics Project by National Science Foundation, USA) was used to analyze the recorded videos for the rotor and the water surface in tank 1 using the auto-tracking tool. All the angular velocity measurements were conducted at an image capturing rate of 240 fps.

To measure the angular speed using the Tracker package, a 4 mm circular-black object is drawn near the tip of one of the rotor's blades; refer to Figure 2d. The circular object is auto-tracked via the Tracker. Using the tracked coordinates of the center of the circular object, the Tracker could determine the instantaneous variations of the angular speed, as shown in Figure 3. The time-averaged values of the angular speed were later calculated by applying the moving window average technique while considering an average time of one second. The reader can refer to [5] for more details about the tracking manipulation.

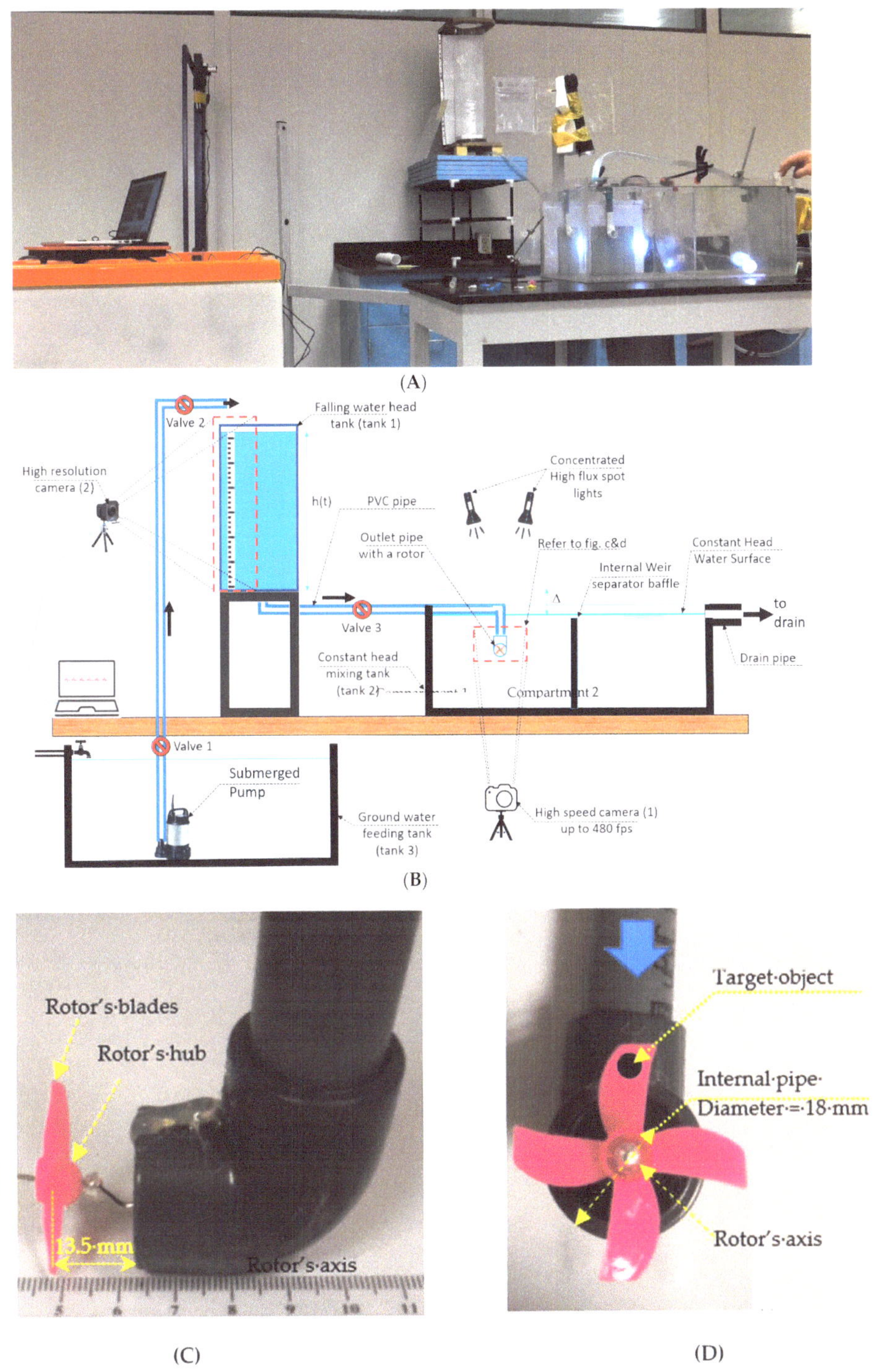

Figure 2. Experimental setup: (**A**) a snapshot of the experimental setup, (**B**) Elements of the experimental setup, (**C**) side view of the rotor, (**D**) front view of the rotor.

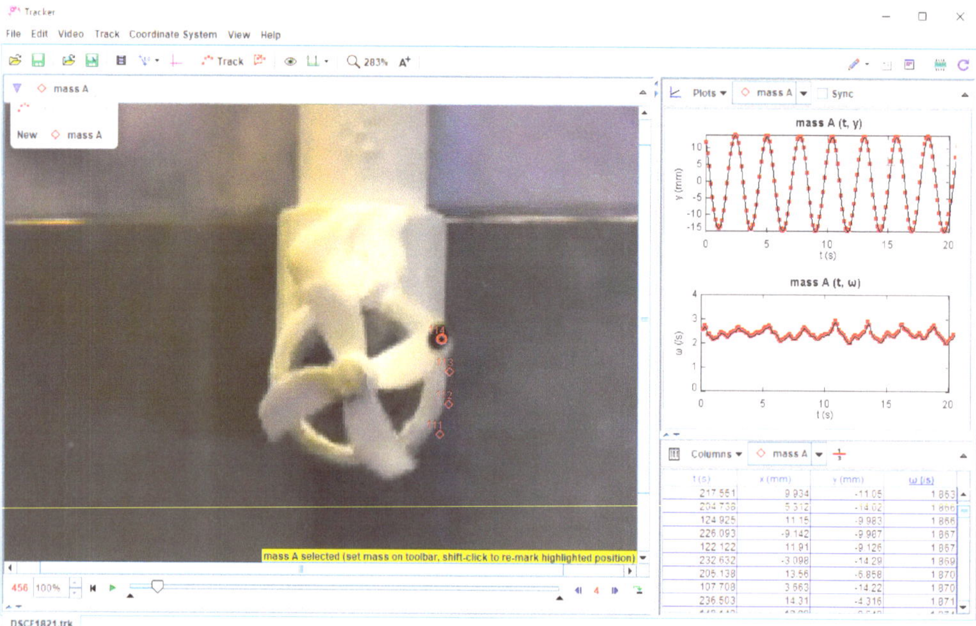

Figure 3. Auto-tracking of the rotor's angular speed via the Tracker (camera capturing rate = 240 fps).

2.4. Accuracy of Measuring the Angular Speed Using Video Auto-tracking Tracking

Two tracking techniques are commonly used in the Tracker. The first is manual tracking, where the user is obliged to manually identify the location of the center point of the tracked "target" object in each frame of the video recorded during the measurement. The second technique is auto-tracking, where the location of the tracked object is automatically identified by the software. Due to the large number of image frames (240 times the duration of the recorded video in seconds) that need to be processed in each video measurement, auto-tracking is adopted in this study. The auto-tracking is based on creating one template image of a selected feature of interest (target object) and then searching each frame for the best match for that template. The best match is the one with the highest match score.

Despite that, auto-tracking, on the one hand, is more convenient and faster and consumes less time than manual tracking. On the other hand, it could result in less accurate measurements. The expected reduction in accuracy due to the auto-tracking is due to what is called template evolving and drifting, in which "visual" or "apparent" changes might take place regarding the shape and or the colors of the target object (for instance, due to the changes in the light intensity). These changes could lead to a small drifting or small errors in the "looking for" coordinates of the center of the tracked object. This means that using auto-tracking to measure the angular speed of a rotor could result in expected small induced or pseudo-perturbations in the angular velocity that are not physically existing in reality. Nevertheless, such small errors are not expected to affect the time-averaged values of the angular speed measurements due to the expected randomness of their occurrence.

To eliminate such errors, a time averaging process (using the marching window technique) is commonly used and applied to all the measurements based on a duration interval of 1 s.

To assess the accuracy of the time-averaged measurements of the angular speed of the rotor using visual tracking, a number of diagnostic experiments (rotor speed from 160 to 1500 rpm) have been conducted to measure the rotor angular speed using two different measurement techniques. The first technique is the digital photo tachometer, and the second is the visual (video) tracking. These experiments are conducted, where the rotor is kept running in the air (not submerged under water) to avoid the dispersion of the reflected laser beam (in the digital photo tachometer) that is expected to take place when the system is set underwater.

The tested rotor is attached to the axis of an overhead stirrer (Dragon Lab Type model OS40 Pro, Dragon Lab, Rowland St. City of industry, CA, USA). A small piece of reflective tape was stuck to the tip of one of the blades of the rotor, and a handheld, non-contact digital tachometer (model: NEIKO 20713A, NEIKO, Wilmington, Delaware, USA) was used to carry out the non-contact diagnostics. The measurements of the tachometer are based on identifying how many times per minute the laser beams (that emerge from the tachometer) are reflected back (from the reflecting tape that is stuck on the surface of the rotor) to the tachometer. Figure 4 shows the setup of this assessment experiment.

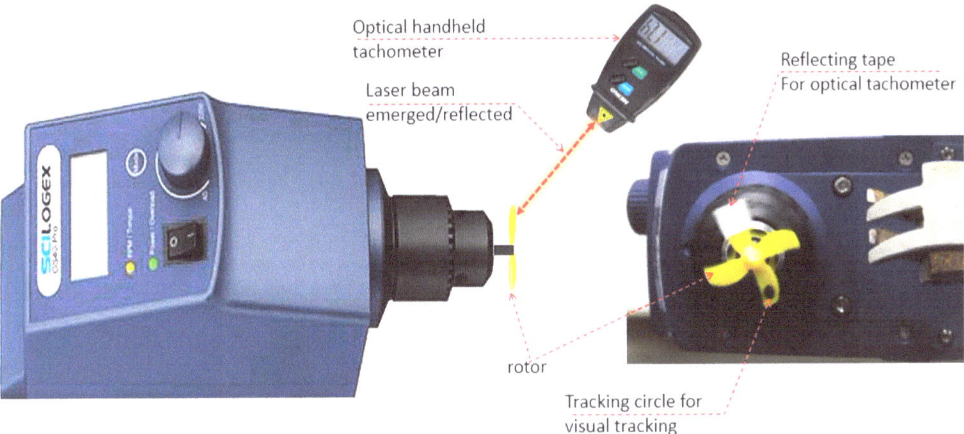

Figure 4. Setup of the assessment of accuracy experiments (using two different methods) for the angular speed measurements: oblique-plan view, side view.

The accuracy of the tachometer is of the order of ±0.05%, and the discrepancies between the measurements from the two techniques were found, in general, to be of the order of +1%.

2.5. Numerical Techniques for Rotating Elements in FLUENT

FLUENT provides three different methods to deal with rotating objects. These methods include the moving reference frames (MRF), the sliding mesh (SM), and the six-degrees-of-freedom (6DOF) [6]. Accordingly, it will be essential to identify the pros and cons of each method, to decide what is the most suitable calculation method for the current application, and to determine the practical numerical workflow that is to be adopted to carry out accurate predictions of the time-varied rotor angular speed subjected to the available computational resources.

The CFD literature reveals that the moving reference frame (MRF) method can be applied for the steady-state simulation, where the rotation speed of the rotor can be considered a constant [7–10]. It should be noted that the MRF has a low computational cost and can accurately predict the characteristics of the constant rotation applications.

In the case of unsteady-state flow, if the rotor's angular speed ω is known as a function in time, then the sliding mesh "SM" technique may be recommended. It should be noted that the sliding mesh technique has been widely implemented to simulate the rotor motion in many turbine applications [11–14]. Nevertheless, it should be mentioned that the computational time resources needed for the sliding mesh technique for unsteady applications are usually 30 to 50 times higher than the steady-state applications [15]. In recent years, an alternative technique known as the overset (chimera) mesh method has caught the attention of the academic community. A comparison between the overset and the sliding mesh methods is used to find their advantages and disadvantages in the two-dimensional simulation of vertical axis turbines [16]. The comparison was established to predict the performance parameters of the turbine in order to check the capabilities of the models to capture the complex flow phenomena in these devices and the computational costs.

The dynamic mesh (6DOF) method is appropriate when the rotor's angular speed as a function of time is not known beforehand; therefore, the subsequent move of the rotor is determined based on the solution at the current time. In this case, the rotor's angular speed ω is computed from the force balance on a solid body via the 6DOF solver. The update of the volume mesh is handled automatically by FLUENT at each time step based on the new positions of the boundaries. The starting volume mesh and a description of the motion of any moving zones in the model have to be provided to use the dynamic mesh technique. It should be noted that the 6DOF method has been used to simulate the rotation of the rotor in different research works, such as [17–20].

Research work [20] compared the experimental results with the results of the CFD model while using the 6DOF technique at a pico scale. They found that the 6DOF method is more accurate than the MS method for predicting the performance of cross-flow turbines. However, the moving mesh requires less computational time and a faster convergence rate than the 6DOF.

In [21], the researchers used a glyph script in POINTWISE-GRIDGEN to generate the domain and mesh of a vertical-axis marine (Water) turbine (straight-bladed Darrieus type), with particular emphasis on the turbine's unsteady behavior. At the same time, the simulations were performed in ANSYS FLUENT v14 (Ansys, San Jose, CA, USA). To simulate the interaction between the dynamics of the flow and the Rigid Body Dynamics (RBD) of the turbine, a User Defined Function (UDF) was generated for the quasi-steady-state.

Research work [5] investigated the effect of adding a passive rotor on the outlet performance, where different sizes and numbers of blades of rotors were considered. In the current research, the CFD simulations have been applied to study the ω of a passive rotor starting from a stationary state under a decelerated cross submersible jet for one size and one shape of a rotor. From the physical model [5], the current numerical study assumes the rotor's recommended size and shape.

In this research, the rotor is assumed to be submersible, not near the water surface; therefore, there is no need to include the free surface and two-phase problem (water and air), such as in [22] and [23]. Moreover, the rotor is also not near the bed, and, thus, the effect of the sediment will not be included [24]. Additionally, the flow is gradually decreased, not rapidly varied, as in [25].

Starting the rotor rotation from a stationary state to reach a constant speed under steady cross velocity has rarely been tackled by researchers, as it has in wind turbines [26]. To the best of the authors' knowledge, the case of a rotor starting to move from a static state under a decelerated cross-water velocity is studied for the first time. This case study is more complicated since ω never achieves a steady state. The rotor's angular velocity rapidly accelerates for a very short time; then, it gradually decelerates. Therefore, the 6DOF technique has been applied, since no known rotating speed is unknown to the computational grid frame. The 6DOF model has three objectives: the determination of ω_{max}, the variation of ω with time, and the equation of ω as a function of υ. The 6DOF results were compared with those obtained from the physical model [5]. The verification includes both ω_{max} and the variation of ω with time. After that, the ω equation as a function of υ was investigated. Then, the sliding mesh technique was used after adding that equation to the User-Defined-Functions (UDF). The accuracy of both CFD methods and the time consumed are discussed. Finally, the results and a discussion of the sliding mesh model are tackled.

2.6. Assumptions and Simplifications

The numerical study on hand considers the following main assumptions:

- There is no eccentricity for the shaft of the rotor to the center of the pipe outlet;
- The shaft of the rotor revolves smoothly around its pivot without resistance;
- The shaft and its rotor have the same angular speed (i.e., no slipping took place);
- No deformation is allowed regarding the blades of the rotor during the simulation;
- The rotor blade is assumed rigid enough, and its deformation is neglected;
- The direction of the water flux from the pipe outlet (before impinging on the rotor) is mainly horizontal, with no inclination;
- For practical reasons, the main focus of the simulation is directed on capturing the decelerated period for the rotor (refer to the calibration section) since this period is the dominating stage, and the accelerating zone lasts less than 1.5% of the whole simulation time.

3. Numerical Model Development

3.1. Workflow Development

Since the rotor is subjected to decelerating water flux and its angular speed is not known a priori and never reaches a fixed value, the dynamic mesh method seems to be the suitable technique to determine the temporal variations in the rotor's angular speed. The downside of adopting the 6DOF technique is that it requires substantial computational and time resources. As a compromise, it is proposed to adopt the following computational workflow:

- Start the simulation using the dynamic mesh method using the 6DOF solver for about 21% of the required total simulation time (for the study on hand, for the first 40 s);
- The k-e turbulence model will be initially selected, and the accuracy of the generated mesh will be checked near the boundaries by assessing the y+ plot;
- The simulation will be repeated many times, and each time, a different turbulence model will be tried, and the rotor's angular speed results will be compared with the measurements;
- Identify the most relevant turbulence model that gives the best match with the measurements;
- Based on the obtained results of the optimal turbulence model, identify the relation between the rotor's angular speed (ω) and the pipe outlet velocity (v) and create a user-defined function (UDF) for it;
- Switch the model to the sliding mesh (SM) technique while adopting the optimal turbulence model, and start the simulation until the end of the simulation time.

Figure 5 presents the block diagram for the abovementioned workflow.

3.2. The Geometry of the CFD Model

The assumed passive rotor is typical of the rotor's recommended size and shape for the previous study [5]. The rotor of four blades that is 31 mm in diameter has been developed in three-dimensional space using the workbench in ANSYS FLUENT v14. The tank's geometry in the CFD model is typical of compartment number 1 in tank number 2 in the experimental study, as shown in Figure 6a. The tank in the CFD model, shown in Figure 6a,b, is 30.5 cm in length and 34.0 cm in width, and the controlled water depth equals 25 cm. The sidewall is 25 m from three sides, and the fourth wall (the baffle wall) is 24.2 cm. The water outlet is located above the baffle wall. The rotor axis is fixed at the center of the inlet pipe. The mesh consists of around 4.7 million tetrahedral cells; the cells on the rotor surface are shown in Figure 7a. Figure 7b illustrates the cells on the surface of one blade and the cells on the surface of the rotor, the inlet, and the fixed walls in Figure 7c. The minimum and the maximum surface areas on the rotor are 2.775×10^{-3} and 1.086×10^{-2} mm^2, with an average of 6.716×10^{-2} mm^2.

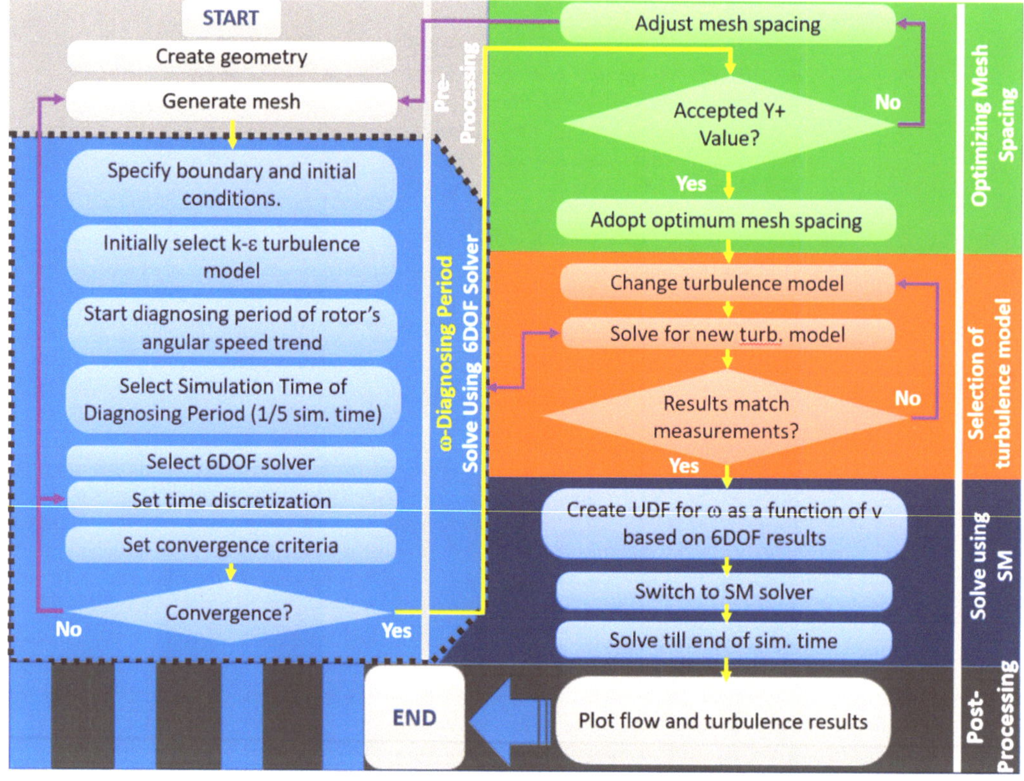

Figure 5. Block diagram for the adopted CFD workflow.

3.2.1. The Boundary Conditions

The inlet boundary is assumed as a mass inlet boundary, and the outlet boundary is considered a pressure outlet. All other boundaries are taken as walls. The inlet boundary and the solver parameters are discussed hereunder.

The Inlet Boundary

Since the inlet velocity in the physical model depends on the falling water head, the CFD's inlet boundary has to precisely simulate the velocity value. The diameter of the inlet is 18 mm at a level of 12.1 cm and is located in the middle of the sidewall, as shown in Figure 6c,d. In the CFD model, the mass flow inlet boundary has been used. The best-fitting curve has been investigated from the experimental results of the flow rate with time. Equation (1) has been assumed using UDF at the inlet. The correlation of the equation is 0.9993.

$$Q = 0.0851 - 0.0002 \times t \tag{1}$$

where:

Q is the mass flowrate (kg/s)

t is the time (s)

This equation is valid only if t < 180 s

The Outlet Boundary

The top of the baffle (side) wall is assumed to have an outlet pressure boundary to match the physical model, as shown in Figure 6c,d. The height of the boundary is 8 mm, and the width is 30.5 cm.

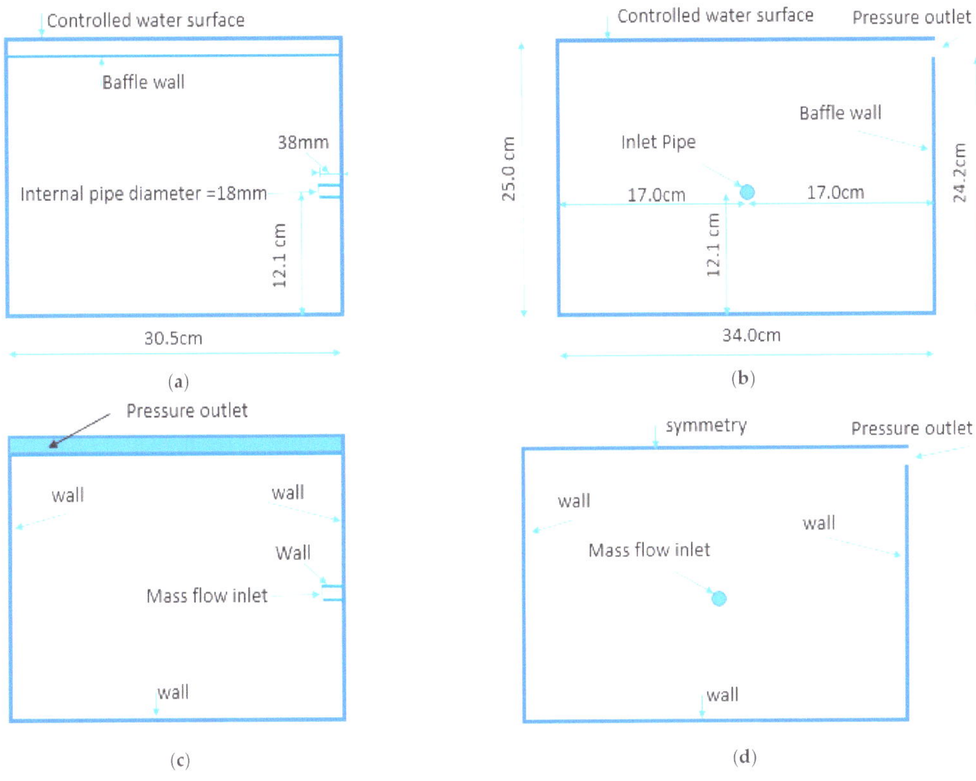

Figure 6. Schematic Figure of the CFD model: (**a**) Side view of the CFD model dimension, (**b**) Front view of the CFD model dimension, (**c**) Side view of the CFD model boundary, (**d**) Front view of the CFD model boundary.

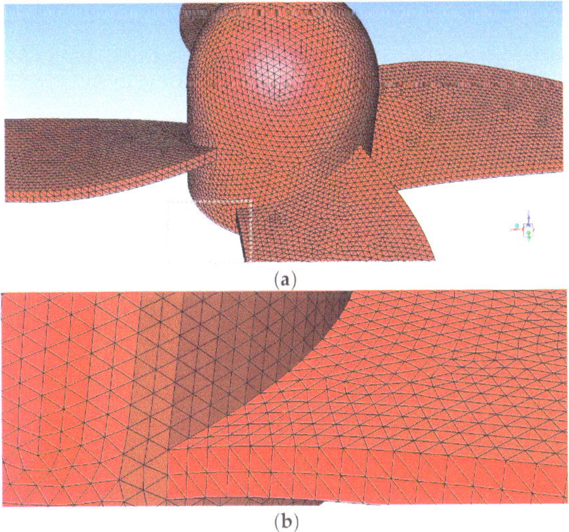

(**a**)

(**b**)

Figure 7. *Cont.*

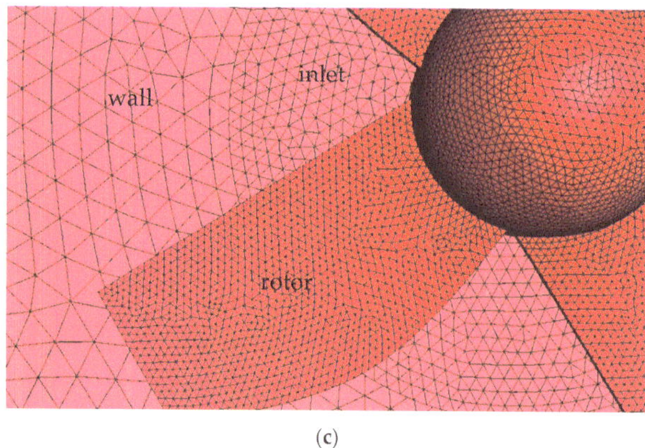

(c)

Figure 7. Passive rotor grid model: (**a**) The mesh on the rotor surface, (**b**) The detail of the mesh on one of the blades (dotted line zone in Figure 5a), (**c**) The detail of the mesh of the rotor, the inlet, and the walls.

The Top of the Model and the Other Boundaries

The model's top is assumed to be the symmetry boundary, as shown in Figure 6d. All other boundaries (the rotor, the tank sidewalls, and the tank bottom) are considered as walls, as shown in Figure 6c,d.

3.2.2. Solver Parameters

The solver parameter settings for the propeller open water simulations were made with a judicious combination of the FLUENT literature recommendations and the trial-and-error evaluations of various solver settings. Combining these options with the domain dependence and grid dependence studies is a vital, though time-consuming, initial effort before any propeller geometry can be investigated for performance. The final solver parameters, including physical constants, are shown below in Table 1.

Table 1. Summary of the boundary conditions and the solver parameters of the CFD models.

Parameters	Settings
Solver	Pressure-based, transient
Velocity formulation	Absolute
Turbulence model	Standard k-ε
Water density	998.2 kg/m^3
Water viscosity	0.001003 kg/m·s
Pressure discretization	Body Force Weighted
Gravity	9.81 m/s^2
The inlet	Unsteady mass flow inlet
The outlet	Pressure outlet
The top of the tank	Symmetry
All other boundaries	Wall

4. Calibration and Accuracy Assessment of the Numerical Model

4.1. Measurements of the Rotor's Angular Speed

Figure 8a shows the measurements of the variation in time of the ensemble average of the angular speed (ω) of the rotor (normalized by the maximum angular speed value ω_{max}). Three distinct stages appear in Figure 8a. The first zone is the acceleration stage (from A to B), where the rotor speed accelerates to achieve its maximum angular speed. Interestingly, the rotor is accelerating at this stage, even though the pipe flow is decelerating. In the second stage (from B to C), the angular speed starts linearly and progressively decreases with time until it reaches point C, where the tank becomes wholly drained. In the last stage (from C to D), the transmission pipe rapidly drains until it reaches point D. This rapid decrease in angular speed may be explained by the substantial difference between the tank and pipe cross-section areas.

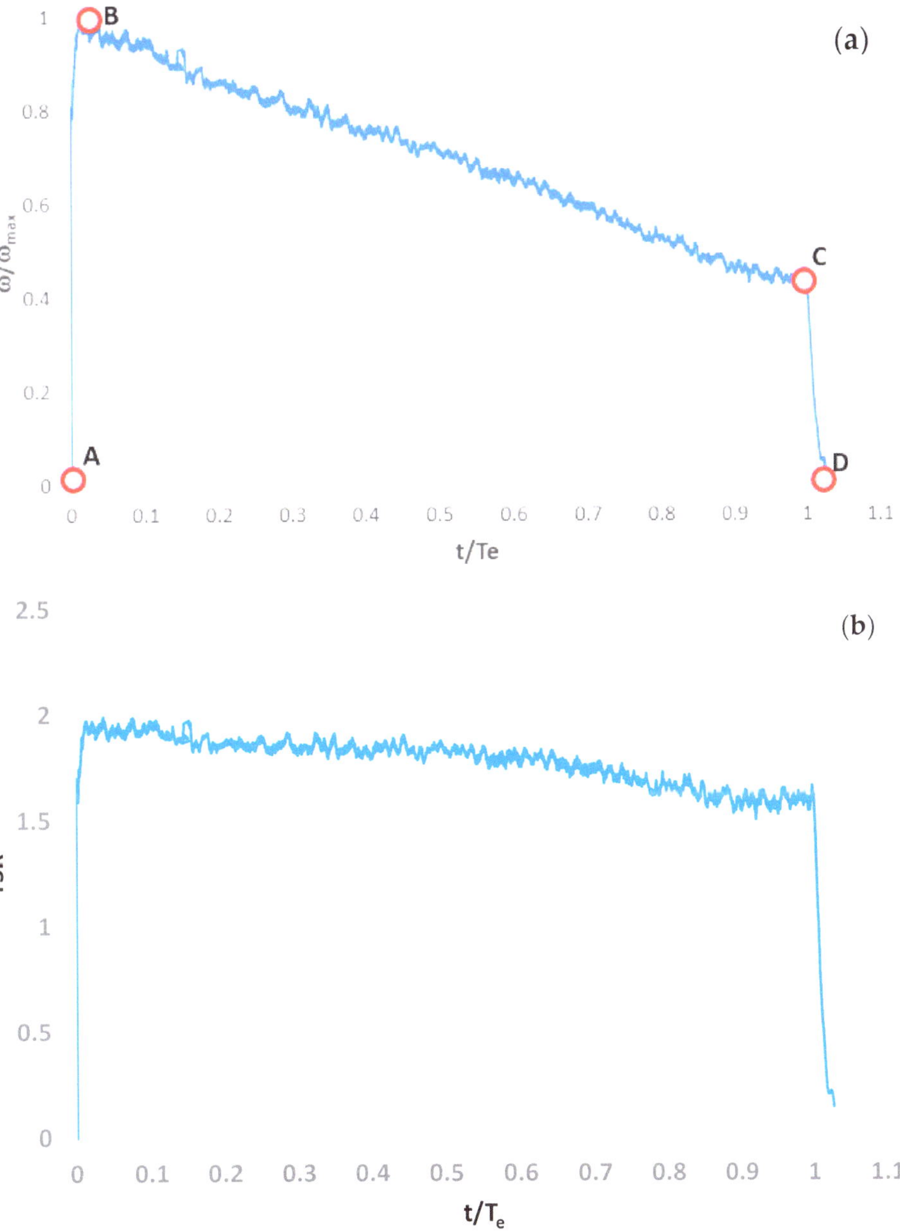

Figure 8. Video analysis measurements of the temporal variations of (**a**) the normalized rotor's angular speed and (**b**) the tip speed ratio (TSR).

The rotor's tip speed ratio (TSR) is expressed using Equation (2)

$$TSR = \frac{V_r}{V_o} \tag{2}$$

where:

The rotor's tip speed could be calculated as $V_r = \omega.d_r/2$, and Vo is the water pipe outlet mean velocity. Figure 8b shows the variations in the rotor's tip speed rotation (TSR) over the course of time. The figure shows that TSR varies from about 1.6 to 1.9, which means that the tip rotor speed is almost 60 to 90% faster than the outlet water velocity.

4.2. Model Discretization and GCI Analysis

Numerical models usually use discrete methods to convert the governing partial differential equations into algebraic equations. All discrete methods introduce discretization errors that might be significant enough to ruin the accuracy of the produced numerical solutions. Therefore, the computational fluid dynamics community and many other CFD-reputable journals require discretization error estimation as a prerequisite for publishing any CFD paper.

The mesh discretization has been studied by applying the grid convergence index (GCI). The generated mesh was assessed by calculating the non-dimensional wall distance for a wall-bounded flow (y+). The validity of the turbulence models is examined by comparing the models' results with the experimental measurements of the rotor's angular speed.

The Grid Convergence Index (GCI) is a relatively new discretization error estimation technique. The GCI is calculated to answer whether the adopted mesh in the simulation is refined enough or not. A minimum of two mesh solutions are required, but three are recommended to calculate the GCI.

In the current study, three groups of mesh sizes with different grid densities are established and are named Low, Med, and High grid density, respectively. The corresponding numbers of mesh cells are almost 1.0, 2.7, and 4.7 million. The three groups of mesh sizes were used to calculate the rotor's maximum angular speed and compare it with the measurements. The GCI was calculated using the method described in [27]. The GCI for the high grid density group was less than 0.8%, and the maximum rotor angular speed error was less than 1.4%. Therefore, the mesh of 4.7 million cells was selected for further analysis.

4.3. Checking Model Discretization near the Boundary

To verify the numerical accuracy near the boundaries, the non-dimensional wall distance for a wall-bounded flow (y+) is calculated for the first layer of the generated grid points throughout the rotor at the instance of maximum angular speed (refer to Figure 9a). The rotor center is presented by the zero point in the figure, as shown in the key figure. Two blades (r = 15.5 mm) are almost horizontal, and the other two are almost vertical. Figure 9b shows the 3-D variation of y+ along the transect A-A. It is noted that the y+ values lie within the acceptable limit of 5 (the maximum Y+ is less than 5 for the rotor, with an average value equal to 2.34, while the maximum Y+ is less than 1 for all other fixed walls, with an average value equal to 0.6 [28,29]).

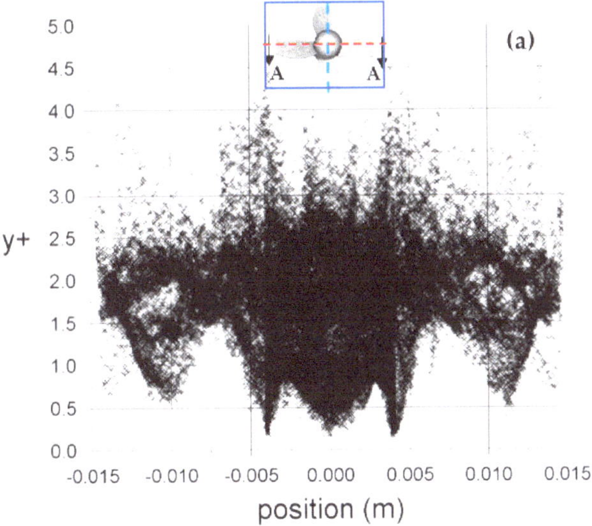

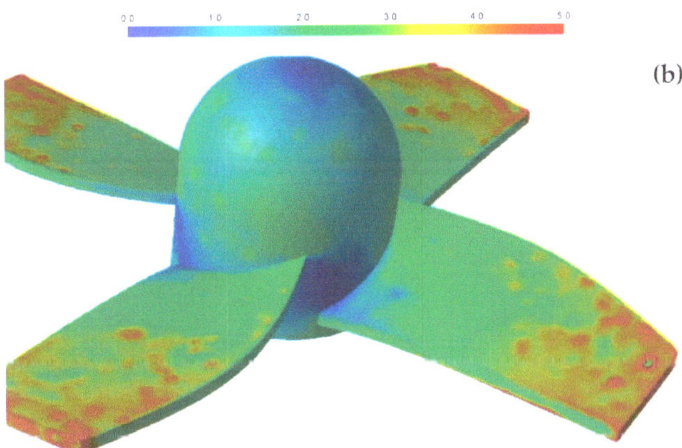

Figure 9. The relationship between the position on the blades and the y+ value at the maximum angular velocity. (**a**) Spatial variations of the wall y+ value along transect A-A, (**b**) Color contours of y+ over the whole rotor.

4.4. The Sensitivity of the Turbulence Model

The validation of the turbulence model is of fundamental importance for evaluating the reliability of the CFD simulations. Several research studies have used FLENUT to study the rotor motion for different applications, where different turbulence models have been adopted. Since the current study topic has not been tackled before, to the best of the researchers' knowledge, the sensitivity of the turbulence models should be studied to select the most appropriate turbulence model with a maximum speed of 0.6 m/s. Almost all of the different turbulence models applied earlier in other research on rotor movement are considered in this study. Eight different turbulence models have been analyzed in this research. A list of these turbulence models considered in the sensitivity analysis is given below:

- Spalart–Allmaras model [30,31],
- The K-ε standard model, [32–35],

- The K-ε RNG model, [36],
- The K-ε realizable model [31,37],
- The K-ω standard model [38],
- The K-ω-SST model [31,39,40],
- The Reynolds Stress Model (RSM) [41],
- The Transition SST (four equations) [18]

The sensitivity study was analyzed in two steps. First, the different turbulence models are compared in terms of how they predict the maximum angular speed of the rotor. The turbulence model that provides the most accurate value of the maximum angular speed is selected. The second step was used to check the accuracy of the chosen turbulence model by comparing the reduction rate of the angular speed with time to the physical model

4.4.1. The Maximum Angular Speed

The maximum rotation speed results for the different turbulence models have been compared directly to those obtained in the physical model. The results are shown in Figure 10a. Figure 10b shows the percentage error in the predicted maximum angular speed for each turbulence model. It is interesting to notice that the three K-ε models yield, in general, the most accurate results compared with the other turbulence models. It is found that the standard K-ε model produces the lowest error (2% error), which is slightly better than the other K-ε models. Therefore, the standard K-ε model has been selected to proceed to the next validation step. It is also of interest to notice that the K-ω standard turbulence model resulted in the most significant error (27% error) among the other models

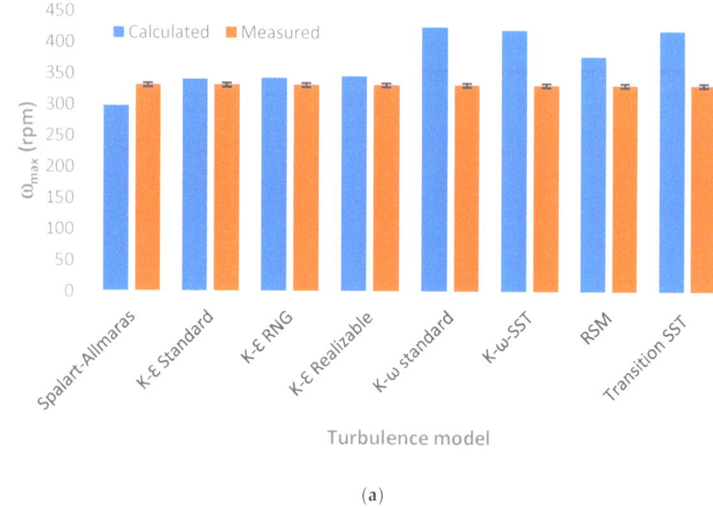

(a)

Figure 10. *Cont.*

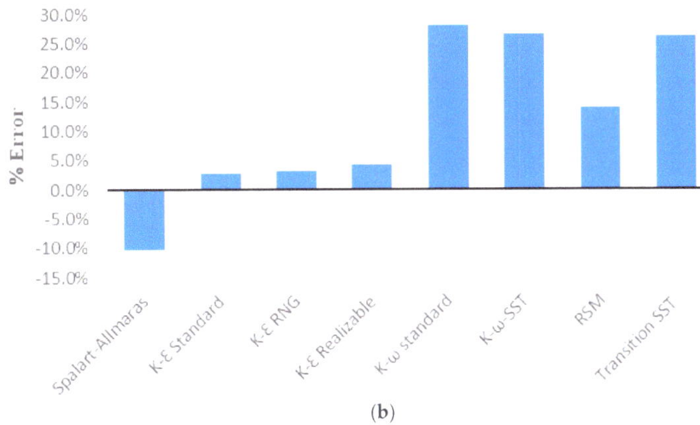

(b)

Figure 10. Predicted rotor's angular speed based on different turbulence models. (**a**) maximum angular speed, (**b**) percentage error in the maximum angular speed.

4.4.2. Comparison of the Temporal Variation in the Rotor's Angular Speed

After selecting the standard K-ε model as the optimal turbulence model for this study (maximum speed = 0.6 m/s), a more comprehensive analysis has been applied by comparing the temporal variation in the rotor's angular speed for both the experimental and 6DOF models. Figure 8a illustrates the time decline trend of the rotor's angular speed for the first 40 s. In general, the measurements and the CFD-6DOF model match reasonably well. Nevertheless, the decline rate of the angular speed of the rotor predicted by the CFD-6DOF model is slightly steeper than the measurements. Some discrepancies exist between the 6DOF model and the measurements, especially within the accelerating phase (from points A to B in Figure 11a,b). Such discrepancies are probably related to the non-homogeneous mechanical resistance experienced by the rotor's shaft in the physical model while initiating the rotor's movement. After the rotor has gained its angular velocity, both trends of the curves are almost identical. Overall, the results of the 6DOF model are very consistent with the experimental results.

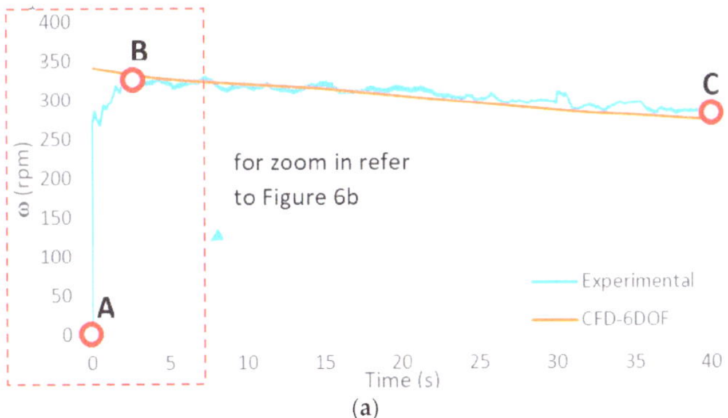

(a)

Figure 11. *Cont.*

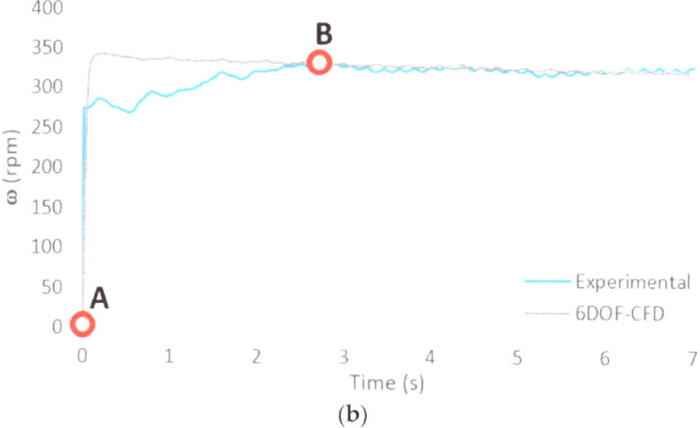

Figure 11. The variation in the rotor's angular speed: (**a**) The time span from rest to 40 s, (**b**) Zoom in for the accelerating zone. (for explanation of points A–C, refer to Figure 8a).

4.5. Computational Resources

The entire CFD study lasted around three months, with a processing speed of 3.1 GHz, 32 used cores, and 15,000 used core-hours.

Using the 6DOF solver for such a case yields accurate results but consumes much time. A total of 11,500 core-hours are required to conduct only forty seconds for the passive rotor with 4.7 million tetrahedral cells. Only forty seconds showed in the 6DOF simulation (out of 180 s, which was the total time for the experimental run), which was sufficient to deduce the mathematical relation between the rotor's angular speed (ω) and the average water velocity at the pipe outlet (υ). This equation has been included in the new UDF for the next sliding mesh simulation step.

4.6. The Relation between the Angular Speed and the Pipe Outlet Water Velocity

The relation between the angular velocity "ω" and the average pipe outlet water velocity "υ" is linear, with a correlation of 0.998, and the mathematical relationship was deduced from the results of the 6DOF simulation. It was given by Equation (3) below.

$$\omega = 12.0425\,\upsilon - 67.785 \tag{3}$$

where
    ω is the angular speed of the passive rotor (rpm)
    υ is the average pipe outlet water velocity (cm/s)
    The equation is only valid for the gradual decelerating zone (from B to C in Figure 10a), where υ lies within the following range (19 cm/s < υ <35 cm/s).

5. Results and Discussion

5.1. Perturbation of Angular Speed

The sliding mesh model, which is a particular case of general dynamic mesh motion wherein the nodes move rigidly in a given dynamic mesh zone, is applied to the case study. Additionally, multiple cell zones are connected through non-conformal interfaces. As the mesh motion is updated in time, the non-conformal interfaces are likewise updated to reflect the new positions of each zone [42,43]. It is essential to note that the mesh motion has been defined using the investigated Equation (3). The model has been used with the same setup, except with the sliding mesh instead of the 6DOF. Additionally, the UDF defines two equations: the relation between the time and the velocity, as stated in Equation (1), and the relationship between υ and ω, as shown in Equation (3). Figure 12 represents the relationship between the angular speed (rpm) and the time (s) for the experimental and sliding mesh-CFD models along the physical test time (180 s). The time consumed is significantly reduced compared to the 6DOF.

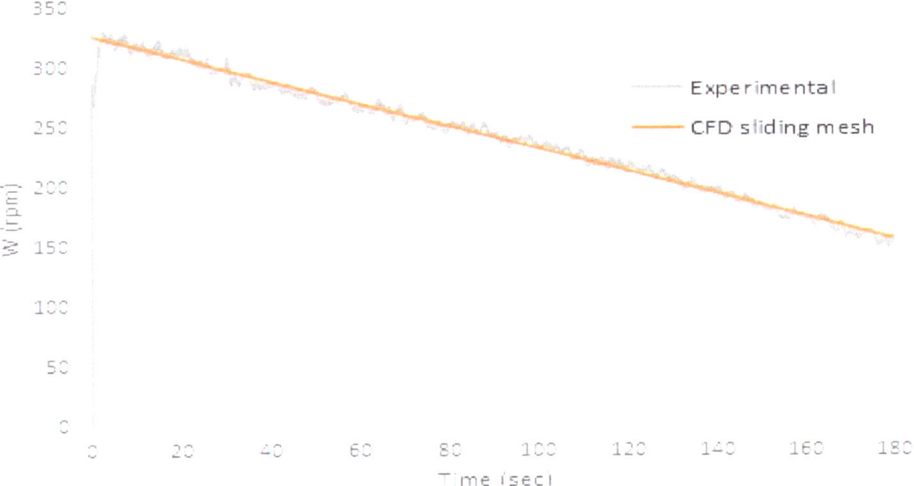

Figure 12. The relationship between the angular velocity (rpm) and the time (s) for the experimental and sliding mesh-CFD models.

Interestingly, both the experimental values and the results of the 6DOF model presented some perturbations for the rotor's angular speed with different amplitudes, as shown in Figure 10b. On the other hand, no perturbations exist in the results of the sliding mesh model, as shown in Figure 12, since the model is based on a given mathematical relation for the angular speed. Nevertheless, the perturbation range of ω in the physical model is more noticeable and considerable than that in the 6DOF CFD model. Such differences could be due to the assumptions stated in the numerical model, which might not 100% comply with the physical situation. Still, the average angular speed values with time for the physical and numerical results are almost identical.

5.2. Effect of the Outlet–Rotor Gap Distance

It is of interest to study the effect of the gap distance (s) between the pipe outlet and the passive rotor on the gained angular speed of the passive rotor. For this, five numerical runs were conducted using the 6DOF model while adopting different pipe outlet–rotor gap distances. Figure 13a shows the temporal variations in the normalized angular speed of the passive rotor for different gap distance ratios. Unsurprisingly, it is observed that as the distance between the rotor and the outlet increases, the amount of emerged jet energy that is received by the rotor decreases, and, therefore, the rotor's maximum angular speed decreases as well. As a result, the rotor's maximum angular speed takes longer to attain, resulting in a longer initial acceleration phase (refer to the dashed red arrow in Figure 13a). The power-law formula fits the reduction trend of the maximum rotor's angular speed with the rotor–pipe outlet gap distance ratio well, as depicted in Figure 13b.

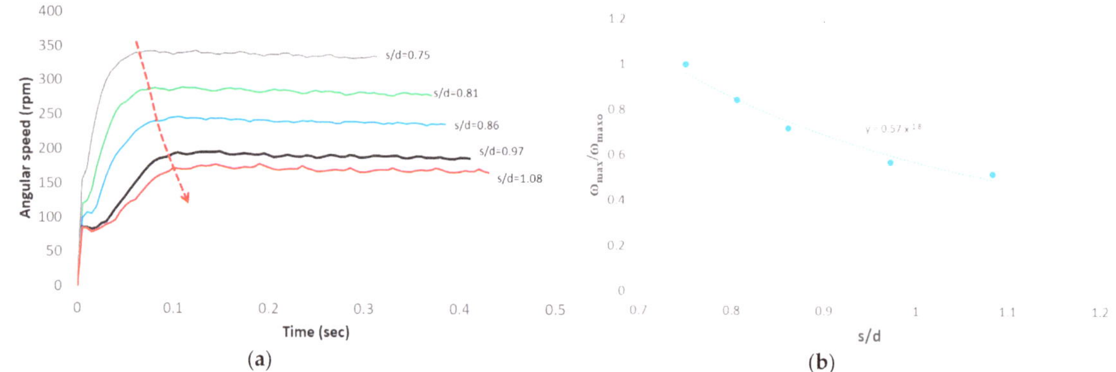

(a)   (b)

Figure 13. Effect of the pipe outlet-rotor gap distance on the gained rotor's angular speed: (**a**) Temporal variation with time, (**b**) Maximum rotor's angular speed.

5.3. Effect of the Passive Rotor on the near Downstream Flow Field

Figure 14 shows an example of the general 3D flow visualization of the flow path-lines downstream of the passive rotor (for the case of $v_{av} = 30$ cm/s and $\omega = 293.5$ rpm). The path lines are also color-coded based on the corresponding values of flow velocity, where a relatively high flow velocity exists near the rotor (within a distance of 2–3 d, where d is the pipe outlet diameter), and the flow velocity significantly decreases downstream. The figure illustrates how a spiral flow is developed by the passive rotor when a rotating wake occupies a flow domain whose average lateral size exceeds the rotor's diameter ($\approx$2.3 D, where D is the rotor's diameter). The rotating wake influence is also observed to extend downward to the front wall of the tank (x $\approx$ 15 d, where d is the pipe outlet diameter).

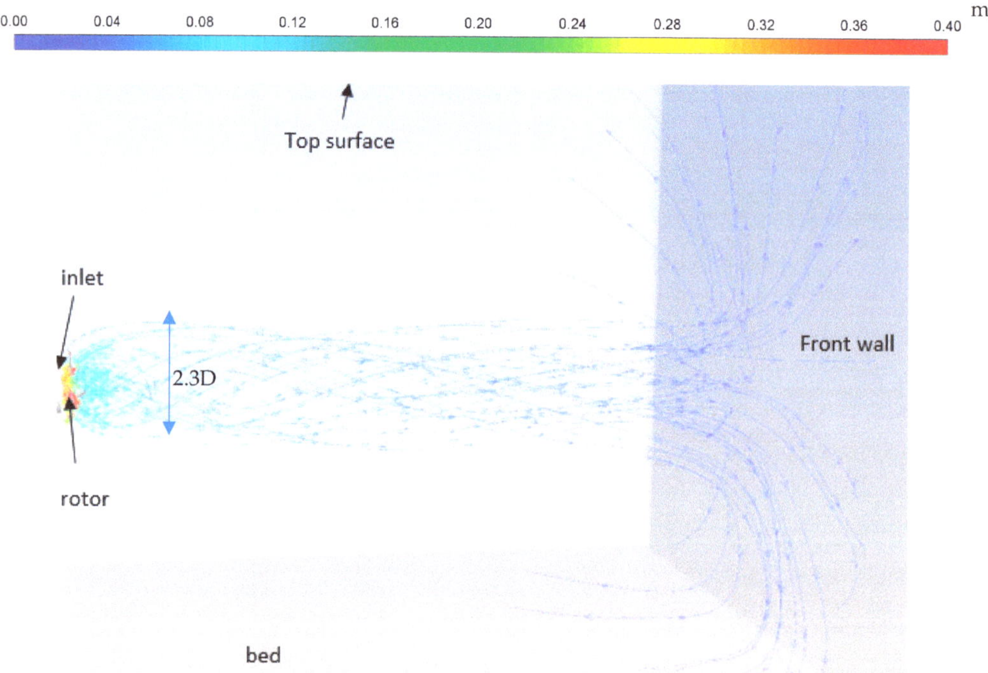

Figure 14. The velocity path-lines at $v_{av} = 30$ cm/s and $\omega = 293.5$ rpm.

5.4. Effect of the Passive Rotor on the Turbulence Intensity

The turbulent intensity (TI) presents the unresolved unsteadiness and the level of turbulence in a flow over time. It is defined as the ratio of the average fluctuating velocity components (u_t') to the time-averaged absolute "resultant" velocity (U) at a certain point in the flow field [44]. Mathematically, the turbulence intensity (TI) can be given as:

$$TI = \frac{u_t'}{U}. \tag{4}$$

where:

$$u_t' = \sqrt{\frac{1}{3}\left(u'^2 + v'^2 + w'^2\right)}. \tag{5}$$

$$U = \sqrt{\left(\bar{u}^2 + \bar{v}^2 + \bar{w}^2\right)} \tag{6}$$

The TI is a vital quantity for many physical phenomena, including the development of the turbulent boundary layer, heat transfer, and mixing.

Figure 15a illustrates the spatial variations in the turbulence intensity values throughout a vertical plane that passes by the centerline of the pipe outlet and the passive rotor (at v_{av} =30 cm/s and ω = 293.5 rpm). The figure shows that the turbulence intensity at the pipe outlet varies between 1.19 and 1.95%, with an average value of 1.3%. Slightly further downstream, the passive rotor causes the turbulence intensity to sharply increase to reach significantly higher values than the values downstream of the pipe outlet. It is also noticed that the turbulence intensity substantially decreases as the flow goes far from the passive rotor.

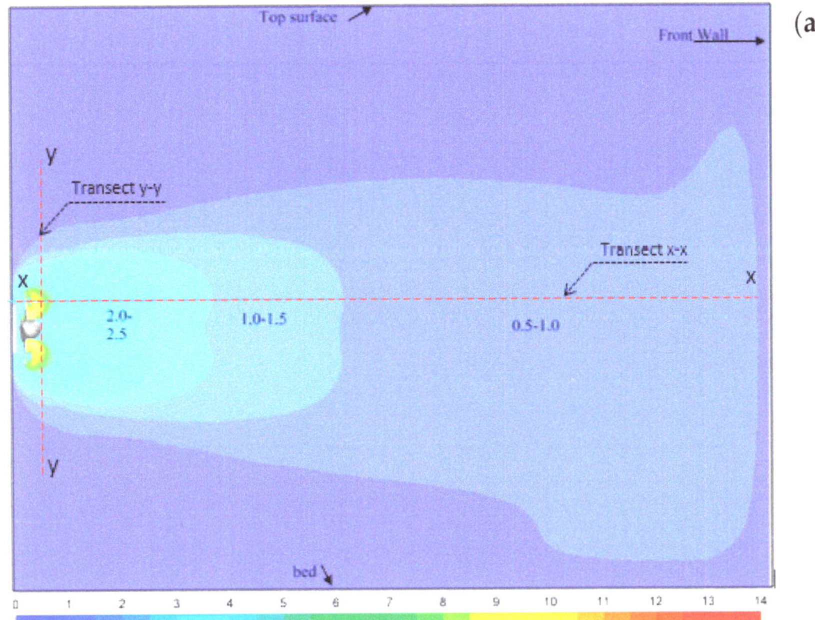

Figure 15. *Cont.*

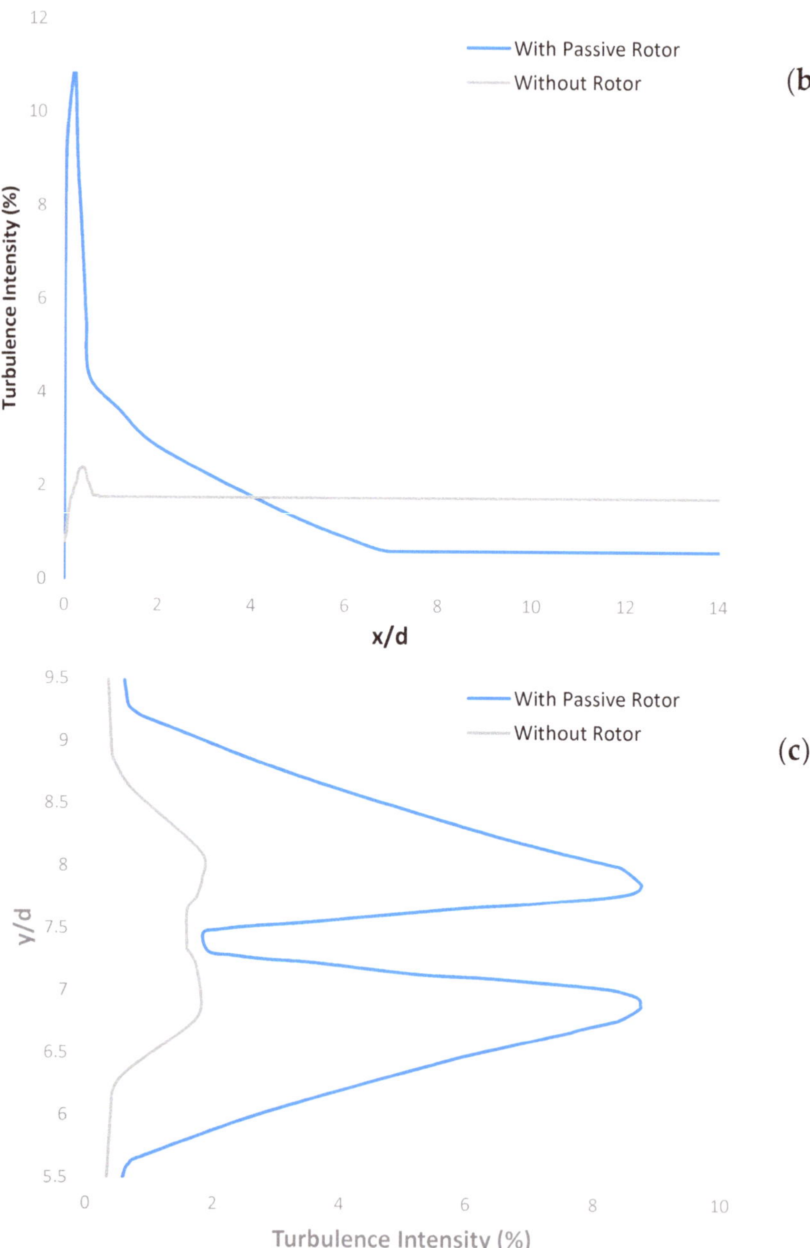

Figure 15. Effect of the passive rotor on the turbulence intensity at v_{av} =30 cm/s and ω = 293.5 rpm. (**a**) Color contour map (vertical plan) of the turbulence intensity downstream of the rotor. (**b**) Spatial variations in turbulent intensity along the x-x transect. (**c**) Spatial variation of TI along the y-y transect.

Figure 15b,c compare the turbulence intensities for the cases of pipe outlets with and without a passive rotor. Figure 15b shows the longitudinal spatial variability throughout the horizontal transect x-x. The vertical position of the transect x-x is selected to pass by the flow zone, where a high turbulence intensity took place. Figure 15c presents the spatial variability in the vertical direction across transect y-y.

It is noticed that the passive rotor resulted in an increase in the turbulence intensity for the near field zone up to a distance of $x \leq 4.0$ d, and the maximum turbulence intensity increased (for the case of "with rotor") to more than five times the corresponding value for the case of "no rotor". However, in the relatively far field zone ($x > 4.0$ d), the turbulence intensity decreased by more than 50 to 60% compared to the "no rotor" case.

The above-mentioned findings mean that the passive rotor has a significant effect on the turbulence in the near field zone of the pipe outlet; however, this effect vanishes after a distance from the pipe outlet equals 4.0 times the diameter of the pipe outlet.

Using the passive rotor creates a spiral motion, which prolongs downstream of the rotor to a long distance. Tt increases the turbulence intensity considerably. Both the increment of the turbulence intensity and the existence of the spiral motion will help mix the outflow with the ambient water.

Therefore, one can conclude that the passive rotors used in pipe outfalls of desalination plants will produce spiral rotating wakes that will significantly increase the intensity of turbulence downstream of the rotor, resulting in improved mixing in the near field of the rotor, which could be beneficial for thermal dissipation and brine dilution applications. It is also important to emphasize that this improvement applies only to the near field area and is not applicable to the far field.

6. Conclusions and Challenges

Passive rotors have been recently introduced and usefully used in a number of water applications. This study proposes a computational modeling workflow based on the ANSYS FLUENT package to simulate the fluid–structure interaction between the emerged water jet from a pipe outlet and a nearby downstream passive rotor. The numerical challenge is to reasonably simulate the temporal variations in the angular speed (ω) of a passive rotor initially at rest and then subjected to time-varying water jet velocity (υ). For the numerical model calibration, a lab experiment for a time-varied water flux that emerged from a pipe outlet of a falling head water tank was set, and a passive rotor was added to the pipe outlet. The temporal variation in the angular speed of the passive rotor was measured using the video analysis tracking method, with the help of the Tracker freeware package.

Two computational techniques were investigated; the first is the six-degrees-of-freedom (6DOF), and the second is the sliding mesh (SM). The 6DOF method was applied first since the rotating speed of the computational grid frame is unknown a priori. The objectives of the 6DOF model were: to define the maximum rotor's angular speed, investigate the variations of ω with time, and deduce a mathematical relation of ω as a function of the water jet velocity (υ). The study has shown that the 6DOF technique is considerably accurate in determining both the maximum and temporal angular speeds, with discrepancies within 3% of the measured values. The 6DOF results indicated a linear relation of the rotor's angular speed (ω) as a function of the average pipe outlet water velocity (υ). The SM technique was used as a second step while adopting the obtained w-υ relation from the 6DOF analysis. It has been found that the SM technique results in temporal variations in the rotor's angular speed compared with the physical model measurements. Additionally, the time consumed using the SM technique is considerably less than that required by the 6DOF. Nevertheless, the main limitation of the SM technique is that it requires the modeler to have a priori knowledge of the relationship between the rotor angular speed and the jet velocity. Such limitation does not exist in the case of the 6DOF technique.

The developed computational workflow was used to investigate the following: (1) the effect of the pipe-outlet-to-the-passive-rotor-gap distance on the angular speed of the passive rotor, and (2) the effect of using the passive rotor on the near flow field zone and the water turbulence. The results indicated that the rotor's maximum angular speed decreases as the gap distance between the pipe outlet and the passive rotor increases, following a decline power-law trend. The numerical simulation also presented that the passive rotor creates a spiral motion of the outlet flow that extends to a relatively long distance (10 times the diameter of the pipe outlet). That spiral motion may help to better mix the incoming water flux inside the water mass. It has also been noticed that using the passive rotor significantly affects the turbulence intensity in the near field zone of the pipe outlet; however, this effect vanishes after the distance from the pipe outlet equals four times the diameter of the pipe outlet.

Using the proposed passive rotors downstream of pipe outlets and outfalls might be challenging in practice. One of the common expected challenges is the sediment and sludges that might be transported with the effluent water, which could break or even remove the rotor from its place. Another challenge could be the expected harm to the fishes swimming nearby the rotor that might take place.

Author Contributions: Conceptualization, M.E. and M.F.; methodology, M.F., K.K. and M.E.; software, M.F.; validation, M.E., K.K. and M.F.; formal analysis, M.F., K.K. and M.E.; investigation, M.F., M.E. and K.K.; resources, M.F., M.E. and K.K.; data curation, M.E., K.K.; writing—original draft preparation, M.F. and M.E.; writing—review and editing, K.K. and M.E.; visualization, M.F., M.E. and K.K.; supervision, M.F.; project administration, M.F.; funding acquisition, M.F. All authors have read and agreed to the published version of the manuscript.

Funding: This research was supported by the Deanship of Scientific Research, Imam Mohammad Ibn Saud Islamic University, IMSIU, Riyadh, Saudi Arabia, Grant No. 20-13-14-003.

Institutional Review Board Statement: Not applicable.

Informed Consent Statement: Not applicable.

Data Availability Statement: Data are available on request from the authors.

Acknowledgments: The authors would like to thank the Deanship of Scientific Research, Imam Mohammad Ibn Saud Islamic University, Saudi Arabia, for supporting and funding this research.

Conflicts of Interest: The authors declare no conflict of interest.

References

1. World Bank. *The Role of Desalination in an Increasingly Water-Scarce World*; World Bank: Washington, DC, USA, 2019.
2. Available online: https://ussaudi.org/water-in-saudi-arabia-desalination-wastewater-and-privatization/ (accessed on 19 June 2022).
3. Lim, J. Consideration of structural constraints in passive rotor blade design for improved performance. *Aeronaut. J.* **2016**, *120*, 1604–1631. [CrossRef]
4. Elgamal, M.; Abdel-Mageed, N.; Helmi, A.; Ghanem, A. Hydraulic performance of sluice gate with unloaded upstream rotor. *Water SA* **2017**, *43*, 563. [CrossRef]
5. Elgamal, M.; Kriaa, K.; Farouk, M. Drainage of a Water Tank with pipe outlet loaded by a passive rotor. *Water* **2021**, *13*, 1872. [CrossRef]
6. ANSYS. *ANSYS FLUENT User's Guide*; ANSYS, Inc.: Canonsburg, PA, USA, 2013.
7. Patil, H.; Patel, A.K.; Pant, H.J.; Vinod, A.V. CFD simulation model for mixing tank using multiple reference frame (MRF) impeller rotation. *ISH J. Hydraul. Eng.* **2018**, *27*, 200–209. [CrossRef]
8. Durkacz, J.; Islam, S.; Chan, R.; Fong, E.; Gillies, H.; Karnik, A.; Mullan, T. CFD modelling and prototype testing of a Vertical Axis Wind Turbines in planetary cluster formation. *Energy Rep.* **2021**, *7*, 119–126. [CrossRef]
9. Lanzafame, R.; Mauro, S.; Messina, M. Wind turbine CFD modeling using a correlation-based transitional model. *Renew. Energy* **2013**, *52*, 31–39. [CrossRef]
10. Carcangiu, C.E. CFD-RANS Study of Horizontal Axis Wind Turbines. Ph.D. Thesis, Dipartimento di Ingegneria Meccanica, DIMeCa, Cagliari, Italy, January 2008.
11. Wang, J.-F.; Piechna, J.; Müller, N. A Novel Design Of Composite Water Turbine Using CFD. *J. Hydrodyn.* **2012**, *24*, 11–16. [CrossRef]
12. McNaughton, J.; Afgan, I.; Apsley, D.D.; Rolfo, S.; Stallard, T.; Stansby, P.K. A Simple Sliding-Mesh Interface Procedure and Its Application to the CFD Simulation of a Tidal-Stream Turbine. *Int. J. Numer. Methods Fluids* **2013**, *74*, 250–269. [CrossRef]
13. Laín, S.; Cortés, P.; López, O.D. Numerical Simulation of the Flow Around a Straight Blade Darrieus Water Turbine. *Energies* **2020**, *13*, 1137. [CrossRef]
14. Mao, X.; Chen, D.; Wang, Y.; Mao, G.; Zheng, Y. Investigation on Optimization of Self-Adaptive Closure Law for Load Rejection to a Reversible Pump Turbine Based on CFD. *J. Clean. Prod.* **2020**, *283*, 124739. [CrossRef]

15. Dick, E.; Vierendeels, J.; Serbruyns, S.; Voorde, J.V. Performance prediction of centrifugal pumps with CFD-tools. *Task Q.* **2001**, *5*, 579–594.
16. Mejia, O.D.L.; Mejia, O.E.; Escorcia, K.M.; Suarez, F.; Laín, S. Comparison of Sliding and Overset Mesh Techniques in the Simulation of a Vertical Axis Turbine for Hydrokinetic Applications. *Processes* **2021**, *9*, 1933. [CrossRef]
17. Bouvant, M.; Betancour, J.; Velásquez, L.; Rubio-Clemente, A.; Chica, E. Design optimization of an Archimedes screw turbine for hydrokinetic applications using the response surface methodology. *Renew. Energy* **2021**, *172*, 941–954. [CrossRef]
18. Warjito; Prakoso, A.P.; Budiarso; Adanta, D. CFD simulation methodology of cross-flow turbine with six degree of freedom feature. *AIP Conf. Proc.* **2020**, *2255*, 020033. [CrossRef]
19. Khanjanpour, M.H.; Javadi, A.A. Experimental and CFD Analysis of Impact of Surface Roughness on Hydrodynamic Performance of a Darrieus Hydro (DH) Turbine. *Energies* **2020**, *13*, 928. [CrossRef]
20. Prakoso, A.P.; Warjito, W.; Siswantara, A.I.; Budiarso, B.; Adanta, D. Comparison Between 6-DOF UDF and Moving Mesh Approaches in CFD Methods for Predicting Cross-Flow PicoHydro Turbine Performance. *CFD Lett.* **2019**, *11*, 86–96.
21. Lopez, O.D.; Meneses, D.P.; Laín, S. Computational Study of the Interaction Between Hydrodynamics and Rigid Body Dynamics of a Darrieus Type H Turbine. In *CFD for Wind and Tidal Offshore Turbines*; Springer: Cham, Switzerland, 2015; pp. 59–68.
22. Chen, F.; Zhu, G.; Jing, L.; Zheng, W.; Pan, R. Effects of diameter and suction pipe opening position on excavation and suction rescue vehicle for gas-liquid two-phase position. *Eng. Appl. Comput. Fluid Mech.* **2020**, *14*, 1128–1155. [CrossRef]
23. Benavides-Morán, A.; Rodríguez-Jaime, L.; Laín, S. Numerical Investigation of the Performance, Hydrodynamics, and Free-Surface Effects in Unsteady Flow of a Horizontal Axis Hydrokinetic Turbine. *Processes* **2021**, *10*, 69. [CrossRef]
24. Hu, J.; Xu, G.; Shi, Y.; Huang, S. The influence of the blade tip shape on brownout by an approach based on computational fluid dynamics. *Eng. Appl. Comput. Fluid Mech.* **2021**, *15*, 692–711. [CrossRef]
25. Li, Y.-B.; Fan, Z.-J.; Guo, D.-S.; Li, X.-B. Dynamic flow behavior and performance of a reactor coolant pump with distorted inflow. *Eng. Appl. Comput. Fluid Mech.* **2020**, *14*, 683–699. [CrossRef]
26. Hsu, C.-H.; Chen, J.-L.; Yuan, S.-C.; Kung, K.-Y. CFD Simulations on the Rotor Dynamics of a Horizontal Axis Wind Turbine Activated from Stationary. *Appl. Mech.* **2021**, *2*, 147–158. [CrossRef]
27. Celik, I.B.; Ghia, U.; Roache, P.J.; Freitas, C.J. Procedure for estimation and reporting of uncertainty due to discretization in CFD applications. *J. Fluids Eng. Trans. ASME* **2008**, *130*. [CrossRef]
28. Balduzzi, F.; Bianchini, A.; Maleci, R.; Ferrara, G.; Ferrari, L. Critical issues in the CFD simulation of Darrieus wind turbines. *Renew. Energy* **2016**, *85*, 419–435. [CrossRef]
29. Maître, T.; Amet, E.; Pellone, C. Modeling of the flow in a Darrieus water turbine: Wall grid refinement analysis and comparison with experiments. *Renew. Energy* **2013**, *51*, 497–512. [CrossRef]
30. Borkowski, D.; Węgiel, M.; Ocłoń, P.; Węgiel, T. CFD model and experimental verification of water turbine integrated with electrical generator. *Energy* **2019**, *185*, 875–883. [CrossRef]
31. Kaniecki, M.; Krzemianowski, Z.; Banaszek, M. Computational fluid dynamics simulations of small capacity Kaplan turbines. *Trans. Inst. Fluid-Flow Mach.* **2011**, 71–84. Available online: https://yadda.icm.edu.pl/baztech/element/bwmeta1.element.baztech-article-BWM8-0039-0004 (accessed on 12 August 2022).
32. Adanta, D.; Nasution, S.B.; Budiarso; Warjito; Siswantara, A.I.; Trahasdani, H. Open flume turbine simulation method using six degrees of freedom feature. *AIP Conf. Proc* **2020**, *2227*, 020017. [CrossRef]
33. Mohamed, A.; Ridha, A.; Boualem, C.; Tayeb, K. Two-Dimensional CFD Simulation Coupled with 6DOF Solver for analyzing Operating Process of Free Piston Stirling Engine. *J. Adv. Res. Fluid Mech. Therm. Sci.* **2019**, *55*, 29–38.
34. Prakash, S.; Nath, D.R. A computational method for determination of open water performance of a marine propeller. *Int. J. Comput. Appl.* **2012**, *58*, 0975–8887.
35. Wu, J.; Shimmei, K.; Tani, K.; Niikura, K.; Sato, J. CFD-based design optimization for hydro turbines. *J. Fluids Eng.* **2007**, *129*, 159–168. [CrossRef]
36. Khunthongjan, P.; Janyalertadun, A. A study of diffuser angle effect on ducted water current turbine performance using CFD. *Songklanakarin J. Sci. Technol.* **2012**, *34*, 61–67.
37. Liu, Y. Development and computational validation of an improved analytic performance model of the hydroelectric paddle wheel. *Distrib. Gener. Altern. Energy J.* **2015**, *30*, 58–80.
38. Kutty, H.A.; Rajendran, P. 3D CFD Simulation and Experimental Validation of Small APC Slow Flyer Propeller Blade. *Aerospace* **2017**, *4*, 10. [CrossRef]
39. Krasilnikov, V.; Sun, J.; Halse, K.H. CFD investigation in scale effect on propellers with different magnitude of skew in turbulent flow. In Proceedings of the First International Symposium on Marine Propulsors, Trondheim, Norway, 22–24 June 2009; pp. 25–40.
40. Thakur, N.; Biswas, A.; Kumar, Y.; Basumatary, M. CFD analysis of performance improvement of the Savonius water turbine by using an impinging jet duct design. *Chin. J. Chem. Eng.* **2018**, *27*, 794–801. [CrossRef]
41. Tog, R.A.; Tousi, A.; Tourani, A. Comparison of turbulence methods in CFD analysis of compressible flows in radial turbomachines. *Aircr. Eng. Aerosp. Technol.* **2008**, *80*, 657–665. [CrossRef]
42. Cao, H.; He, D.; Xi, S.; Chen, X. Vibration signal correction of unbalanced rotor due to angular speed fluctuation. *Mech. Syst. Signal Process.* **2018**, *107*, 202–220. [CrossRef]

43. Sang, L.Q.; Murata, J.; Morimoto, M.; Kamada, Y.; Maeda, T.; Li, Q. Experimental investigation of load fluctuation on horizontal axis wind turbine for extreme wind direction change. *J. Fluid Sci. Technol.* **2017**, *12*, JFST0005. [CrossRef]
44. Basse, N.T. Turbulence Intensity Scaling: A Fugue. *Fluids* **2019**, *4*, 180. [CrossRef]

water

Article

Application of Three-Dimensional CFD Model to Determination of the Capacity of Existing Tyrolean Intake

Aslı Bor [1,2,*], Marcell Szabo-Meszaros [3,4], Kaspar Vereide [1,5] and Leif Lia [1]

[1] Department of Civil and Environmental Engineering, Norwegian University of Science and Technology (NTNU), S.P. Andersens veg 5, 7491 Trondheim, Norway; kaspar.vereide@ntnu.no (K.V.); leif.lia@ntnu.no (L.L.)
[2] Department of Civil Engineering, Izmir University of Economics, Izmir 35330, Turkey
[3] SINTEF Energy Research, Sem Sælandsvei 11, 7048 Trondheim, Norway; szabo-meszaros.marcell@emk.bme.hu
[4] Department of Hydraulic and Water Resources Engineering, Faculty of Civil Engineering, Budapest University of Technology and Economics, Muegyetem Rakpart 3, H-1111 Budapest, Hungary
[5] Sira-Kvina Kraftselskap, Stronda 12, 4440 Tonstad, Norway
*   Correspondence: asli.b.turkben@ntnu.no

Abstract: CFD models of intakes in high-head hydropower systems are rare due to the lack of geometric data and cost of modeling. This study tests two different types of software to see how modeling can be performed in a cost-effective way with scarce input data and still have sufficient accuracy. The volume of fluid (VoF) model simulations are conducted using both ANSYS Fluent and OpenFOAM. The geometry is modelled from Google Earth satellite images, drone scanning data, and design drawings from the construction period and supported by field observations for extra quality control. From the model, both capacity parameters and flow pattern are calculated. For capacity, the C_d factor is calculated and compared with the literature. The simulations are conducted for a Tyrolean weir with rectangular bars (flat steel) in the rack. Simulated flow patterns through the rack with ANSYS Fluent and OpenFOAM are compared. OpenFOAM simulations yielded 15% to 20% higher water levels compared to the VOF model applied in Ansys Fluent. Also, when the flow rate was high, the water capture capacity calculated with ANSYS Fluent was 10% higher than that obtained with OpenFOAM. However, considering the total simulation times, modeling with OpenFOAM offered approximately 11% faster results.

Keywords: intake; ANSYS Fluent; OpenFOAM; rectangular bars; water capture capacity; flow profile; discharge coefficient

Citation: Bor, A.; Szabo-Meszaros, M.; Vereide, K.; Lia, L. Application of Three-Dimensional CFD Model to Determination of the Capacity of Existing Tyrolean Intake. *Water* **2024**, *16*, 737. https://doi.org/10.3390/w16050737

Academic Editor: Charles R. Ortloff

Received: 12 January 2024
Revised: 17 February 2024
Accepted: 21 February 2024
Published: 29 February 2024

1. Introduction

Secondary intakes, also known as brook intakes, are constructed in addition to the main intake to take in additional rivers or brooks to increase the inflow to hydropower plants. Hydropower schemes in alpine regions are more complex than in large rivers and have sophisticated systems, including long tunnels and many brook intakes. Such tunnels with several intakes along their length act as roof-gutters over large mountain areas. The intakes, which only catch the discharge and not the hydraulic pressure, are called brook intakes. The most frequent intake type is the Tyrolean weir, and all of these are designed as self-cleaning and fixed structures. Secondary intakes can be a highly efficient method of increasing the inflow, especially in alpine regions with steep hills and many tributaries. They are also efficient for those hydropower plants with long headrace tunnels that may cross below several river systems, and those located remotely without any grid connection or other communication. More thorough inspections during excessive flooding, which has become more frequent with the changing climate, may indicate a substantial loss of water for power production. Looking beyond theoretical insufficient capacity, it is timely to reassess the capacity of secondary intakes, and retrofit them if necessary.

The efficiency of the intake structure depends on several factors, such as the shape of the bars, net spacing between them (void ratio), amount of flow and flow conditions, initial flow depth, and angle and length of the rack. Many researchers have conducted studies to design the minimum rack area required to transmit the maximum flow rate (Garot [1]; Bouvard [2]; Kuntzmann and Bouvard [3]; Noseda [4]; Noseda [5]; Noseda [6]; Mostkow [7]; Brunella et al. [8]). There are some assumptions in these studies; the flow on the rack is one-dimensional, the flow gradually decreases, the hydrostatic pressure distribution acts on the rack in the flow direction, and the energy level or energy head is constant along the rack. There are two approaches to the constant energy head; the energy level is either parallel to the river surface or to the slope of the rack. The orifice effect is one of the most important mechanisms for managing water withdrawal in the intakes, and the proportion of directed flow, q_{intake}, can be expressed by Equation (1):

$$q_{intake} = \frac{dq}{dx} = Cm\sqrt{2gH},\tag{1}$$

where q_{intake} is the diverted discharge through the bottom rack per unit of length x, C is the discharge coefficient, m is the void ratio (the ratio of the opening area of the screen), and H is the hydraulic head. Many researchers' experimental work has further developed equations for the discharge coefficient (Garot [1]; Noseda [6]; De Marchi [9]; Frank and Von Obering [10]; Dagan [11]; García [?]), which plays an important role in intake design. Researchers have shown that the value of the discharge coefficient depends on certain parameters; in particular, the shape, geometry, and spacing between bars are significant dimensions of the discharge coefficient (Righetti et al. [13]).

Another parameter that plays an important role in intake design is the minimum required wetted rack length. There are studies in the literature that consider the theoretical application of 1D energy and momentum conservation equations for wetted rack length calculation (Kuntzmann and Bouvard [3]; Noseda [5]; Frank and Von Obering [10]). More recently, based on the aforementioned studies, researchers performed laboratory experiments for wetted rack length and produced empirical correlations (Brunella et al. [8]; Castillo et al. [14]). In these studies, the researchers carried out experiments with differently shaped bars and defined the discharge coefficient. Table 1 summarizes the equations in the literature and those used in this study.

Table 1. Intake unit discharge equations taken for the study from the literature.

Reference	Orifice Equation	Wetted Rack Length $L(m)$	Discharge Coefficient C
Garot [1]	$dq/dx = C_d m\sqrt{2gQ}$		$C_d = \sqrt{\dfrac{1}{(\frac{b_1}{b_1+b_w})^2\beta(\frac{b_w}{b_1})^{\frac{4}{3}}}}$
Noseda [6]	$dq/dx = C_d m\sqrt{2gh}$	$L = \frac{1.848h}{C_d m}$	$For\,subcritical\,flow:$ $C_d(h) = 0.66\,\mathrm{m}^{-0.16}(\frac{h}{l})^{-0.13}$ $For\,critical\,flow:$ $C_d(h) = 0.78(\frac{h}{l})^{-0.13}$
Frank et al. [10]		$L = \frac{2.561q}{C_d\sqrt{h}}$	$C_d(h) = 1.22C_{qh}(h_0)$
Drobir [15]		$L = \frac{0.846h}{C_d m}$	
Brunella et al. [8]	$dq/dx = C_{d0} m\sqrt{2gh\cos\theta}$	$L = \frac{0.83h}{C_d m}$	
Bekkeinntak Report [16]		$L = \frac{Q}{C_d h^{3/2}}$	$C_d = 1.7 \sim 2.1$

Laboratory experiments or adequate computational fluid dynamics (CFD) simulations for intakes are rare because of the complex geometry and high cost of accurate numerical investigations. Studies in the literature consider a two-dimensional perspective on the vertical plane of the rack. However, in reality, the flow is three-dimensional, and in this case, CFD models, once validated against experimental values, can help to better understand the phenomenon (Bombardelli [17]; Blocken and Gualtieri [18]). While most researchers modelled the intake structure in one or two dimensions, only a few researchers examined the study in three dimensions (Bombardelli [17]; Castillo et al. [19]; Carrillo et al. [20]). However, there are very few studies in the literature modeling the water intakes. In addition, different simplifications were used in these studies. No reported study has modelled the brook intake together with the natural riverbed topography.

This study aimed to determine the behavior of the three-dimensional flow over the existing water intake structure under various flow rates using the CFD-VOF method. A further aim was to investigate measures to improve and optimize existing Tyrolean intake. Two different CFD finite volume codes were used for the simulations, ANSYS Fluent and OpenFOAM, and their performance was compared in terms of time and cost, as well as the effort spent. Previous studies numerically simulated with ANSYS CFX were conducted with circular bars or T-shaped bars (García [?]; Castillo et al. [19]; Carrillo et al. [20]). In our current contribution, we focused our attention on a rack consisting of rectangular bars. The physical behavior of the flow profile on the wetted rack was determined and the rack length values were compared with those in previous studies.

2. Project-Related Background

A case study has been selected, namely the Stigansåni brook intake, which is constructed as a Tyrolean weir, located in Southern Norway with coordinates Euref89 UTM33 6528477N 23422E. The elevation of the inflow weir is at 523.5 m.a.s.l. and the winter season generally lasts from December to March. This brook intake diverts water into the headrace tunnel of the 960 MW Tonstad HPP and is owned and operated by Sira-Kvina kraftselskap. Figure 1 includes a site map that shows the location of Tonstad HPP. This brook intake is found to have insufficient capacity and periods of water spilling every year. This intake type was selected as being among the most common intake designs in Norway. The Stigansåni brook intake diverts water from a 5 km^2 catchment with an annual inflow of 11 million m^3 with mean discharge $Q_m = 5$ m^3/s. The water is directed into the headrace tunnel of the 960 MW Tonstad power plant, which is also diverted towards the Åna-Sira power plant further downstream. The annual production of the diverted water equals about 12 GWh. As can be seen in Figure 1, this intake experiences both extremes of dry and wet conditions. The hydrograph of the catchment area is characterized by flash floods from extreme participation or sudden snow-melting events. Figure 1b shows a situation with water loss, where the intake is saturated, and water spills over both the intake rack and also over the concrete weir. This situation occurred during the winter, as heavy rainfall coincided with a temperature rise and the resulting snow melt. Such events are typical in this region and occur several times annually at this intake. The power plant owner is interested in investigating potential measures to increase the capacity of this intake.

Figure 1. Downstream view of Stigansåni brook intake during (**a**) dry conditions and (**b**) wet conditions.

3. Numerical Modeling

The subject of this research is the simulation of an intake with racks set in a natural riverbed topography under multiphase flow conditions, which represents a complex problem in fluid mechanics. A key challenge is modeling the relatively large river flows with larger mesh sizes, with the added intricacy of modeling the intake with racks, which feature relatively small and particularly narrow rack spacings within the riverbed. In order to obtain a highly accurate simulation, it is important to select the appropriate number of cells in the fine network created in the rack region of the water intake structure.

Simulations were performed using the finite-volume method (FVM) and commercially licensed CFD solver ANSYS Fluent with open-source platform OpenFOAM program used to compare the reliability and performance of the analyses. Both programs are widely used in research institutions. In this study, the current state of the existing Tyrolean intake was analyzed with the minimum amount of data from the study site necessary to perform 3D hydrodynamic modeling in several flow conditions. For this purpose, easy-to-access remote sensing photos were used as modeling input. First, a high-resolution Digital Elevation Model (DEM) from Google Earth satellite images was used to create the upstream and downstream river terrain of the Stigansåni intake area. For detailed topography of the inside of the riverbed, these images were supported with 3D scanning drone images. The DEM file extracted from the satellite images was transferred to the Autocad CIVIL 3D environment, and used to create a 3D triangular model of the riverbed as an .stl file. Secondly, the .stl file was further used in the ANSYS Space Claim software and also in OpenFoam. In ANSYS Space Claim, when created from remote sensing photos, this type of file is generally of good quality, but due to the nature of the topography, there may be frequent spikes or missing faces and holes in the terrain. The knit was checked with ANSYS Space Claim and the .stl file was repaired in the process of creating the geometry. The spikes in the field were softened by 40% shrink-wrap and the holes were closed. This approach also facilitates the meshing step, performed in the advanced stages. Thirdly, the existing structural project of the Stigansåni brook intake was redrawn in 3D in the CAD environment and the .stl file of the structure was saved as a separate file. Finally, the structure was added to the river topography prepared in the ANSYS Space Claim and the 3D full-scale geometry of the area to be modelled was created. A "fluid domain" was defined to represent the riverbed, and the intake body was added. The meshing process simply involved the combination of the two .stl files from the terrain file and from the intake structure in the OpenFoam environment.

In order to model river flows in 3D, the three-dimensional Reynolds-averaged Navier–Stokes equations involving conservation of mass and momentum need to be solved, assuming the flow is incompressible. Continuity and momentum equations for x, y, and z are given in Equations (2) and (3):

$$\nabla.u = 0, \tag{2}$$

$$\rho\left(\frac{\partial u}{\partial t} + u.\nabla u\right) = -\nabla P + \mu\nabla^2 u - \rho\nabla.\left(\overline{u'u'}\right) + F, \tag{3}$$

where u is flow velocity, ∇ is divergence, ρ is density of water, P is pressure, μ is dynamic viscosity, and F is gravity force.

In this study, the Eulerian multiphase approach was used to solve the air and water two-phase flow. Air was defined as the primary phase, and water as the secondary phase, representing the open channel flow. The material properties of both phases were introduced separately, and their densities were taken as 1.225 kg/m^3 and 998.2 kg/m^3, respectively. The surface tension modeling considers the surface tension force as a volume force concentrated at the interface. A surface tension coefficient of 0.072 N/m was specified. No wall adhesion was considered. In the ANSYS Fluent model, the viscous flow model was activated with the $k - \Omega$-based Shear-Stress Transport (SST) model (Menter [21]), which solved the Reynolds-averaged Navier–Stokes (RANS) equations. On the other hand, with OpenFOAM, the

$k - \varepsilon$ turbulence model was used and the interFoam solver was chosen to characterize the free surface flooding at the Stigansåni inlet. Both phases in the Eulerian flow model were considered as continuous fluids. The sum of the volume fractions $(r\alpha)$ of the air and water phases was 1 in each control volume. Cells with 0.5 air volume fraction were considered as the free water surface. The time-dependent volume fraction formulation was used, and the volume fraction was obtained with an explicit formulation, such that the Courant number was 0.25 in each time step. Second-order discretization schemes were used to solve the divergence and gradient. Table 2 shows a comparison of model applications and simulation times used for the two solvers.

Table 2. Comparison of ANSYS Fluent and OpenFOAM simulation times and model implementations.

	ANSYS Fluent	**OpenFOAM**
Domain Size:	10 m × 10 m	10 m ×15 m
Elements:	2.8×10^6	$1\sim3 \times 10^6$
Turbulence Model:	RANS $k - \Omega$ (SST) model	RANS $k - \varepsilon$ model
Wall function:	Standard wall function [22]	Standard wall function [22]
Solution Methods Gradient: Divergence: Turbulence Kinetic energy: Time derivative:	Second-order least squares Second-order upwind Second-order upwind First-order implicit	Second-order least squares Second-order upwind Second-order upwind First-order implicit
Multiphase model:	VoF	VoF
Interface capturing method:	SIMPLEC	SIMPLEC
Pressure–velocity coupling:	PISO	PISO
Simulation Time:	300 s	100 + 20 s
Parallel Computing:	32 Core	32~48 Core
Computation Time:	72 h	24 h

Simulations were performed with four specific flow rates for Stigansåni intake that were selected based on reports from NVE (the Norwegian Water Resources and Energy Directorate). All flow rates were operated under steady flow conditions of 300 s for ANSYS Fluent and 100 + 120 s for OpenFOAM simulations (for initial flow development with low-resolution domain and for high-resolution simulations, respectively). The river velocity inlet boundary condition values for each flow rate are given in Table 3 below.

Table 3. Inlet boundary conditions.

Test No	Q **(m³/s)**	v **(m/s)**
Q1	3.7	0.18
Q2	5	0.25
Q3	5.5	0.27
Q4	7	0.35

Q is the flow rate; v is the river inlet velocity.

3.1. Domain Model with ANSYS Fluent

After connecting the water intake structure in the CAD and the river topography prepared in the ANSYS Space Claim, a 10 m × 10 m "fluid area" was defined to represent the riverbed. These consist of two parts, the flow area and the water intake structure, as seen in Figure 2.

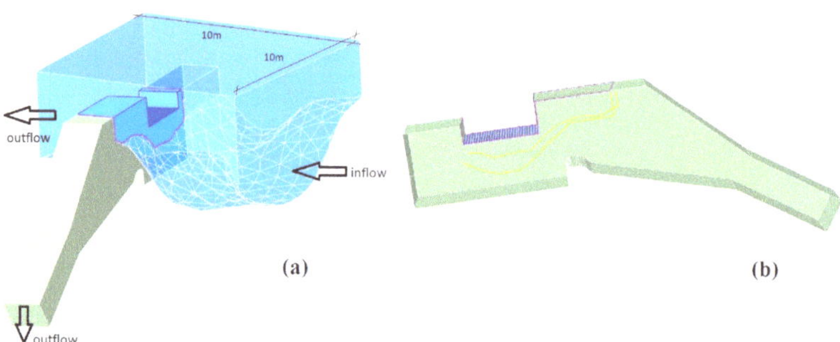

Figure 2. (**a**) Fluid Domain. (**b**) Intake structure.

The meshing of the domain was performed via the 'sweep' method for the bottom rack to the brook intake, and the brook intake to riverbed, and the "tetrahedral" meshes for the fluid domain to obtain mesh cells due to the complex geometry (ANSYS [23]). The generated mesh of the fluid domain consisted of approximately 2.8×10^6 cells, 13×10^6 faces, and 8.1×10^6 nodes, which means each cell size is sufficiently small relative to the approximate rack grids assigned in simulations. In the simulations, the minimum curvature mesh element size was taken as 1 mm, which is smaller than the rack grid in the intake, and this cell size was applied to the denser mesh at the level of the bottom rack. The denser mesh was applied with smaller cell sizes in areas where the intake was. The bottom rack and intake into the riverbed were densely dispersed through the domain, and the cell sizes were increased towards the edges, to avoid unnecessary mesh density. Regarding the mesh quality of the prepared geometry, the maximum aspect ratio, average skewness, and averaged orthogonal quality criteria were calculated as 24, 0.025, and 0.71, respectively. According to the obtained values, the model is considered to have good mesh quality (Figure 3).

Figure 3. The computational mesh used in the full-scale modeling using ANSYS Fluent.

As for the boundary conditions, the upper river-free surface of the meshed fluid area was simplified as the no-slip wall conditions. The outlet of the river fluid domain and the outlet of the intake structure were defined as the "pressure outlet" boundary condition, with a relative pressure of 0 Pa. This is based on previous flow observations, as the river inlet boundary condition (NVE) modeling was conducted with four different flow rates at the entrance to the velocity inlet. In addition, the water volume fraction was taken as 1 and the air volume fraction as 0 in the inlet conditions, and this was simulated. The grid spaces of the intake structure were defined by the "interior" boundary condition between the river fluid domain and the intake fluid domain. Grids, as well as all other outer surfaces (riverbed, riverside surfaces, bottom, up and side surfaces of intake structure), were defined as the no-slip wall conditions (Figure 4).

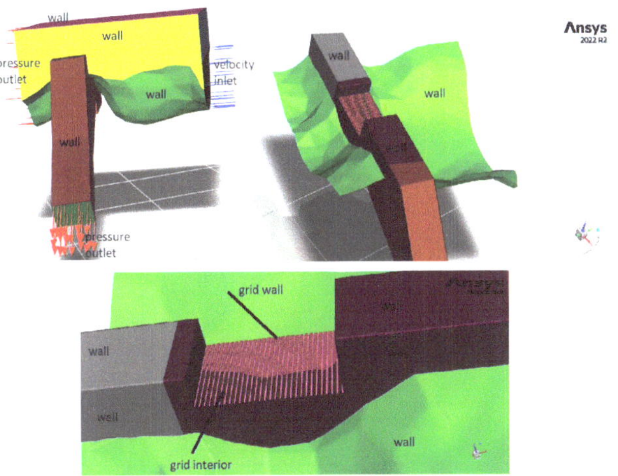

Figure 4. Boundary conditions used in the full-scale modeling using ANSYS Fluent.

The simulations were performed in double precision, and as transient flow using the Eulerian multiphase flow model. The solution method of pressure velocity coupling was chosen, combined with the SIMPLEC Scheme. Spatial discretization methods for solving the equation were applied as a second-order upwind scheme to increase the accuracy of the solution. Another crucial parameter influencing the accuracy and stability of numerical solutions is the length of the time step (dt), calculated using the LAX Finite Difference Scheme. The initial water level in the riverbed was defined within the domain's fluid regions using the "patch" method in Fluent. Thus, at the start of the simulation, the water was positioned at the beginning of the water intake structure. Consequently, electing the time step for different Courant numbers involved a grid spacing of ($dx = 0.001$ m), representing the minimum mesh size at the water intake structure. Test simulations were conducted with reasonable time steps and a Courant number in the range ($C \in [0.2, 1.0]$) and a maximum speed of (0.35 m/s), with time steps ($dt \in [0.001, 0.01]$). As a result of these tests, the time interval with Courant number is 0.25, which requires approximately 10 iterations for convergence at each time step, and this was selected as ($dt = 1.25 \times 10^{-3}$ s). This selection aligns with values reported in the literature for numerical modeling of full-scale hydraulic structures (Torres et al. [24]). Given the small time step and the large number of meshes, it was necessary to use a workstation capable of parallel computing for Computational Fluid Dynamics (CFD) simulations. The model was simulated over a total of 300 s, and it took approximately 72 h to model one scenario, running on 32 cores on an Nvidia RTX A5000 workstation with 1024 GB of memory.

3.2. Domain Model with OpenFOAM

The simulation domain was defined differently as for modeling with ANSYS Fluent. The upstream end of the OpenFOAM domain was set approximately 15 m upstream of the intake structure, but only a few meters downstream (in both directions: from the intake tunnel towards to the HPP, and from the downstream part of the watercourse) (Figure 5).

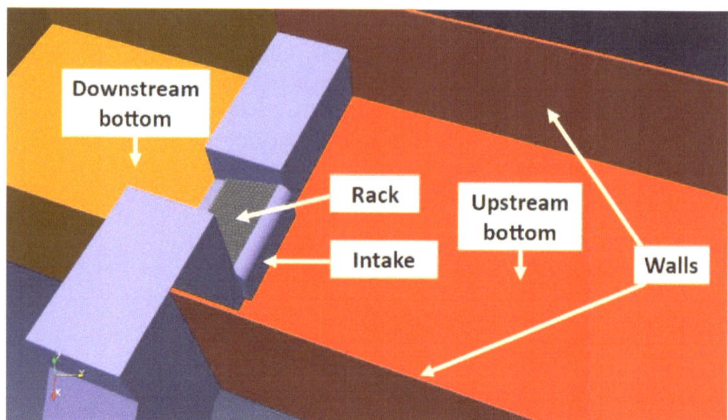

Figure 5. Defined computational and modelled structures for OpenFOAM simulations. Flow direction is from right to left.

A high-resolution topological survey was available at the area around Stigansåni. At the original bathymetry, various polyhedra cell shapes would be generated during the meshing process that increase the size of the final computational grid (in terms of cell numbers), and the computation effort (in terms of resources needed to finish a simulation) necessary to achieve the converged hydraulic state. However, instead, a simplified topology was used. The flow condition over the intake rack structure is estimated to be free-flowing towards the downstream direction without any impact propagated towards upstream (i.e., backed hydraulic jump is not expected to form immediately above or downstream of the intake structure, neither is any blockage found at the downstream vicinity of the intake). At the upstream side, there is a need to test how the natural topology defines the approaching flow towards the intake compared to simplified surroundings. Thus, a comparison analysis was performed between two cases under the same conditions, but with different topology used upstream. The isometric views of the two topologies are presented in Figure 6.

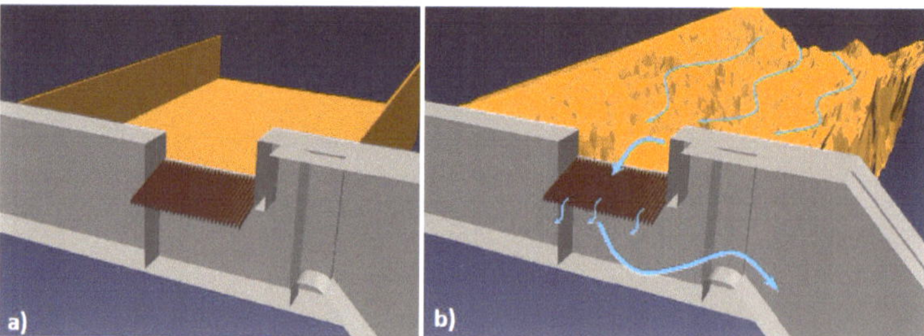

Figure 6. (**a**) Modelling domain with simplified surroundings of Stigansåni river intake (**b**) Modelling domain with measured topology. Intake structure is presented at the front with the weir inside.

The computational grid was generated with the snappyHexMesh utility of Open-FOAM. It generates hexahedra-dominant mesh on a user-defined domain. Finer mesh resolution was applied at the intake opening and near the inside weir (Figure 7). Depending on the different setups, the mesh size varied between 1×106 and 3×106 elements.

The modeling strategy with OpenFOAM was set on two levels. On the first one, the domain was meshed with moderate resolution. In addition, there was no rack structure defined at the intake opening in this case. Such simplifications spared significant computational effort in initializing the hydraulic conditions at the simulated domain. With a "cold-start" the simulation was set to fill up the upstream reach of the domain with a predefined flowrate (e.g., $Q = 4.0 \text{ m}^3/\text{s}$) and let it overflow the intake opening. Flowrates were constantly monitored at three patches: inlet, outlet towards the HPP, outlet towards downstream. The first level of simulation was continued until flowrates through the outlets become constant over time. Simulated time varied between 80 and 100 s at the first level for the different simulated flowrates. Once hydraulic conditions were developed, it was used as an initial condition on the second level of modeling.

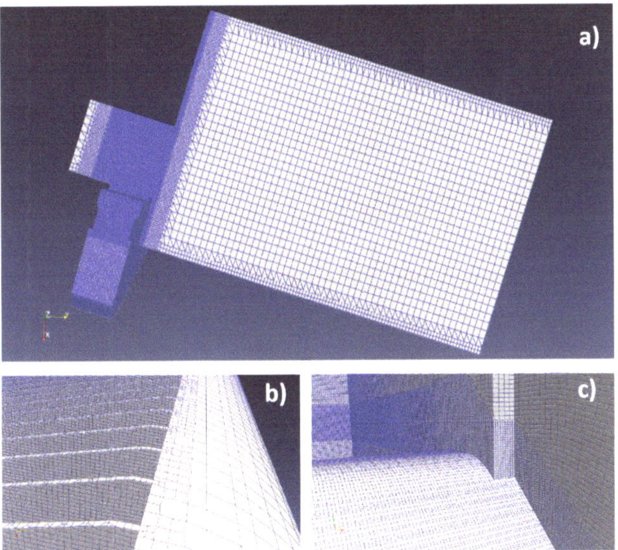

Figure 7. Generated grid with fine resolution that are applied in OpenFOAM. (**a**) Grid resolution at the domain from the top. (**b**) Grid resolution at the upstream end of the rack bars. (**c**) Grid resolution inside the intake with the weir.

On this level, the domain that included the intake rack bars was meshed with a finer resolution. The simulation was run further on the finer setup following the same strategy with monitoring the flowrates through the outlets. The solution was monitored, showing that the intake overflow fluctuates over time. Upon reaching a hydraulic developed state, the simulation was run long enough to determine the statistical average of flowrates through the outlets. That time was set to 20 s and with that it takes 100~120 s in total for simulation of a single case on the two levels. The length of the time step, (dt), calculated using the LAX Finite Difference Scheme, was selected as $dt = 2.0 \times 10^{-3}$ s.

4. Results and Discussion

In Figures 8 and 9, the longitudinal free surface depths inside the intake and results can be observed for four specific flows (Q1, Q2, Q3, and Q4), depending on the water volume fraction at 300 s for ANSYS Fluent and 100~200 s for OpenFOAM simulations. The 0.5 level of the water volume fraction shows the water levels.

The results obtained from OpenFOAM reveal that the water levels both inside of the intake and over the rack are approximately 15% to 20% higher for each inlet boundary condition. In Figure 9, the water levels inside the brook intake for four different flow rates are compared, and it is seen that the values are in harmony.

4.1. Wetted Rack Length and Flow Profile over the Rack

Figure 10 provides a comparison of the water levels over the rack for four different flow rates. It is seen that OpenFOAM simulations gave approximately 15% to 20% higher results.

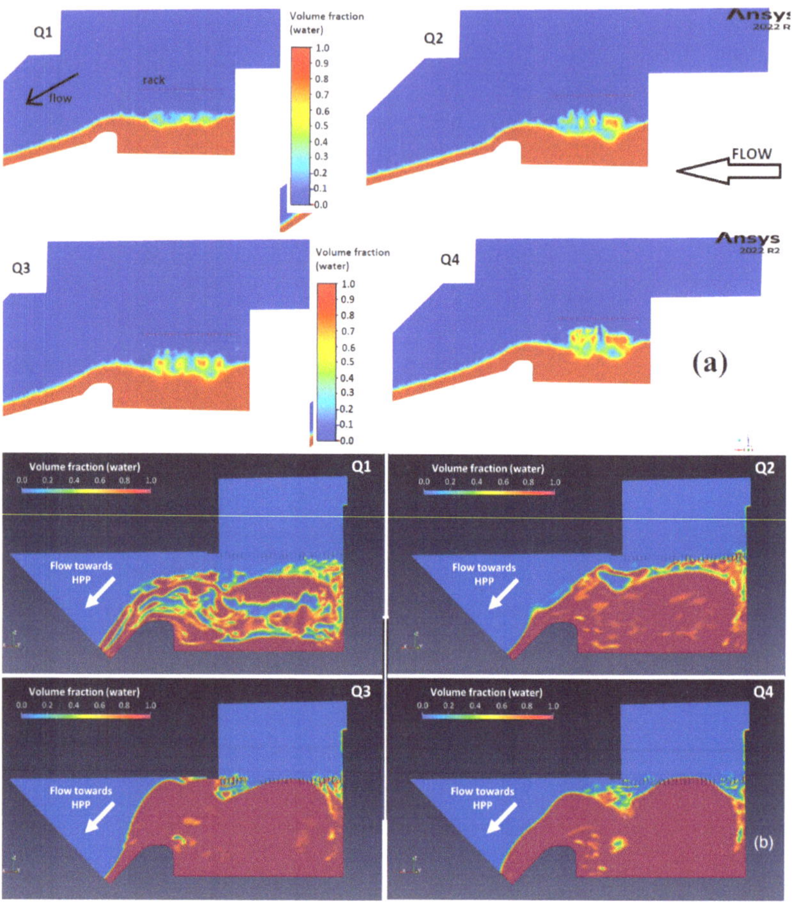

Figure 8. Variation of water volume fraction inside of the brook intake at (**a**) 300 s for ANSYS Fluent. (**b**) 100~200 s for OpenFOAM simulations. (water level is $(r\alpha) = 0.5$) (The figure from ANSYS Fluent is taken the beginning side of the inclined rack, while the figure from OpenFOAM is taken the end side of the inclined rack).

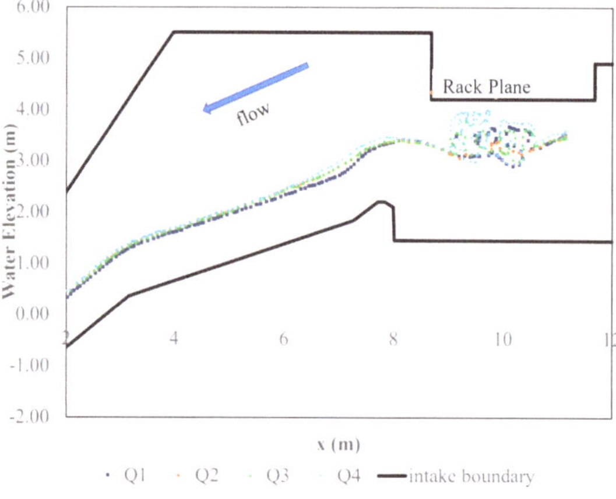

Figure 9. Flow profiles inside of the brook intake for different flow rates at 300 s with ANSYS Fluent.

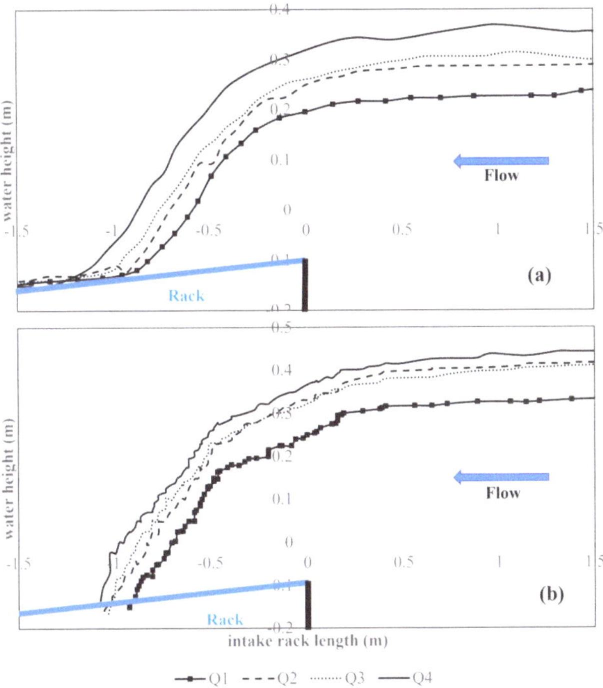

Figure 10. Flow profiles over the rack at 300 s with (**a**) ANSYS Fluent. (**b**) OpenFOAM.

To compare the flow profiles, the water depths were dimensioned using the critical depth h_c, the void ratio m, and the discharge coefficient C_d (Brunella et al. [8]). Figure 11 shows the dimensionless flow profiles over the rack.

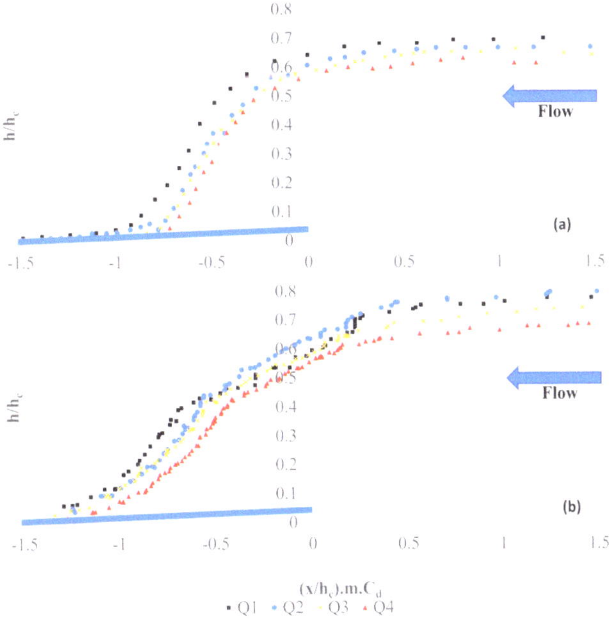

Figure 11. Dimensionless flow profiles over the rack (**a**) with ANSYS Fluent. (**b**) with OpenFOAM.

Considering the dimensionless profiles, if divided into two regions (i.e., pre-rack and post-rack) as $x/h_c mC_d > -0.75$ and $x/h_c mC_d < -0.75$, in general, similar behavior is seen for all flow rates. Although the two regions tend to show differences, their behavior can be explained by the following polynomial curve for $R^2 > 0.97$, which is similar to that described by Carrillo et al. [20]:

$$\frac{h}{h_c} = a\left(\frac{x}{x_c}mC_d\right)^4 + b\left(\frac{x}{x_c}mC_d\right)^3 + c\left(\frac{x}{x_c}mC_d\right)^2 + d\left(\frac{x}{x_c}mC_d\right) + e, \tag{4}$$

Unlike [20], model results with OpenFOAM showed a logarithmic behavior over bar spacing:

$$\frac{h}{h_c} = aln\left(\frac{x}{x_c}mC_d\right) + e, \tag{5}$$

Since the pre-rack behaviors of ANSYS Fluent and OpenFOAM results were similar, coefficients with $R^2 > 0.91$ curve fit were calculated by averaging values. On the other hand, although their behavior over bar and over bar spacing shows the same mathematical expression, the coefficients are calculated separately because R^2 decreases. Curve fits are $R^2 > 0.99$ in this state. The coefficients of the fourth-degree polynomial expression obtained for the regions as $x/h_c mC_d > -0.75$ and $x/h_c mC_d < -0.75$ and over bar spacing are presented in Table 4.

Table 4. Coefficients of the mathematical expressions for Equation (4).

$(x/h_c)mC_d$		a	b	c	d	e
>−0.75		0.153	0.97	2.06	1.91	0.70
<−0.75 Over Bar		−0.071	0.30	0.42	0.24	0.61
ANSYS Fluent	<−0.75 Over Bar spacing	−0.02				0.11
OpenFOAM	<−0.75 Over Bar spacing	0.17				0.31

In order to create a design which optimizes the diversion of water from the river, the minimum wetted rack length should be accurately estimated (García et al. [?]). The minimum wetted rack length was defined by Drobir [15] as the distance from the beginning of the rack to the section where the nappe enters directly through the racks (measured between the bars). Figure 12 shows how the discharge coefficient C_d is estimated for the brook intakes in Norway, according to the Bakkeinntak Report [16]. Table 5 also shows how the minimum wetted rack length is calculated according to the Bakkeinntak Report [16]. When the discharge coefficient C_d is chosen as 2.0 on the safe side, the minimum wetted rack length is calculated as 2.2 m, though it is 2.0 m in the current project. Table 5 shows the comparison of the L_1 values obtained from the numerical simulations and the values calculated from the three different equations available in the literature. It can be seen that Frank et al.'s [10] equation gives overestimated results compared to those from other methods. Very similar results were obtained from the methods of ANSYS Fluent, Brunella et al. [8], and Drobir [15]. It can be seen that the relationship between flow rate and wetted rack length exhibits a logarithmic behavior (Figure 13).

4.2. Discharge Coefficient and Water Capture Capacity

Although, theoretically, the discharge coefficient varies along the rack, formulas presented in the literature generally give average values for each rack. Table 6 shows the discharge coefficient values obtained by Frank and Von Obering [10], Noseda [6], and Garot [1]. Garot's [1] formula is not flow parameter-dependent, and therefore, not affected by flow rate changes, but for the wetted rack length in the Noseda [6] formula, ANSYS Fluent values were used (all formulas in Ref. Table 1).

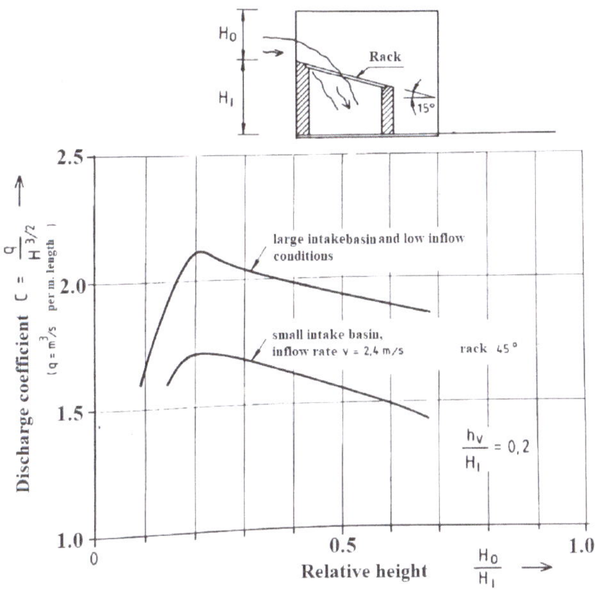

Figure 12. Variation in overflow coefficient for Tyrolean weir.

Table 5. Computed wetted rack lengths with various formulas from the literature.

Q (m³/s)	$L_{1-ANSYS}$ (m)	$L_{1-OpenFOAM}$ (m)	$L_{1-[10]}$ (m)	$L_{1-[8]}$ (m)	$L_{1-[15]}$ (m)	$L_{1-[16]}$ (m)
3.7	0.68	1.07	3.70	0.63	0.64	2.21
5	0.86	1.07	4.64	0.80	0.81	2.21
5.5	0.95	1.38	4.99	0.86	0.87	2.21
7	1.03	1.54	5.98	1.03	1.04	2.21

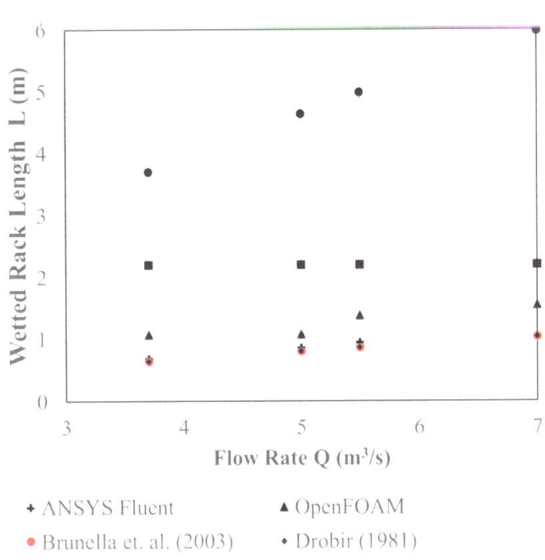

Figure 13. Change of Wetted Rack Length under different flow rates ([8,10,15,16]).

Table 6. Computed discharge coefficients with various formulas from the literature.

Q (m³/s)	$C_{d-ANSYS}$	$C_{d-OpenFOAM}$	$C_{d-[1]}$	$C_{d-[6]}$	$C_{d-[26]}$
3.7	1.76	1.55	1.22	0.83	1.311
5	1.63	1.49	1.22	0.83	1.271
5.5	1.57	1.43	1.22	0.84	1.26
7	1.51	1.40	1.22	0.84	1.24

In Figure 14, discharge coefficient values and those obtained by Frank [26], Noseda [6], and Garot [1] equations and numerical results for different specific flows are compared.

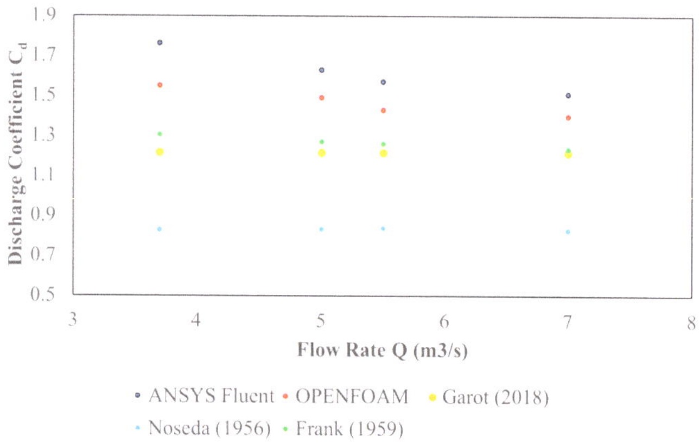

Figure 14. Comparison of discharge coefficients obtained from variable expressions under different flow rates ([1,6,26]).

Figure 15 shows the dimensionless variation in discharge coefficient across the rack under various flow rates for Stigansåni intake. The values obtained from both ANSYS Fluent and OpenFOAM were used for the current flow depths. The discharge coefficient changes before and after the rack, and along the rack. In Figure 15, the zero point is the start of the rack, and the change after this point is shown for rectangular profiles for Stigansåni intake. In the literature, remarkable differences were observed between the behaviors of T-shaped and circular bars (Castillo et al. [19] and Carrillo et al. [20]). The rack with circular bars has higher discharge coefficients than that with T-shaped bars. The results of the current research show that rectangular-shaped bars have higher discharge coefficients than both circular and T-shaped bars. It should be noted that the studies were carried out in clear water conditions without sediments.

The behavior of the non-dimensional discharge coefficient in Figure 15 can be mathematically expressed by the following Equation (6) for rectangular bars. The a and b coefficients found for this study are presented in Table 7.

$$C_d = \frac{ae^{b\left(\frac{x}{h_c}m\right)}}{1 + tan\alpha},\tag{6}$$

In the analyses with OpenFOAM, it was observed that the water levels on the rack upstream and on the rack were higher than the results obtained from ANSYS Fluent. Therefore, more water passed downstream of the structure and the amount of water entering the intake decreased by approximately 10%, especially when the flow rate increased ($Q = 7$ m³/s). Flow rates taken into intake calculated with numerical models are given in Table 8.

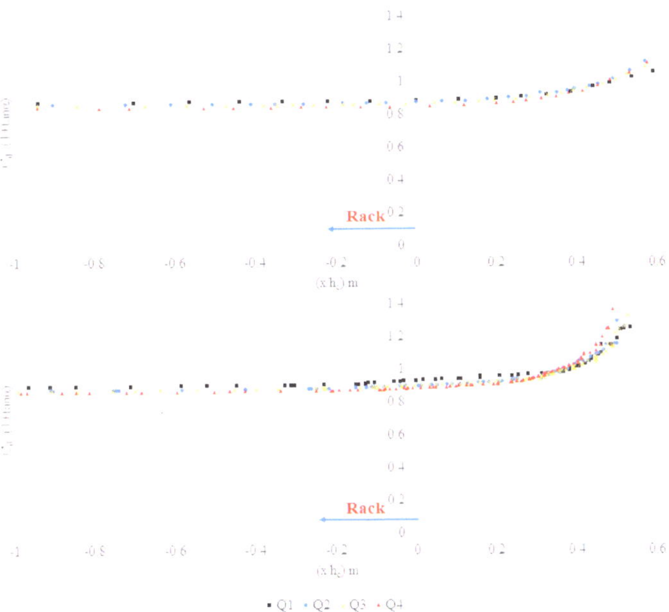

Figure 15. Variation of the discharge coefficient along the rack for rectangular bars ($m = 0.429$).

Table 7. Coefficients of the mathematical expressions for Equation (6).

C_d	a	b
$C_d = \frac{ae^{b\left(\frac{x}{h_c}m\right)}}{1+tan\alpha}$	1	0.0463

Table 8. Flow rates calculated with numerical models.

Q (m³/s)	$Q_{ANSYSFluent}$ (m³/s)	$Q_{ANSYSFluent}$ (%)	$Q_{OpenFOAM}$ (m³/s)	$Q_{OpenFOAM}$ (%)
3.7	3.23	87	3.61	98
5	4.48	90	4.36	87
5.5	4.68	85	4.84	88
7	6.08	87	5.23	75

The marginal difference was spotted in simulations with simplified and with natural topology tested with OpenFOAM. With $Q = 5.5$ m³/s inlet flow, the modelled flowrate averaged over time at the outlet patch towards the HPP shows a 3% difference between the two cases in Figure 16. The case with simplified surrounding slightly underestimates the flow rate at the intake outlet, that has been accepted for performing further model simulations on the simplified domain.

The water capture capacity factor of the Tyrolean weir is calculated according to Yilmaz [27] and Yildiz et al. [28]:

$$WCC = Q_{wi}/Q_{wT},\qquad(7)$$

where Q_{wi} is the diverted discharge flow through the racks, and Q_{wT} is the total discharge of the approaching main river flow. Figure 17 presents the water capture capacity curve of Stigansåni intake for ANSYS Fluent and OpenFOAM numerical model results under different flow rates. In the simulation with OpenFOAM, the higher water capture capacity factor was calculated at lower discharges. An increase of approximately 50% from the average value of $Q = 5 \ \mathrm{m^3/s}$ increases the water passing downstream of the intake, and the amount of water entering the structure decreased by 10% compared to the ANSYS Fluent results. It is seen that while the discharge Q_{wT} increases, the water capture capacity factor gradually decreases (Q_{wi}/Q_{wT}). On the other hand, this may be due to the greater number of meshes in the ANSYS Fluent model compared to OpenFOAM.

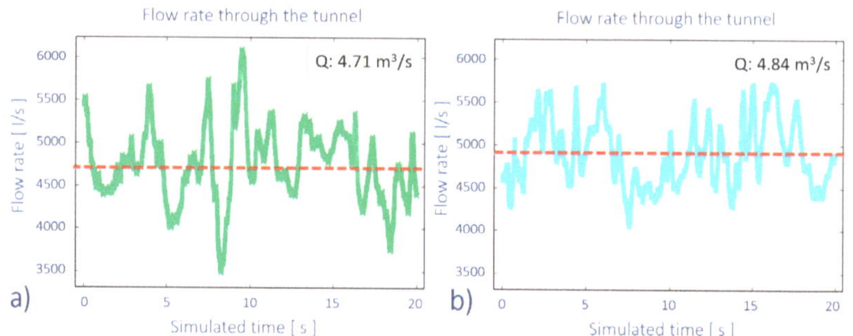

Figure 16. Simulated flow rate over time at the intake outlet with (**a**) simplified surroundings and (**b**) with natural topology.

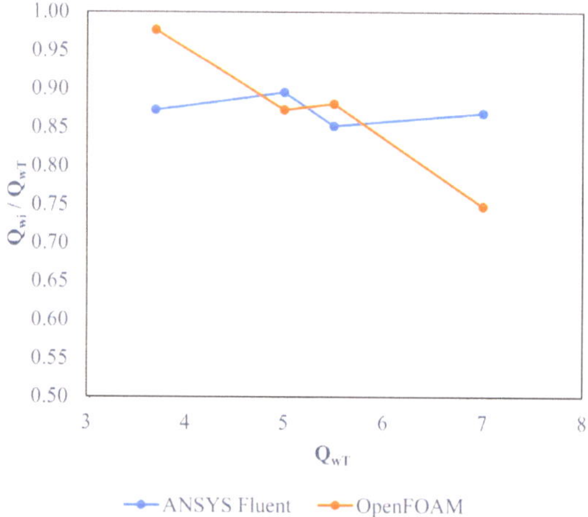

Figure 17. Water capture capacity for the Stigansåni intake ($m = 0.429$).

5. Conclusions

This study aimed to fill the literature gap in numerical estimation of the hydraulic flow over existing Tyrolean intake using the existing natural riverbed topography. In addition, the study has included a comparison of results from ANSYS Fluent (commercial) and OpenFOAM (freeware), which were both used in the project. The simulation results showed the following:

- CFD methods offer fast, practical, and inexpensive solutions for investigating the efficiency of existing systems. Modeling can be cost-effective with limited input data, and still have sufficient accuracy.
- Simulations for four different flow rates yielded 15–20% higher water levels in the VOF model applied in OpenFOAM compared to the simulations in the VOF model applied in ANSYS Fluent. Consequently, while higher water capture capacity was calculated at lower discharges according to the OpenFOAM simulations, 10% less water capture capacity was found at higher discharges compared to ANSYS Fluent

results. On the other hand, the current behaviors of the VOF model applied in ANSYS Fluent were found to be compatible with the literature.

- According to the studies, the estimation of the flow rate coefficient (C_d), which is the design criterion, and accordingly the wet rack length, is the most important parameter that affects the flow rate entering the intake structure. Good calculation of this parameter will provide both economical solutions and optimum flow rate input.
- The turbulence models in flow separation have different behavior. In this study, the $k - \varepsilon$ turbulence model and the $k - \Omega$-based Shear-Stress Transport (SST) model, which are the most widely used RANS turbulence models in the literature, were used. The results obtained with the tests are almost the same when considering the behavior of the flow curves and the total amount of water taken.
- In this study, different formulations were applied for rectangular bars. Regarding the coefficient of discharge, rectangular bars have been found to show larger maximum values than T-shaped bars in the literature, and thus, will require relatively shorter rack length.

This study examined the usability of 3D CFD modeling for old structures with complex and limited data, such as secondary intakes, and in this regard, the study's recommendations may be useful for consulting companies dealing with similar issues. Considering the total simulation times for domains containing approximately the same number of elements and in the simulations with the same parallel computation, modeling with OpenFOAM yielded approximately 11% faster calculations. However, it should be noted that this advantage is increased by the availability of a "cold-start" simulation technique for OpenFOAM. The results are in agreement with the studies in the literature. The intake is expected to collect the same amount of water as clean water with an increase in the required wet rack length. As such, experiments with sediments are required to understand the behavior of the rack under these conditions. For future studies, different turbulence models should be tested, and the results should be compared.

Author Contributions: A.B., M.S.-M., K.V. and L.L., methodology; A.B. and M.S.-M., numerical simulations; A.B., formal analysis; K.V., resources and data curation; A.B., writing and original draft preparation; A.B., M.S.-M., K.V. and L.L., writing, review and editing. All authors have read and agreed to the published version of the manuscript.

Funding: This project was initiated through FME HydroCen, which is a Centre for Environment-friendly Energy Research (FME) dedicated to research on hydropower. The project was initiated in 2022 through the open calls process, where the main initiative came from Statkraft Energy. The project has had one year of funding with a total budget of NOK 500000. The project has been conducted in cooperation between NTNU and SINTEF.

Data Availability Statement: The data that support the findings of this study are available from the Statkraft Energy but restrictions apply to the availability of these data, which were used under license for the current study and are not publicly available. Data are, however, available from the authors upon reasonable request and with permission of the Statkraft Energy.

Acknowledgments: The researchers would like to thank Sira-Kvina hydropower company for their great help with providing data to the project. We also would like to thank Statkraft, who initiated the project. Thanks also go to the 'Fagutvalg/Technical committees' in HydroCen, who supported and voted for the project 'CFD model simulations for the purpose of optimizing secondary intakes' in competition with many others. The authors would also like to thank Simon Edward Mumford for his help with language editing and proofreading.

Conflicts of Interest: The authors declare no conflicts of interest. The funders had no role in the design of the study; in the collection, analyses, or interpretation of data; in the writing of the manuscript; or in the decision to publish the results.

Abbreviations

The following abbreviations are used in this manuscript:

CAD	Computer-Aided Design
CFD	Computational Fluid Dynamics
DEM	Digital Elevation Model
FVM	Finite Volume Method
NVE	Norwegian Water Resources and Energy Directorate
SST	Shear-Stress Transport Model
VOF	Volume of Fluid Method
3D	3 Dimensional
b_1	spacing between the rack bars;
b_w	distance between the middle of two bars (For Norway, it is normally around m $\cong$ 0.9);
b_1	space between bars;
b_w	bar width;
C	discharge coefficient;
C_{d0}	discharge coefficient calculated under static conditions;
C_{qh}	discharge coefficient for flow depth;
F	gravity force;
h	local flow depth;
H	hydraulic head;
h_0	specific flow depth that is approaching to the rack;
h_c	critical flow depth;
L	wetted rack length;
m	void ratio (the ratio of the opening area of the screen);
P	pressure;
Q	diverted discharge;
Q_{wi}	diverted discharge flow through the racks;
Q_{wT}	total discharge of the approaching main river flow;
q	incoming unit flow discharge;
q_{intake}	intake unit flow discharge;
dq/dx	diverted discharge for unit width for length dx;
ρ	density of water;
$r\alpha$	volume fractions;
u	time-averaged velocity vector (for x, y and z direction);
WCC	Q_{wi}/Q_{wT} water capture capacity factor;
x	streamwise coordinate;
v	mean velocity;
μ	contraction coefficient;
θ	angle of rack;

References

1. Garot, F. De Watervang met liggend rooster. *Ing. Ned. Indie* **1939**, *6*, 115–132.
2. Bouvard, M. Debit d'une grille par en dessous. *Houille Blanche* **1953**, *3*, 290–291. [CrossRef]
3. Kuntzmann, J.; Bouvard, M. Theoretical study of bottom type water intake grids. *Houille Blanche* **1954**, *40*, 569–574. [CrossRef]
4. Noseda, G. Operation and design of bottom intake racks. In Proceedings of the VI General Meeting IAHR, The Hague, The Netherlands, 1955; Volume 3, pp. C17-1–C17-11.
5. Noseda, G. Correnti permanenti con portata progressivamente decrescente, defluenti su griglie di fondo. *Energ. Elettr. 1* **1956**, *33*, 41–51.
6. Noseda, G. Correnti permanenti con portata progressivamente decrescente, defluenti su griglie di fondo. *Energ. Elettr. 6* **1956**, *33*, 565–581.
7. Mostkow, M. Sur le calcul des grilles de prise d'eau. *Houille Blanche* **1957**, *4*, 569–576. [CrossRef]
8. Brunella, S.; Hager, W.; Minor, H.E. Hydraulics of Bottom Rack Intake. *J. Hydraul. Eng.* **2003**, *129*, 2–10. [CrossRef]

9. De Marchi, G. *Profili Longitudinali della Superficie Libera delle Correnti Permanenti Lineari con Portata Progressivamente Crescente o Progressivamente Decrescente Entro Canali di Sezione Constante*; Consiglio Nazionale Delle Ricerche: Rome, Italy, 1947; pp. 203–208.
10. Frank, J.; Von Obering, E. *Hydraulische Untersuchungen für das Tiroler Wehr*; Der Bauingenieur: Prague, Czech Republic, 1956; pp. 96–101.
11. Dagan, G. Notes sur le calcul hydraulique des grilles par-dessous. *Houille Blanche* **1963**, *1*, 59–65. [CrossRef]
12. García, B. Estudio Experimental y Numérico de los Sistemas de Captación de Fondo. Ph.D. Thesis, Escuela Internacional de Doctorado de la Universidad Politécnica de Cartagena, Universidad Politécnica de Cartagena, Cartagena, Spain, 2016.
13. Righetti, M.; Rigon, R.; Lanzoni, S. Indagine sperimentale del deflusso attraverso una griglia di fondo a barre longitudinali. In Proceedings of the XXVII Convegno di Idraulica e Costruzioni Idrauliche, Geneva, Italy, 12–15 September 2000; Volume 3, pp. 112–119.
14. Castillo, L.; Bermejo, J.T.G. Estudio Experimental y Numérico de los Sistemas de Captación de Fondo. Ph.D. Thesis, Universidad Politécnica de Cartagena Departamento de Ingeniería Civil, Cartagena, Spain, 2016.
15. Drobir, H. *Entwurf von Wasserfassungen im Hochgebirge*; Österreichische Wasserwirtschaft: Mondsee, Austria, 1981; pp. 243–253.
16. Stokkebø, O. *Bekkeinntak på Kraftverkstunneler—Sluttrapport fra Bekkeinntakskomiteen*; Vassdragsregulantenes Forening: Norway, 1986.
17. Bombardelli, F.A. Computational multi-phase fluid dynamics to address flows past hydraulic structures. In Proceedings of the Symposium on Hydraulic Structures, Porto, Portugal, 9–11 February 2012.
18. Blocken, B.; Gualtieri, C. Ten iterative steps for model development and evaluation applied to Computational Fluid Dynamics for Environmental Fluid Mechanics. *J. Environ. Model. Softw.* **2012**, *33*, 1–22. [CrossRef]
19. Castillo, L.G.; García, J.T.; Carrillo, J.M. Influence of Rack Slope and Approaching Conditions in Bottom Intake Systems. *Water* **2017**, *9*, 65. [CrossRef]
20. Carrillo, J.M.; García, J.T.; Castillo, L.G. Experimental and Numerical Modelling of Bottom Intake Racks with Circular Bars. *Water* **2018**, *10*, 605. [CrossRef]
21. Menter, F.R. Two-equation eddy-viscosity turbulence models for engineering applications. *AIAA J.* **1994**, *32*, 1598–1605. [CrossRef]
22. Launder, B.E.; Spalding, D.B. The numerical computation of turbulent flows. *Comput. Methods Appl. Mech. Eng.* **1974**, *3*, 269–289. [CrossRef]
23. Ansys Inc. ANSYS Fluent Workbench Totorial Guide. ANSYS2021. Available online: https://forum.ansys.com/uploads/846/SCJEU0NN8IHX.pdf (accessed on 20 February 2024).
24. Torres, C.; Borman, D.; Sleigh, A.; Neeve, D. Three dimensional numerical modelling of full-scale hydraulic structures. In Proceedings of the 37th IAHR World Congress, Kuala Lumpur, Malaysia, 13–18 August 2017; pp. 1335–1343.
25. García, J.T.; Castillo, L.G.; Haro, P.L.; Carrillo, J.M. *Diseño de Sistemas de Captación de Fondo*; Las Jornadas de Ingeniería del Agua (JIA): Coruña, Spain, 2017. (In Spanish)
26. Frank, J. *Fortschritte in der Hydraulic des Sohlenrechens*; Der Bauingenieur: Prague, Czech Republic, 1959; pp. 12–18.
27. Yilmaz, N.A. *Hydraulic Characteristics of Tyrolean Weirs*; Orta Doğu Teknik Üniversitesi: Ankara, Turkey, 2010.
28. Yildiz, A.; Marti, A.I.; Gogus, M. Numerical and experimental modelling of flow at Tyrolean weirs. *Flow Meas. Instrum.* **2021**, *81*, 102040. [CrossRef]

Article

Numerical Modelling on Physical Model of Ringlet Reservoir, Cameron Highland, Malaysia: How Flow Conditions Affect the Hydrodynamics

Safari Mat Desa [1], Mohamad Hidayat Jamal [2,3,*], Mohd Syazwan Faisal Mohd [1], Mohd Kamarul Huda Samion [1], Nor Suhaila Rahim [2,3], Rahsidi Sabri Muda [4], Radzuan Sa'ari [2], Erwan Hafizi Kasiman [2,3], Mushairry Mustaffar [2], Daeng Siti Maimunah Ishak [2,3] and Muhamad Zulhasif Mokhtar [2]

[1] National Water Research Institute of Malaysia (NAHRIM), Ministry of Environment and Water (KASA), Lot 5377, Jalan Putra Permai, Seri Kembangan 43300, Selangor, Malaysia
[2] Faculty of Civil Engineering, Universiti Teknologi Malaysia (UTM), Johor Bahru 81310, Johor, Malaysia
[3] Centre of River and Coastal Engineering (CRCE), Research Institute for Sustainable Environment (RISE), Universiti Teknologi Malaysia (UTM), Johor Bahru 81310, Johor, Malaysia
[4] TNB Research Sdn Bhd, Lorong Ayer Itam, Kawasan Institusi Penyelidikan, Kajang 43000, Selangor, Malaysia
* Correspondence: mhidayat@utm.my

Abstract: The relative impacts of changes in the storage capacity of a reservoir are strongly influenced by its hydrodynamics. This study focused mainly on predicting the flow velocities and assessing the effectiveness of groynes as control mitigation structures in changes in the water depth and velocity distributions in Ringlet Reservoir. Initially, the physical model of the Habu River (the main part of Ringlet Reservoir) was fabricated, and flow velocities were measured. Then, a two-dimensional HEC-RAS was adapted to numerically simulate the hydrodynamics of the annual recurrence intervals of 1, 5, and 100 years in the Ringlet Reservoir. Experimental data acquired at the Hydraulic and Instrumentation Laboratory of the National Water Research Institute of Malaysia (NAHRIM) was used to calibrate and validate the numerical models. The comparison of simulation and experimental results revealed that the water levels in all simulations were consistent. As for the velocity, the results show a comparable trend but with a slight variation of results compared to the experiments due to a few restrictions found in both simulations. These simulation results are deemed significant in predicting future sediment transport control based on hydrodynamics in this reservoir and can be of future reference.

Keywords: numerical modelling; physical modelling; Ringlet Reservoir; flow velocity; hydrodynamics; HEC-RAS 2D

Citation: Mat Desa, S.; Jamal, M.H.; Mohd, M.S.F.; Samion, M.K.H.; Rahim, N.S.; Muda, R.S.; Sa'ari, R.; Kasiman, E.H.; Mustaffar, M.; Ishak, D.S.M.; et al. Numerical Modelling on Physical Model of Ringlet Reservoir, Cameron Highland, Malaysia: How Flow Conditions Affect the Hydrodynamics. *Water* **2023**, *15*, 1883. https://doi.org/10.3390/w15101883

Academic Editor: Charles R. Ortloff

Received: 15 March 2023
Revised: 19 April 2023
Accepted: 9 May 2023
Published: 16 May 2023

1. Introduction

Dams and reservoirs are essential for impounding or storing water for many uses, such as flood control, water supply, hydropower, irrigation, navigation, and recreation. A reservoir typically has two storage features. First, active storage, or the volume used for storing water, generating power, supplying water, or as reserves for flood management; and second, dead storage, which refers to the volume below the minimum operational elevation [1]. There are different types of reservoirs; for instance, run-of-river type reservoirs. These reservoirs are commonly used for generating hydropower due to their small-volume active and large-volume dead storage [2]. Worldwide, hydropower continues to be the most significant renewable energy source for producing electricity [3]. It has generated more energy than all other renewable sources, with 4418 TWh in 2020 [4].

The Ringlet Reservoir in Cameron Highlands, Malaysia, is an essential water source for domestic, agricultural, and hydropower purposes. However, the increasing demand for water supply and activities such as agriculture and logging have resulted in significant changes in the hydrodynamics of the reservoir. Sediment movement also leads to deposition or erosion, and its movement is primarily influenced by flow velocity and direction. In general, sediment deposition can reduce water storage capacity by up to 0.8%, resulting in a shorter project life [5]. This situation makes it difficult for reservoirs to cater to the rising need for water [6]. Another issue is sediment buildup around power plant gates, reducing a dam's ability to operate effectively.

Furthermore, a significant amount of sediment in the water intake might harm or lessen the effectiveness of the hydro turbines and other electromechanical components [7,8]. Dredging activities have been carried out since the dam's initial operation to provide adequate reservoir storage volume. Unfortunately, the cost of dredging has risen dramatically and is anticipated to increase even further as the rate of silt deposition rises [9]. While the projected life for the Ringlet Reservoir in the Cameron Highlands is 80 years with a gross storage of 6.7 million m^3, after barely 35 years of operation, 52 percent of its storage was already utilised, with 34 percent filled with sediment [10]. Hence, further research and predictions are required for the continuous operation of the current power generation plant. One of the solutions for this is to study the hydrodynamics of the flow velocity in those reservoirs.

Flow velocity is a crucial parameter in understanding the behavior of fluids in reservoirs. Physical models and numerical simulations are two common methods used to study flow velocity. Physical models involve constructing a scaled-down physical model that mimics the behavior of the reservoir, while numerical simulations use mathematical equations to simulate the behavior of the reservoir. Several studies have compared the accuracy of physical models and numerical simulations in predicting reservoir flow velocity. For example, a study by Kositgittiwong et al. found that physical models provided more accurate predictions of flow velocity than numerical simulations due to the many uncertainties in the real application [11]. However, a study by Zhang et al. found that numerical simulations were more accurate than physical models in predicting flow velocity [12]. Research conducted by McCoy et al. used a physical model to investigate the effect of groynes on flow velocities in a reservoir. The study found that installing groynes reduced flow velocities and increased water retention time in the reservoir [13]. Numerical simulations have also been used to predict flow velocities in reservoirs. Research conducted by Liang et al. used a numerical simulation model to investigate the effect of groynes geometry and arrangement on flow velocities, erosion, and sedimentation in reservoirs. The study found that the installation of groynes led to a reduction in flow velocities, the formation of vortices, and scouring, which was consistent with the findings from their laboratory experiment [14].

The primary objective of this investigation is to determine the effect of hydrodynamics on the flow regime of the reservoir by assessing the rates and directions of flow in Ringlet Reservoir using a physical and 2D numerical model. The present work aims to characterise the Ringlet reservoir's velocity patterns for up to a 100-year return period through physical and numerical modelling, the common practices used by several researchers [15–17]. First, the influences of hydrodynamics on the flow pattern from the reservoir were investigated, and a numerical simulation model was compared with physical observations better to understand the flow pattern along the Ringlet Reservoir. The results of this physical model were then compared with the numerical simulation to validate the accuracy of the simulation. The simulation used a 2D numerical model based on the Reynolds-Averaged Navier-Stokes (RANS) equations. Finally, the numerical model was calibrated and validated using experimental data from the physical model. Further, this study examined the construction of groynes or access rams to obtain the hydrodynamics for this prediction [18].

The results from the physical model and numerical simulation revealed that the flow conditions in Habu (Ringlet Reservoir) are affected by various factors such as velocity, bathymetry, and the presence of hydraulic structures. The findings also showed agreement in the flow pattern between the physical and numerical models, indicating that the numerical model can predict future sediment transport patterns. Furthermore, the results showed agreements in the reservoirs' flow pattern between the physical and numerical models. In the meantime, the installation of groynes significantly affects the hydrodynamics of flows at the installation area. Overall, the study provides valuable insights into the hydrodynamics of Ringlet Reservoir, which can be used to predict future changes and implement effective management strategies.

2. Methodology

The methodology adopted for numerical modelling of the physical model of Ringlet Reservoir (Habu), located in Cameron Highland, Malaysia, was based on a four-step procedure, as depicted in Figure 1. The first step was a desktop study, which involved gathering relevant data about the reservoir, including aerial images, topography, and hydrology. The second stage was the small-scaled model setup, which comprised two significant components: experimental work and numerical model setup using HECRAS. The experimental work involved the physical construction of a small-scale model of the Ringlet Reservoir. In contrast, the numerical model setup involved using HECRAS software to simulate the hydraulic behaviour and performance of the reservoir. The third step of the study was the comparison of the physical and numerical models. This involved comparing the observed data from the physical model with the results obtained from the HECRAS numerical model setup. The final step involved the presentation of the results, a discussion of the findings, and the study's overall conclusions. Overall, this methodology incorporates physical and numerical modelling that provides a robust approach to assessing the hydrodynamic performance and prospects of the Ringlet Reservoir in Cameron Highland, Malaysia.

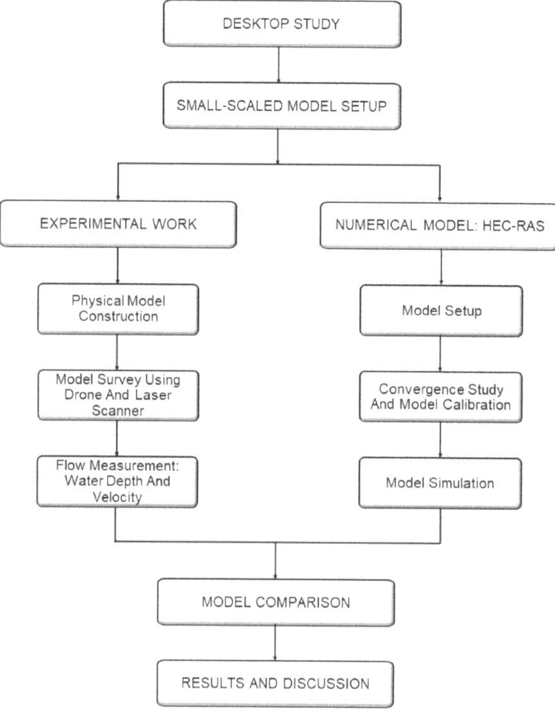

Figure 1. Flowchart of the study. The major components of the study are the experimental works and validation of the model using a numerical approach.

2.1. Study Area

Ringlet Reservoir, in the central highlands of West Malaysia, approximately 160 km north of Kuala Lumpur, is a part of the Cameron Highlands-Batang Padang hydroelectric scheme. It comprises the Sultan Abu Bakar Dam, which was constructed in 1963. It was initially estimated that the sediment loading to the reservoir was m^3/annum [19]. Tenaga National Berhad (TNB) is responsible for developing and operating most major hydropower projects. At present, TNB is the main operator for three of the largest hydroelectric schemes in Sungai Perak (1249.1 MW), Kenyir (665 MW), and Cameroon Highlands-Batang Padang (622 MW) [20]. A map of Habu is depicted in Figure 2. The Ringlet Reservoir is located within the Bertam catchment, with a combined area of 70.4 km^2. The Bertam Catchment is divided into six sub-catchments: Upper Bertam, Middle Bertam, Lower Bertam, Habu, Ringlet, and Reservoir. Bertam River, Habu River, and Ringlet River are the principal rivers supplying the Ringlet Reservoir. Teh et al. state that most sediment loaded into the Ringlet Reservoir comes from the Habu end [21].

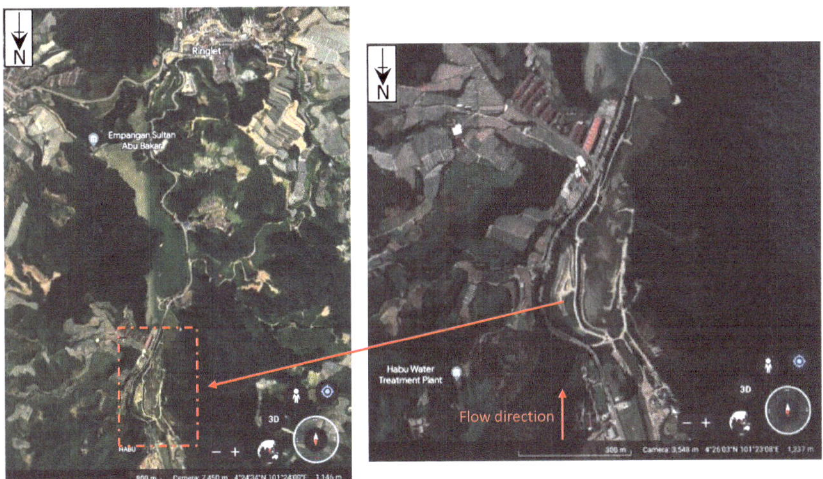

Figure 2. Map area of Habu. Habu is part of the Ringlet Reservoir, Cameron Highland, Malaysia. Hence, the name 'Habu' was used throughout.

An area of around 183 km^2 comprises the upper watershed that feeds the Ringlet Reservoir [22,23]. Electricity is generated from headwater from two main rivers, Sungai Telom and Sungai Bertam. In this regard, even though the Ringlet reservoir had an initial water storage capacity of 6.7 million m^3, it has experienced a loss of operational volume over the years because of accelerated sedimentation [19,24]. This phenomenon is caused by millions of tons of additional sediment mobilized by rapid developments in the upper catchments area. This situation has gradually decreased the reservoir's capacity for hydro generation and led to a higher risk of downstream flooding [25]. The higher sediment deposition rate would significantly reduce the projected useful life of the reservoir [26]. Additionally, it has a negative impact on the dam's stability and risks the ability to store water for flood control [27,28].

2.2. Experimental Works

2.2.1. Physical Model Construction

Typically, physical models are used to investigate, evaluate, and formulate solutions to various sedimentation issues in hydraulic structures like reservoirs [29,30]. A physical hydraulic model is often a scaled-down version of the research site and was essential for data calibration and validation of the HEC-RAS numerical model. The dynamic similarity makes it possible to scale results from model tests to predict corresponding results for the full-scale prototype. The dynamic similarity makes it possible to scale results from model tests to predict related results for the full-scale prototype. Dynamic similarities can be found by comparing the ratios of various significant forces acting on the system. Because gravitational force dominates fluid motion in free surface flow, Froude scaling was used. Laboratory tests have been conducted in the Hydraulic and Instrumentation Laboratory, National Water Research Institute of Malaysia (NAHRIM). A model of Habu River (also known as Habu), a significant part of Ringlet Reservoir in the Cameron Highlands, was constructed with a geometric scale of 1:30. The model was constructed using a fixed bed setup; hence, no erosion or accretion occurred during the tests. The model's bed was levelled to its original or existing condition, equivalent to its prototype.

2.2.2. Model Surveying

The model's bathymetry was determined using three-dimensional surveying techniques (3D), Terrestrial Laser Scanning (TLS), and Digital Close-Range Photogrammetry (DCRP). A medium-range GLS 2000 TLS model (Topcon, Livermore, CA, USA) was used for the laser scanning research. The output of laser scanning is an array of points known as a point cloud, a set of vertices defined in a 3D coordinate system (x, y, and z) capable of reconstructing a highly detailed 3D physical model. The high-density point cloud from TLS will support a numerical model as input parameters (base data). In the meantime, DCRP was implemented to gather 3D information about features from two or more photos of the same object. This technique was used to create accurate 2D orthophoto and 3D surface models in the form of point cloud data using a sequence of overlapping digital photos acquired by a drone. The drone (Phantom 4 Pro, DJI, Tokyo, Japan) flew manually along the designated flight line approximately 5 m above the model. A series of digital images were captured at intervals of three seconds to enable a 3D model and orthophoto of the physical surface model to be generated.

2.2.3. Flow Measurements

Water was pumped through the inlet pipe upstream of Habu. The 350 SZ (EBARA, Subang Jaya, Malaysia) external pumps, with a capacity of 1000 L/min and a head of 7 m, were employed for this study. Water was poured until the model or waterway reached its maximum level. The pump was then turned on until the specified flow rate for each test was reached. Dynaflox Series DMTFP Portable Transit Time Ultrasonic Flow Meters (Emin Group, Singapore) were put near the stabilizer tank to measure the inlet flow rate. Steady flow conditions for all tests were established by manipulating the setting of the pump rate to achieve the designated flow discharges, refer to Table 1. H1 and H2 denote Habu with Existing condition and Groynes, respectively, while the last numerical value after the dash symbol represents their respective ARIs.

Table 1. Test conditions for Habu (Ringlet River).

Location	Conditions	ARI (Year)	Test Series	Flow Rate Prototype (m^3/s)	Flow Rate Model (l/s)
Habu	H1 Existing	1	H1-1	23.5	4.7
		5	H1-5	34.5	7.0
		100	H1-100	55.5	11.3
	H2 with Groynes	1	H2-1	23.5	4.7
		5	H2-5	34.5	7.0
		100	H2-100	55.5	11.3

Flow velocity is a crucial hydraulic and hydrological component utilised in the velocity–area approach to evaluate discharge [31]. Flow velocities were measured using a 1D AEM electromagnetic current meter (ECM). With a minimum depth of 3 cm and a range of 0 to 5 m/s, this ECM can precisely detect water speeds in shallow water storage. The flow was measured at the observation point located at every CH, marked with black dots in Figure 3. An average of five readings were measured for every point at roughly 0.5 of the depth.

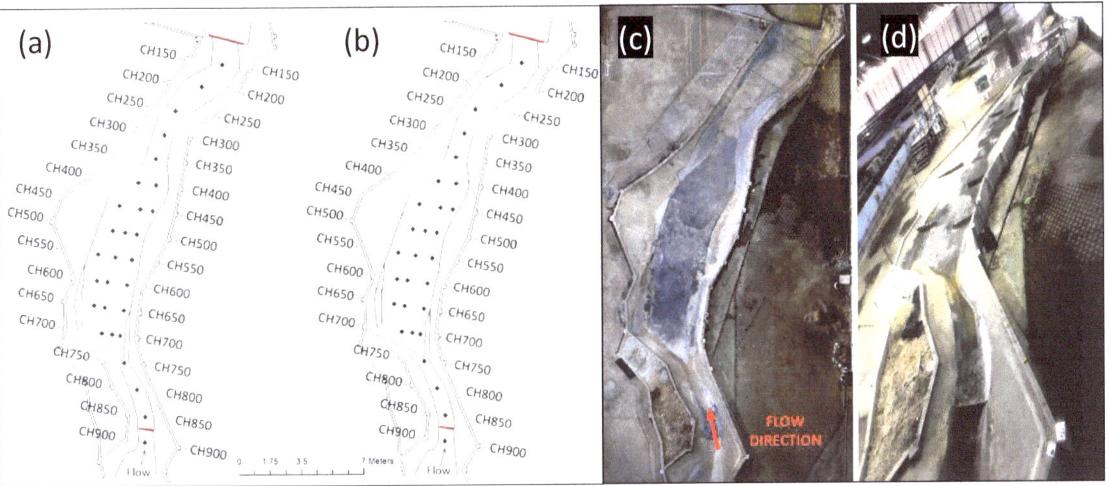

Figure 3. Diagram of the physical model in Habu. (**a**) existing and (**b**) with groynes conditions. (**c,d**) live images of the constructed physical models of (**a,b**).

Groynes were constructed after completing the measurements for the existing condition. Figure 3b shows the arrangement of eight groynes with a width-to-length ratio, w/l, approximately equal to 0.1. The groynes were constructed approximately perpendicular to the channel. These impermeable groynes are designed to be non-submerged [32–34]. Therefore, the groynes were approximately built perpendicular to the channel. The same procedures used during the existing condition were repeated for mitigation (with groynes condition).

2.3. Numerical Model (HEC-RAS)

2.3.1. Model Setup

HEC-RAS software was used to model the flow of water in this project. The 2D Shallow Water Equations (SWE) are the basis for hydrodynamic two-dimensional (2D) models in HEC-RAS. The SWE is found by integrating the Reynolds-averaged Navier-Stokes equations over the flow depth [35]. In this integration process, a hydrostatic pressure distribution is assumed. The 2D finite volume method is used for the solution of the SWE, allowing greater stability (due to shock capturing capability) and accuracy compared to the finite element or finite difference-based method. Two types of models are available in HEC-RAS 2D, the full SWE and the Diffusion Wave Approximation (DWA). The DWA model was used as an initial condition for the SWE model to provide stability for the unsteady model. The point cloud taken from the 3D scanning of the model was processed into Digital Elevation Model (DEM) and exported as a GeoTIFF file format. This was then imported into HEC-RAS as terrain for the simulation model.

In all cases, an initial water level was prescribed. This means at the start of the simulation. The domain contains water at a certain level. This initial condition proved more effective than starting the simulation with the domain completely dry. Boundary conditions were imposed at the inlet and outlet of the computational domain. The inlet is located at the upstream part of the domain, while the outlet is located precisely on the check dam. A constant flow rate based on the Annual Recurrence Interval (ARI) of 1-, 5- and 100 years was prescribed at the inlet. At the outlet, a constant water level was used. The level was initially assumed as no water level data was observed during the experiment. The value was then changed until the water level just upstream of the outlet matched the experimental observation. This is one of the calibration steps used to set the current value for the unknown boundary conditions.

2.3.2. Convergence Study and Model Calibration

A computational mesh was generated for the domain of interest. Three mesh sizes of 7680, 30,065, and 185,338 were used for 15 cm, 10 cm, and 5 cm, respectively. The water surface elevation and velocity plots at CH500 for different mesh sizes were plotted and compared. Although the coarse mesh can provide an equally accurate result, for the simulation that follows, a mesh size of 5 cm is used throughout. Constant flow rates were imposed at the inlet, while constant water levels were used at the outlet. An adaptive time step was used in the simulation with a minimum and maximum Courant number between 1 and 3. Two of Manning's values of 0.0025 and 0.01 were used for calibration. The Manning value was selected based on the closest agreement of water level and velocity between simulation and experimental results. The simulation period for all cases is 30 min.

2.3.3. Model Simulation

The wetted boundary of the model is considered through the Manning coefficient. Because the shallow water equation assumes averaged vertical velocity, the effect of the boundary layer cannot be modelled directly. The influence of the boundary layer due to surface roughness is represented as additional source terms. This source term includes, amongst others, the Manning coefficient. For the current work, it is assumed that the bed is quite rough. A different value for Manning was tested. The best Manning value was selected based on the closest agreement for water level and velocity between simulation and experimental results. A Manning value of 0.0025 was used throughout the simulation. With the mesh generated and the initial and boundary conditions imposed, the hydrodynamics of the model was simulated for 15 computational minutes based on a steady state obtained during the preliminary study. The 2D HEC-RAS model used in the current work solves the unsteady SWE. HEC-RAS uses an explicit formulation for the solution of the SWE. Due to the explicit formulation, the time step needed to keep the solution stable needs to be smaller than the critical time step calculated by the Courant number [36]. Two schemes for time marching are available in HEC-RAS: fixed and adaptive time stepping. Initially, the fixed time step was used. However, as the mesh size decreased, the time step selected violated the critical time step and resulted in an unreliable solution. The adaptive time step was then selected. The use of adaptive time-stepping requires the setting of a minimum and maximum Courant number. The minimum and maximum Courant numbers 1 and 3 were selected for all simulations. The selection was based on the best practices outlined in the HEC-RAS manual [37].

3. Results and Discussions

3.1. Physical Model

Flow velocity measurements in Habu were plotted as velocity points for the existing and mitigation (groynes). Velocity points are illustrated in Figure 4 for existing conditions and mitigation (groynes), while Figure 5 shows the scatter plot comparisons between different ARIs for both conditions. The original velocity points are used to show the distribution of data points on the model and to demonstrate the accuracy of the gridding methods used for an accurate comparison with the simulation.

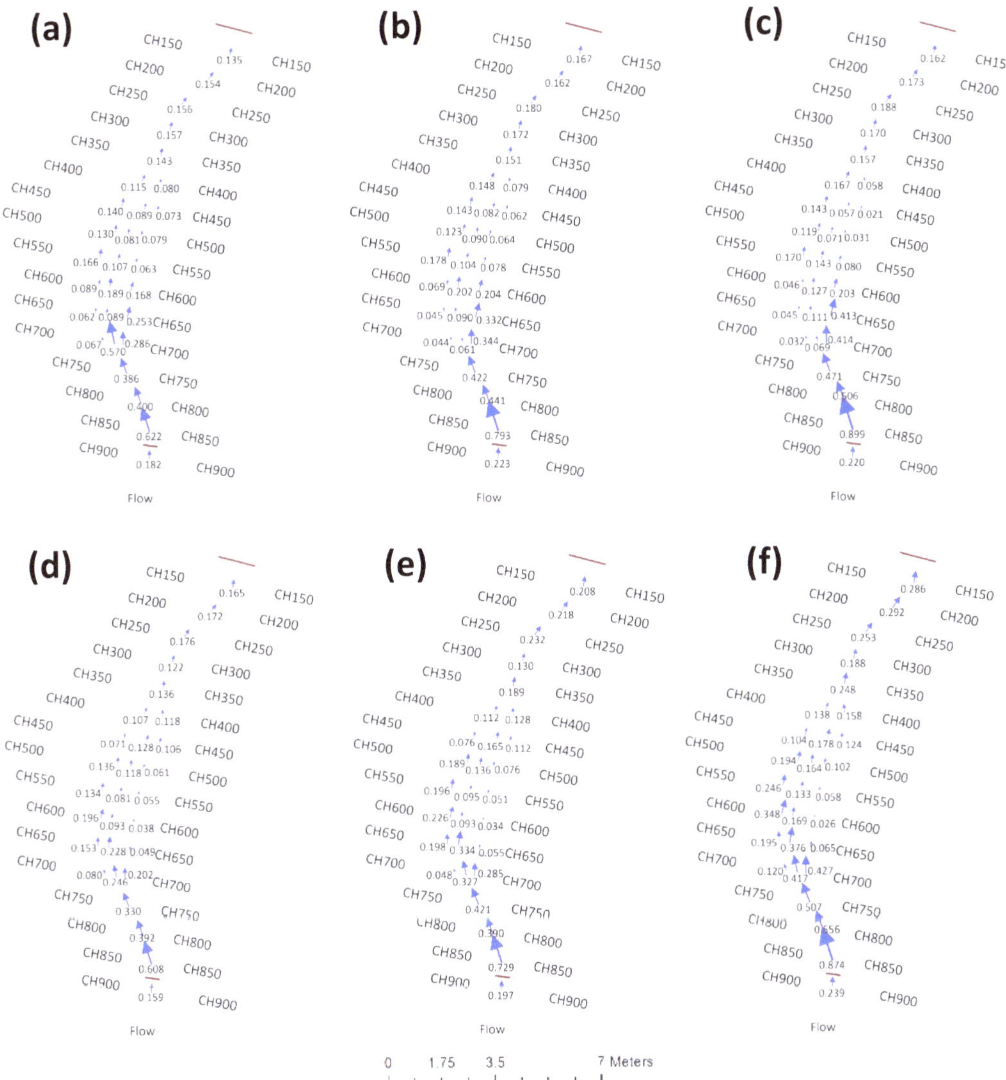

Figure 4. Velocity points at Habu during; existing condition: (**a**) 1-year, (**b**) 5-years and (**c**) 100-years ARI, Groynes condition: (**d**) 1-year, (**e**) 5-years and (**f**) 100-years ARI, respectively. Display vectors were present with the vector arrow tail at the data location.

Generally, the local velocity in the reservoir or channel varies with its location. The velocity is lower near the boundary (side). It was found that as the flow rate, Q increased, the velocity also increased for all tests in both existing and mitigation conditions. This indicated that the velocities in the reservoir increased when the water depths and flow rates increased. The summary of flow velocities at Habu for existing and during mitigation (groynes) is summarized in Table 2.

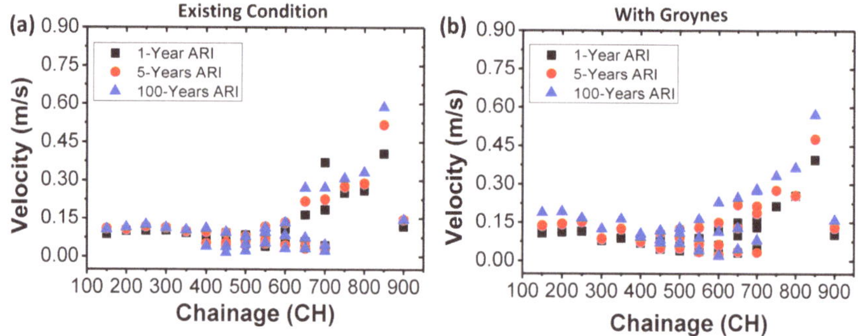

Figure 5. Velocity distribution comparisons between different ARIs (1-year, 5-years, and 100-years) for (**a**) existing conditions and (**b**) with Groynes.

Table 2. Summary of velocities measured at Habu.

Condition	ARI (Years)	Flow Rate, Q (L/s)	U_{min} (m/s)	U_{max} (m/s)	U_{ave} (m/s)
H1 Existing	1	4.7	0.062	0.622	0.180
	5	7.0	0.044	0.793	0.181
	100	11.3	0.021	0.899	0.188
H2 Groynes	1	4.7	0.038	0.608	0.161
	5	7.0	0.034	0.729	0.195
	100	11.3	0.026	0.874	0.248

Most of the flow in Habu takes place in the middle of the channel for both existing and mitigation conditions. The highest velocity occurred at CH850, right after the first check dam in Habu for existing and mitigation conditions. The velocity decreases after CH650 downstream in Habu for existing conditions. Meanwhile, the velocity decreases after CH600 downstream in Habu for mitigation conditions. This indicates the influence of groynes to reduce the flow velocity in the area close to their location and focus or concentrate the flow velocity in the middle of the channel [38]. Therefore, the second groyne effectively conveys the flow to the middle of the channel. Based on the experimental results, the flow velocity patterns for all return periods tested are the same; the only difference is the magnitude of the velocity. This can also be observed by the existence of groynes in the channel. Groynes also changes the pattern of the velocity, with lower velocity observed on the side of the channel and higher velocity being more concentrated in the middle of the channel.

3.2. HEC-RAS

3.2.1. Convergence Study

This section discusses the convergence study for simulations with different mesh sizes. The convergence study aims to determine the best mesh size that needs to be used to minimise computational expenses and, at the same time, ensure that the solution is not sensitive to the changes in mesh size (mesh independent solution) [13]. Therefore, a convergence study was carried out for these mesh sizes. Figure 6a–c compares mesh sizes used throughout the simulation.

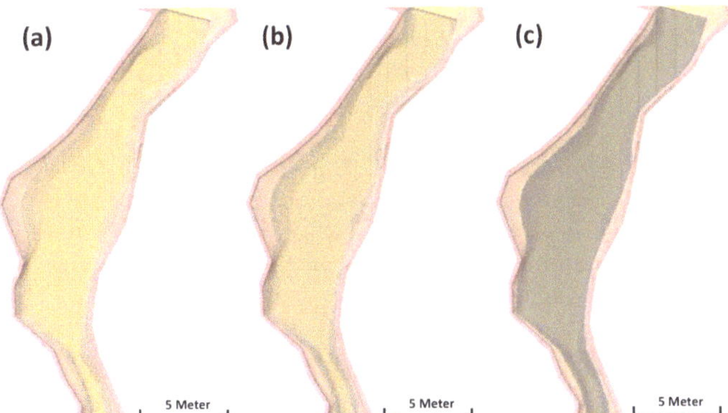

Figure 6. Mesh with sizes of (**a**) 15 cm, (**b**) 10 cm and (**c**) 5 cm.

It should be noted that the solution produced by the DWA might be different from SWE depending on the corresponding Reynold number, with higher values representing turbulence flow. As DWA neglect the nonlinear effect caused by the convective acceleration, the effect of turbulence on the flow cannot be accounted for. For the case considered here, the result is mostly turbulence. The use of SWE is more appropriate and was used for this study. A cross-section at CH500 was chosen for the water surface elevation and velocity evaluation. To simplify the analysis of the result, only the DWA model was considered. Figure 7a shows the cross sections in which the values for the water surface elevation (WSE) and velocity were evaluated. Figure 7b–d shows the plot of WSE along the centre line of Habu for different mesh sizes. From the figure, it is shown that the WSE is almost identical for all mesh cases. From the figure, it can be concluded that even with the coarsest mesh used here (15 cm), HEC-RAS can provide accurate prediction for the WSE.

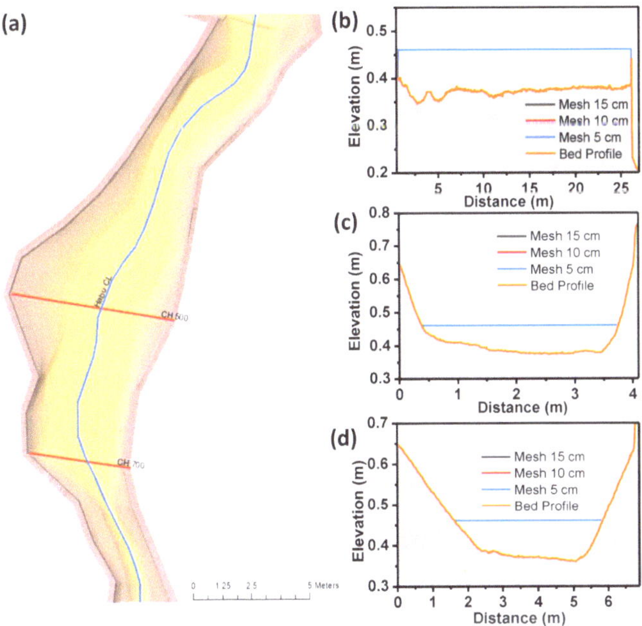

Figure 7. (**a**) River centreline (blue line) and CH (red line) for Habu. WSE plot along the Habu end centreline for different mesh sizes (**b**) centreline, (**c**) CH700, (**d**) CH500.

The velocities along CH700 and CH500 are presented in Figure 8. The result shows that mesh convergence is also achieved for velocity. Although the coarse mesh can provide an equally accurate result, for the simulation that follows, a mesh size of 5 cm was used throughout. It should be noted that the solution produced by the DWA might be different from the SWE depending on the corresponding Reynolds number, with a higher value representing turbulence flow [39]. As DWA ignores the nonlinear effect caused by the convective acceleration, the impact of turbulence on the flow cannot be accounted for. For the case considered here, where the result is mostly turbulence, the use of SWE is more appropriate and was used for this study. The DWA method was adopted just to simplify the comparison of the convergence study.

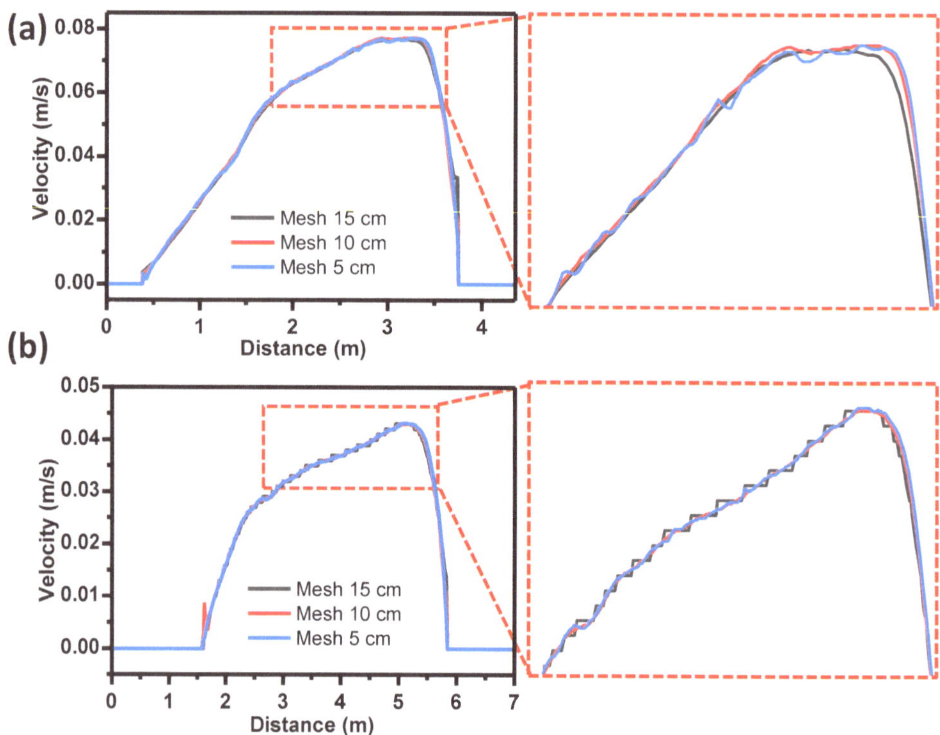

Figure 8. Sensitivity plot of velocity along Habu (**a**) CH700, (**b**) CH500 for different mesh sizes. The insets show the enlarged curves to aid comparisons.

3.2.2. Model Calibration

Calibration is finding the most appropriate parameter that must be included in the model to produce results representing the physical condition. For hydraulic modelling involving SWE, an important calibration factor that needs to be determined is the Manning coefficient. Manning's coefficients represent the influence the bottom roughness has on the solution. As the physical model involves a bottom boundary that is relatively smooth, the influence Manning has on the solution might not be that critical. Therefore, two Manning values were used for calibration: 0.0025 for a smooth surface and 0.01 for a slightly rough one. Normal concrete has a roughness of 0.01. A smaller value increases velocity but reduces water surface elevation, while a larger value has the opposite effect. However, since the physical model's surface is assumed to be uniformly smooth, a single Manning value was used for the entire domain.

Figure 9a shows the observation points from the experimental work. The point is located along the channel's CH. The red dots are the points where a comparison between simulation and experimental observation is carried out. To determine the influence of Manning on the result, a comparison is carried out for the water depth and velocity at the observation points. In addition, a comparison was made between DWA and SWE, the former assuming negligible flow acceleration while the latter includes the full nonlinear solution. Please note that the calibration should only be carried out for the 100-year ARI scenario, as this represents the worst-case scenario and provides the largest velocity changes compared to the rest of the scenarios.

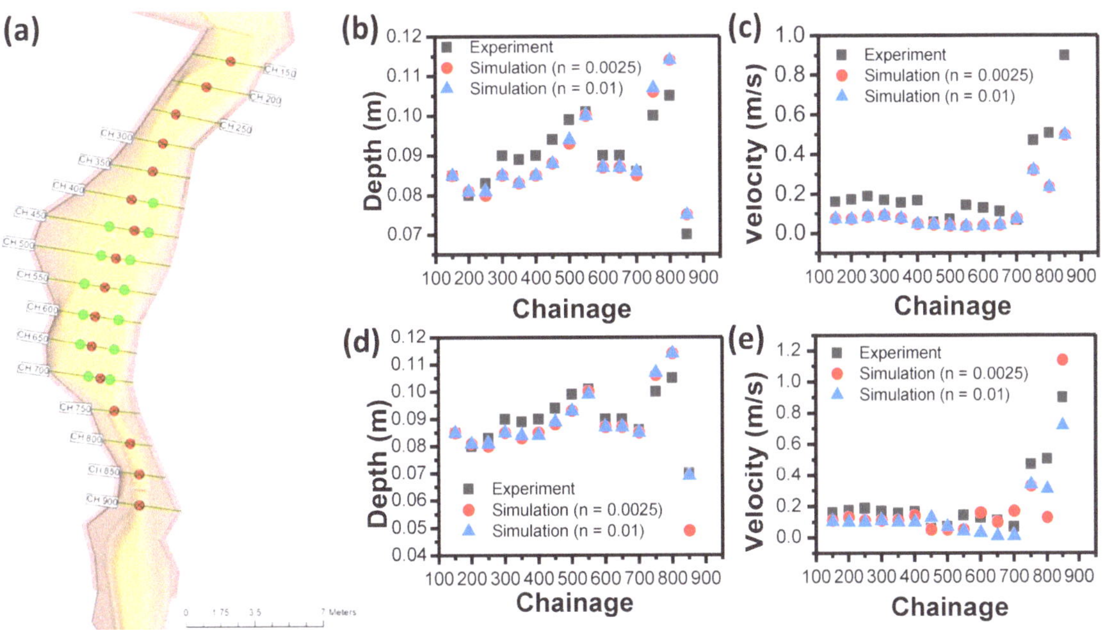

Figure 9. (**a**) Observation points from experimental work. Red dots denote the location of sampling used for comparison between experiment and simulation, while green dots denote the additional experimental sampling locations. Water depth and velocity comparison for DWA (**b**), (**c**) and SWE (**d**), (**e**) between experimental observation and simulation results.

Figure 9b,d show the comparison of the water depth between experimental observation and simulation for DWA and SWE, respectively. The plot shows that the water depth can be predicted accurately by both DWA and SWE. A closer inspection shows that SWE provides overall better accuracy than DWA, except for CH850, due to the presence of the check dam where the turbulence and hydraulic jump occurred. The experimental and simulation values differed by 25–65%. The differences between the Manning results were quite small for this comparison. Figure 9c,e compares the velocity at the observation points between experimental and simulation results for DWA and SWE, respectively. For this comparison, much larger differences can be seen. For DWA, the velocity predicted by the simulation model is consistently lower than those obtained from the experiment, with the result from Manning varying slightly. However, for the SWE, the result shows better prediction by the simulation model, with a closer agreement for the Manning value of 0.0025 compared to 0.01. Apart from CH800 and CH850, the Manning value of 0.0025 shows good agreement with the observed data.

It is important to note that to compare the velocity of the simulated model against the physical model, the velocity for SWE is taken as the average velocity over some time.

This is because the simulation does not reach a true steady state globally [40]. The velocity fluctuation is presented in Figure 10a,b. At CH750 in Figure 10a, where the channel is relatively narrow, the flow is predominantly one-dimensional. The velocity at the centre of the channel fluctuates rapidly at the beginning of the simulation but stabilizes over time. However, at CH500, where the channel is wider, the flow becomes two-dimensional, and vortices form as the simulation progresses. As the vortices develop, the velocity taken at the middle of the cross-section fluctuates, as shown in Figure 10b. The velocity fluctuation becomes less pronounced once the vortices are fully formed. To compare with the experimental data, the velocity magnitude is the average over a period during which the vortices have fully formed. The root mean square of velocity fluctuation is small.

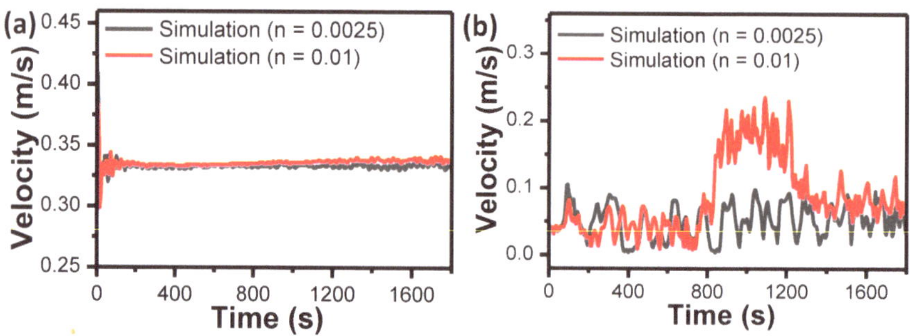

Figure 10. Velocity time history at (**a**) CH750, (**b**) CH500 for simulation with different Manning's values.

A comparison with observation data for the Habu catchment with groynes installed has also been carried out. Figure 11a shows the location of the groynes along the Habu catchment. Figure 11b,c shows the water depth and velocity comparison between the experiment and simulation result at the observation location. The result shows good agreement for the water depth and a slight variation for the velocity. These small variations were due to uncertainties in model parameters. The numerical model relies on specific input parameters, such as friction coefficients or roughness values, which may be uncertain or difficult to estimate accurately and may not perfectly capture the exact physics of the flow in the physical system. However, comparing both chosen Manning values shows a small influence of the value on the hydrodynamics of the flow.

3.3. Simulation

3.3.1. Habu Existing Conditions

This section presents the simulation results and analysis for Habu without the groynes and focuses on the hydrodynamics of the flow. At the start of each simulation, a fixed water level was set for the entire domain. The simulation period for all cases was 30 min. Figure 12a shows a plot of the water surface elevation (WSE) along the Habu centreline. Although the WSE in all cases does not vary significantly along the channel, the corresponding water depth is not constant due to the undulating bathymetry. Figure 12b–d displays the simulated water depth at the end of the simulation for 1-year, 5-years, and 100-years ARI. The figures reveal that the water depth throughout the domain varies significantly at several locations due to variations in the bathymetry. These variations, coupled with differences in the channel width, are expected to lead to a spatially varying velocity distribution.

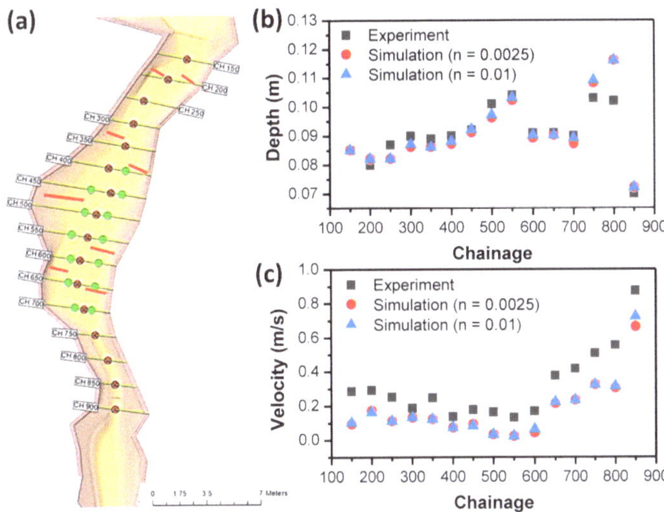

Figure 11. (**a**) Location of the groynes along Habu catchment and experimental sampling locations. Red dots denote the location of sampling used for comparison between the experiment and simulation, while green dots denote the additional experimental sampling locations. (**b**) water depth (**c**) velocity comparison between experimental observation and simulation results (SWE) for a model with groynes installed.

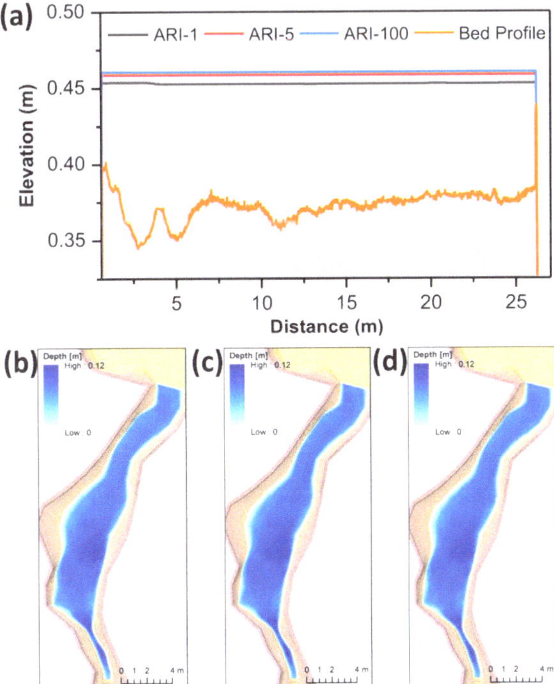

Figure 12. (**a**) Water Surface Elevation (WSE) along the Habu centreline. Simulated water depth at the end of the simulation for (**b**) 1 year (**c**) 5 years, and (**d**) 100 years ARI.

The velocity magnitude for different ARI at time steps of 10 min, 20 min, and 30 min is presented in Figure 13. In all cases, the highest velocity can be observed at the upstream end of the channel. At this upstream part, the channel is narrow, with a very steep bottom. This combination forces the water to flow at a relatively high velocity within this constricted area.

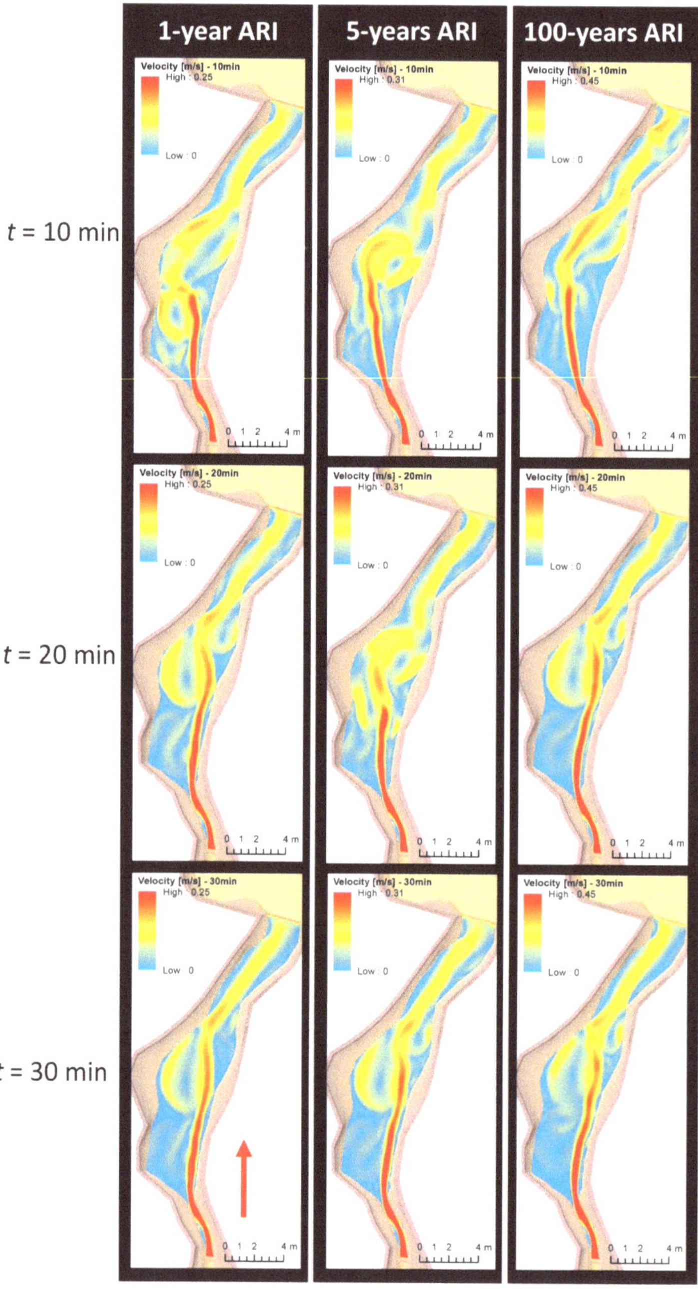

Figure 13. Velocity magnitude at a time step, *t* = 10, 20 and 30 min, for 1-year, 5-years and 100-years ARI. The arrow represents the flow direction of the water and is applied to all cases.

For 100 years of ARI, a comparison with the experiment shows that the simulated velocity is underpredicted. At CH 850, experimental observation recorded a value of around 0.9 m/s, whereas simulation predicted a value of about 0.37 m/s. The differences can be attributed to how the inlet boundary is applied. Due to stability problems at the upstream check dam, where rapid wetting and drying occur, the inlet has been moved to just after the check dam. This reduces the high velocity due to free fall. Unfortunately, a velocity boundary condition cannot be specified in HEC-RAS, as only flow rate is allowed; hence, the velocity at the boundary is calculated from the flow rate. As the water enters a broader part of the channel, the variation of the velocity magnitude becomes increasingly large. Flow recirculation or rollers can be observed as the channel enlarges. The same roller-like flow patterns were observed in the physical model; see Video S1. This was due to a turbulent flow at the inlet area. Turbulence occurs when the flow of water becomes irregular, forming vortices and eddies that cause the water to roll. This rolling motion can also be caused by the irregular surface beneath the water. The velocity reduces in the downstream direction due to a downstream check dam. The particle tracing plots in Figure 14a–c show that the flow is highly unsteady, especially at the start of the simulation. By observing the evolution of the magnitude of the velocity at increasing time steps, one can observe significant changes in the velocity distribution. At time step t = 5 min, the high-velocity region impinges on the left side of the channel. However, as time progresses, this pattern changes. At the next step, t = 20 min onward, the high-velocity region after the expansion changes sides, impinging to the right. This change is caused by the formation of vortices at around CH 450. As the simulation reaches a steady state, the velocity shows an identical pattern for all ARI cases.

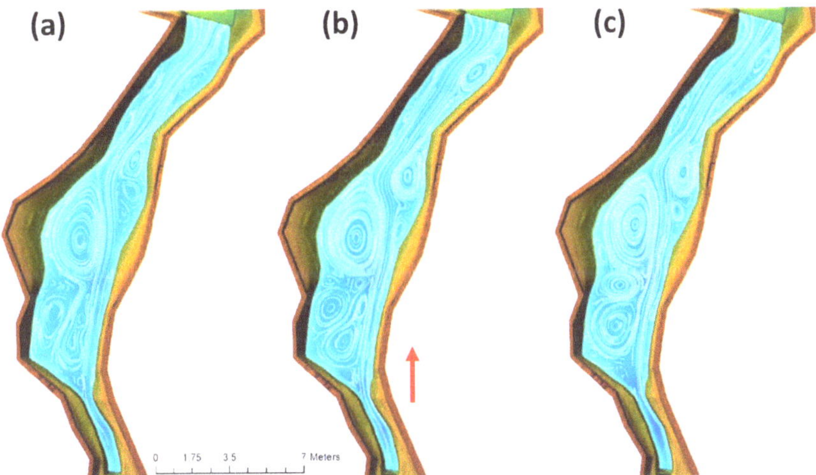

Figure 14. Particle tracing at time step, t = 30 min, for (**a**) 1-year, (**b**) 5 years and (**c**) 100-years ARI. The arrow represents the flow direction of the water and is applied to all cases.

Figure 15a–c displays the cross-sectional velocity profiles for CH650, CH500, and CH350. These plots compare the velocity across the chainage at t = 30 min (steady state) for different ARI cases. The distance between banks is measured from left to right when looking downstream. It should be noted that the plots only show the magnitude of the velocity and do not consider its direction. With higher ARI, the inflow increases, resulting in an increase in velocity magnitude for all cases. The velocity profiles at CH650 and CH500 reveal that the right side has a higher velocity than the left. This observed velocity profile may be explained by the eddy that forms on the left side of the channel around CH650 and CH500, as shown in Figure 14. As this eddy forms due to low-velocity flow, a high-pressure region is developed, which pushes the high-velocity flow to the right. At CH350, the eddy forms on the right side and pushes the flow to the right, creating a high-velocity region, as shown in Figure 15c.

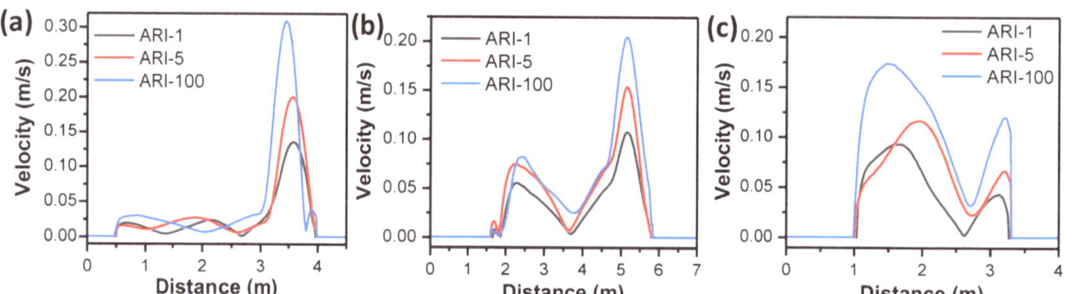

Figure 15. Velocity comparison between 1-year, 5-years and 100-years ARI during steady-state conditions at (**a**) CH650, (**b**) CH500 and (**c**) CH350.

The differences in the flow profile across the domain have large implications for sediment transport dynamics. The movement of sediment is highly dependent on the hydrodynamics of the flow. For example, the formation of vortices can entrain sediment and force it to settle.

3.3.2. Habu with Groynes Installed

This section presents the results and analysis of the simulation of different ARI events with groynes installed. Figure 16a–c shows the plot of water depth at the end of the simulation for all cases. Despite the presence of the groynes, no appreciable difference in the water depth can be observed compared to the existing condition. With the installation of the groynes, the "unsteadiness" of the flow pattern has somehow been suppressed.

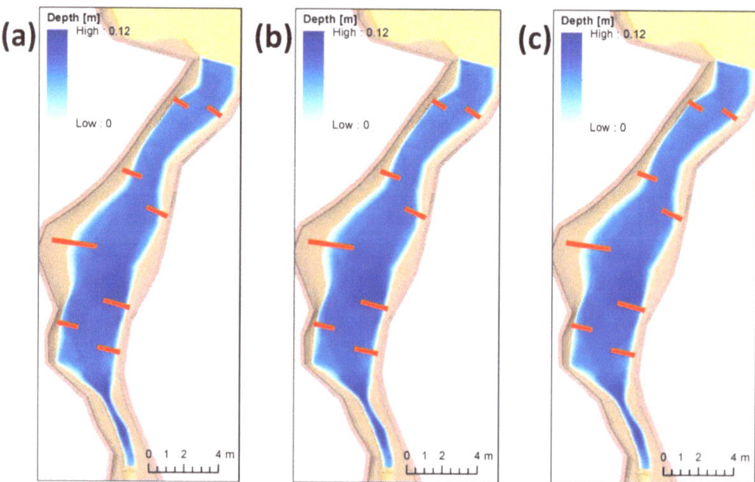

Figure 16. Water depth at Habu with groynes installed for (**a**) 1 year (**b**) 5 years, and (**c**) 100 years ARI.

This is evidenced by the velocity magnitude plot at time $t = 10$ min, $t = 20$ min, and $t = 30$ min, given in Figure 17. The plot shows almost identical patterns for all cases and time steps. High-magnitude velocities can be observed to flow at the centreline of the channel. In addition, vortices can be observed forming on the front side of the groynes (looking downstream). These vortices have a lower velocity than the main flow and could be significant in trapping sediment. The number of vortices has also increased as compared to the existing condition.

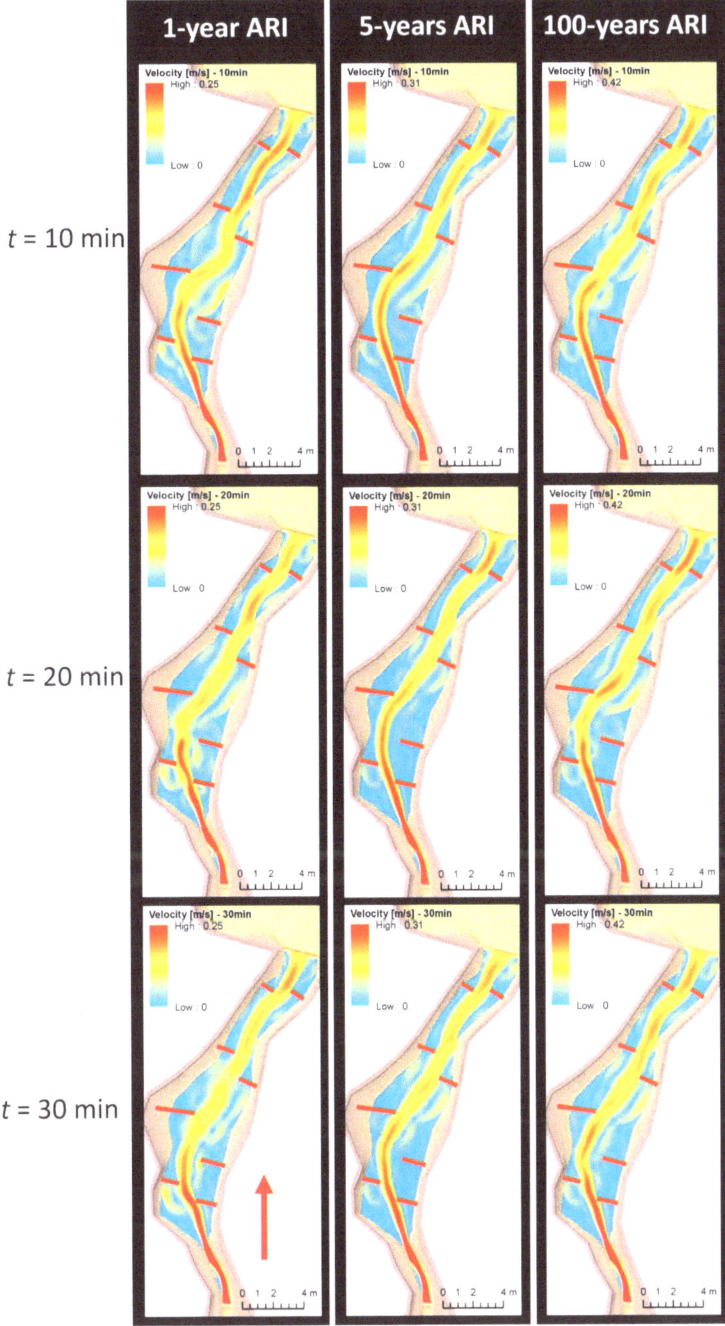

Figure 17. Velocity magnitude comparison between 1-year, 5-years and 100-years ARI for Habu with groynes installed at a time step, t = 10, 20 and 30 min. The arrow represents the flow direction of the water and is applied to all cases.

The particle tracing plot in Figure 18 shows the dominant flow pattern and the vortices generated. The effect of groynes on the flow pattern and hydrodynamics of water can vary depending on the groynes' size, shape, and placement. When water flows towards the groyne, it is forced to slow down and divert around the structure. This creates eddies and vortices on the downstream side of the groyne, resulting in areas of slower-moving water that will lead to sediment deposition. On the upstream side of the groyne, water accelerates and creates faster-moving currents. This can cause erosion at the base of the groyne and create a scour hole. The placement of groynes can also affect the flow pattern and hydrodynamics of water. Too many groynes placed too closely together can disrupt the natural flow of water and create potential stagnant areas, e.g., around the CH600, while too few groynes can lead to increased erosion, e.g., in the area between the CH600 and the CH400. The optimal placement and number of groynes to minimise the impact on the flow pattern and hydrodynamics must be determined to mitigate erosion effectively.

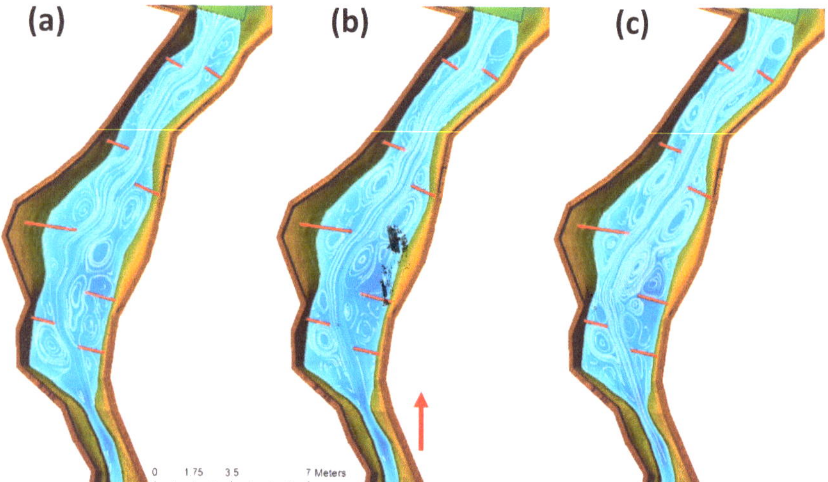

Figure 18. Particle tracing comparison between (**a**) 1-year (**b**) 5-years, and (**c**) 100-years ARI for Habu with groynes installed. The arrow represents the flow direction of the water and is applied to all cases.

A comparison between velocity along CH600, CH500, and CH400 between the existing condition and with groynes installed is presented in Figure 19. These plots show the difference in magnitude and location of the dominant flow. At CH600 to CH500, the high-velocity magnitude is located on the opposing side of the channel for cases with and without groynes. The location change is caused by installing groynes at the upstream part of the channel. Furthermore, the magnitude of the dominant velocity at the upstream part (CH600 and CH500) is reduced with the installation of groynes and increases at the downstream part (CH400). In this scenario, the existence of groynes can change the flow pattern. To trap sediment, groynes can be built to generate flow that directs sediment to a specific area within the domain [41]. The ability of a numerical model to predict sediment transport and migration in Ringlet Reservoir will be explored and discussed in our future study using Flow-3D software that solves tougher free-surface flow problems.

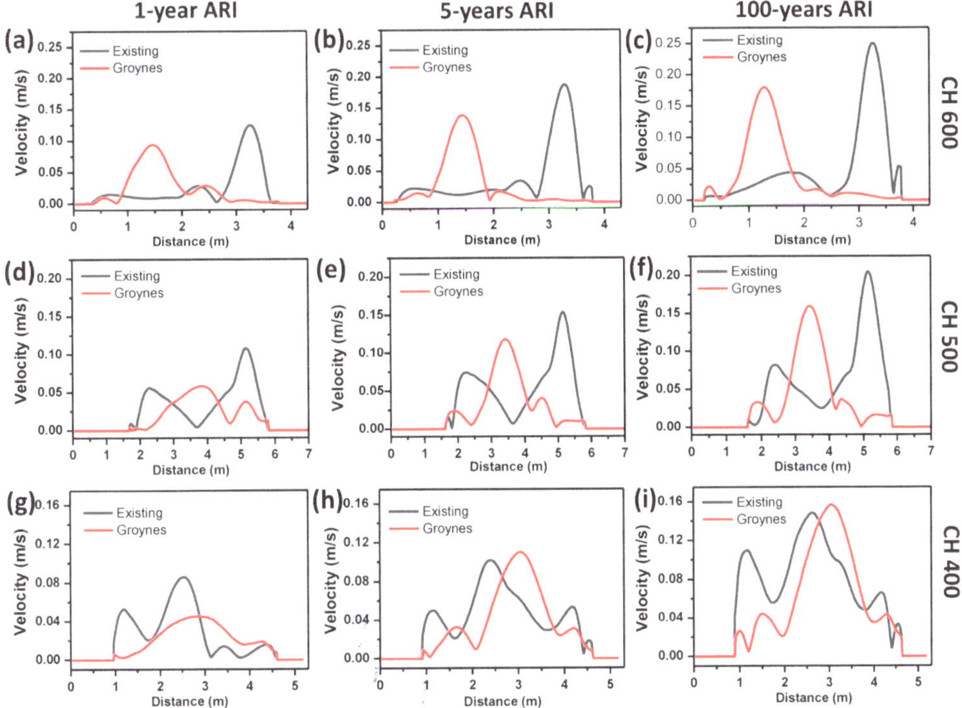

Figure 19. Velocity comparison between existing condition (black line) and groynes installed (red line) at CH 600 (**a**–**c**), CH 500 (**d**–**f**), CH400 (**g**–**i**) for 1-year, 5-years and 100-years ARI.

4. Conclusions

Flow velocity measurements in Habu (Ringlet Reservoir) were plotted in velocity points for existing and mitigation (groynes) conditions. Based on the experimental results, the flow velocity patterns for all return period tested is very similar but with different magnitude of velocities. This also can be observed in the existence of groynes in the channel. Groynes also change the pattern of the velocity. Lower velocity was observed on the side of the channel, and higher velocity was more concentrated in the middle. An unsteady shallow water module in HEC-RAS has been used to investigate the flow hydrodynamic of the problem. Comparison with experimental results shows that HEC-RAS can provide an excellent prediction for the water surface elevation but falls slightly off when predicting the velocity. The unsteady shallow water equation allows HEC-RAS to simulate complex flow phenomena involving highly energetic free surface turbulence flow. The output can provide further insight into the flow behaviour, which provide essential information for the experimental work and future flow predictions. Detailed flow information such as velocity and free surface flow provided in this study will give valuable information that can be used to design better mitigation options. Understanding flow characteristics in reservoirs is crucial for designing effective mitigation options because it helps identify potential risks and predict behavior. For example, knowledge of flow characteristics can help determine the likelihood of sedimentation or erosion and design appropriate measures to prevent them.

Supplementary Materials: The following supporting information can be downloaded at: https://www.mdpi.com/article/10.3390/w15101883/s1, Video S1: Physical observation of vortex formation using dye tracer.

Author Contributions: S.M.D. and M.H.J. Conceptualization, and project supervision; M.S.F.M., M.K.H.S., R.S. and R.S.M. methodology and field investigation; E.H.K. and M.M. performed software modelling and data validation; N.S.R., D.S.M.I. and M.Z.M. performed data curation, general analysis, and manuscript preparation. All authors have read and agreed to the published version of the manuscript.

Funding: The National Water Research Institute of Malaysia (NAHRIM), Ministry of Natural Resources, Environment and Climate Change, spearheaded and funded this research.

Data Availability Statement: The data can be obtained from the corresponding author upon reasonable request.

Acknowledgments: The author would like to thank TNB Research Sdn Bhd (TNBR) for collaborating on this project's research and exchanging data. This paper's contents are based solely on research and are intended solely for research purposes; they do not reflect the views of TNBR or its subsidiaries. The National Water Research Institute of Malaysia (NAHRIM) provided the physical model experiment facilities necessary for this work. The authors are grateful for the help.

Conflicts of Interest: The authors have declared there are no competing interests. The present paper is an original work, and they are not associated with any profitable organization or company with any financial interest.

References

1. Maass, A.; Hufschmidt, M.M.; Dorfman, R.; Thomas, J.H.A.; Marglin, S.A.; Fair, G.M. *Design of Water-Resource Systems*; Harvard University Press: Cambridge, MA, USA, 1962.
2. Annandale, G.W.; Morris, G.L.; Karki, P. *Extending the Life of Reservoirs: Sustainable Sediment Management for Dams and Run-of-River Hydropower*; The World Bank: Washington, DC, USA, 2016; p. 188.
3. Branche, E. The multipurpose water uses of hydropower reservoir: The SHARE concept. *Comptes Rendus Phys.* **2017**, *18*, 469–478. [CrossRef]
4. IEA. Hydroelectricity. 2022. Available online: https://www.iea.org/reports/hydroelectricity (accessed on 13 March 2023).
5. Huang, C.-C.; Lai, J.-S.; Lee, F.-Z.; Tan, Y.-C. Physical Model-Based Investigation of Reservoir Sedimentation Processes. *Water* **2018**, *10*, 352. [CrossRef]
6. Basson, G. Reservoir sedimentation—An overview of global sedimentation rates and predicted sediment deposition. In Proceedings of the Oral Contribution to the International CHR Workshop-Expert Consultation: Erosion, Transport and Deposition of Sediments, Bern, Switzerland, 28–30 April 2008; pp. 74–79.
7. Garcia, M. *Sedimentation Engineering: Processes, Measurements, Modeling, and Practice*; ASCE: Reston, VA, USA, 2008; pp. 2–12.
8. Padhy, M.K.; Saini, R.P. A review on silt erosion in hydro turbines. *Renew. Sustain. Energy Rev.* **2008**, *12*, 1974–1987. [CrossRef]
9. Sidek, L.M.; Luis, I.J. Sustainability of Hydropower Reservoir as Flood Mitigation Control: Lesson Learned from Ringlet Reservoir, Cameron Highlands, Malaysia. 2014. In Proceedings of the 2nd International Conference on Water Resources (ICWR2012), Langkawi, Malaysia, 5–7 November 2012.
10. Sidek, L.M.; Luis, I.J.; Desa, M.N.M.; Julien, P.Y. Challenge In Running Hydropower As Source Of Clean Energy: Ringlet Reservoir, Cameron Highlands Case Study. In Proceedings of the National Graduate Conference 2012 (NatGrad2012), Kajang, Malaysia, 8–10 November 2012.
11. Kositgittiwong, D.; Chinnarasri, C.; Julien, P. Numerical simulation of flow velocity profiles along a stepped spillway. *Proc. Inst. Mech. Eng. E J. Process Mech. Eng.* **2013**, *227*, 327–335. [CrossRef]
12. Zhang, Y.; Han, L.; Chen, L.; Wang, C.; Chen, B.; Jiang, A. Study on the 3D Hydrodynamic Characteristics and Velocity Uniformity of a Gravity Flow Circular Flume. *Water* **2021**, *13*, 927. [CrossRef]
13. McCoy, A.; Constantinescu, G.; Weber Larry, J. Numerical Investigation of Flow Hydrodynamics in a Channel with a Series of Groynes. *J. Hydraul. Eng.* **2008**, *134*, 157–172. [CrossRef]
14. Choufu, L.; Abbasi, S.; Pourshahbaz, H.; Taghvaei, P.; Tfwala, S. Investigation of Flow, Erosion, and Sedimentation Pattern around Varied Groynes under Different Hydraulic and Geometric Conditions: A Numerical Study. *Water* **2019**, *11*, 235. [CrossRef]
15. Desa, S.M.; Samion, M.K.H.; Hamzah, A.F.; Manan, E.A.; Noh, M.N.M. Physical Modelling Study on Sediment Removal of Sungai Habu and Sungai Ringlet, Cameron Highlands. In *ICDSME 2019—Water Resources Development and Management*; Springer: Singapore, 2020; pp. 184–196.
16. Al-Qaisi, D.A.; Alqaisi, M.; Abdulridha, R. Analysis of the hydraulic characteristics of al mahawil stream using hecras: A field study. *Pollut. Res.* **2020**, *39*, 856–863.
17. Mojtahedi, A.; Basmenji, A.B. Numerical and Field Investigation of the Impacts of the Bank Protection Projects on the Fluvial Hydrodynamics (Case Study: Ghezel Ozan River). *Int. J. Eng. Technol.* **2017**, *9*, 492–497. [CrossRef]
18. Walker, D.J.; Dong, P.; Anastasiou, K. Sediment Transport near Groynes in the Nearshore Zone. *J. Coast. Res.* **1991**, *7*, 1003–1011.
19. Luis, J.; Sidek, L.M.; Desa, M.N.M.; Julien, P.Y. Hydropower Reservoir for Flood Control: A Case Study on Ringlet Reservoir, Cameron Highlands, Malaysia. *J. Flood Eng.* **2013**, *4*, 87–102.

20. Berhad, T.N. TNB's Nenggiri Dam to Play Multi-Fold Role. 2021. Available online: https://www.thesundaily.my/local/tnb-s-nenggiri-dam-to-have-multifold-roles-YG7811573 (accessed on 23 March 2023).
21. Teh, S.H. Soil Erosion Modeling Using RUSLE and GIS on Cameron Highlands, Malaysia for Hydropower Development. Master's Thesis, The School for Renewable Energy Science, Akureyri, Iceland, 2011.
22. Luis, J.; Sidek, L.M.; Desa, M.N.M.; Julien, P.Y. Sustainability of hydropower as source of renewable and clean energy. *IOP Conf. Ser. Earth Environ. Sci.* **2013**, *16*, 012050. [CrossRef]
23. Raj, J.K. Land use changes, soil erosion and decreased base flow of rivers at Cameron Highlands, Peninsular Malaysia. *Bull. Geol. Soc. Malays.* **2002**, *45*, 3–10. [CrossRef]
24. Razad, A.Z.A.; Abbas, N.A.; Sidek, L.M.; Alexander, J.L.; Jung, K. Sediment management strategies for hydropower reservoirs in active agricultural area. *Int. J. Eng. Technol.* **2018**, *7*, 228–233.
25. Vangelis, H.; Zotou, I.; Kourtis, I.M.; Bellos, V.; Tsihrintzis, V.A. Relationship of Rainfall and Flood Return Periods through Hydrologic and Hydraulic Modeling. *Water* **2022**, *14*, 3618. [CrossRef]
26. Morales-Marin, L.A.; French, J.R.; Burningham, H.; Battarbee, R.W. Three-dimensional hydrodynamic and sediment transport modeling to test the sediment focusing hypothesis in upland lakes. *Limnol. Oceanogr.* **2018**, *63*, S156–S176. [CrossRef]
27. Kang, T.; Jang, C.-L.; Kimura, I.; Lee, N. Numerical Simulation of Debris Flow and Driftwood with Entrainment of Sediment. *Water* **2022**, *14*, 3673. [CrossRef]
28. Ma, C.; Xu, X.; Yang, J.; Cheng, L. Safety Monitoring and Management of Reservoir and Dams. *Water* **2023**, *15*, 1078. [CrossRef]
29. El Kadi Abderrezzak, K.; Die Moran, A.; Mosselman, E.; Bouchard, J.-P.; Habersack, H.; Aelbrecht, D. A physical, movable-bed model for non-uniform sediment transport, fluvial erosion and bank failure in rivers. *J. Hydro-Environ. Res.* **2014**, *8*, 95–114. [CrossRef]
30. Isaac, N.; Eldho, T.I.; Gupta, I.D. Numerical and physical model studies for hydraulic flushing of sediment from Chamera-II reservoir, Himachal Pradesh, India. *ISH J. Hydraul. Eng.* **2014**, *20*, 14–23. [CrossRef]
31. Sun, Y.; Zhang, L.; Liu, J.; Lin, J.; Cui, Q. A Data Assimilation Approach to the Modeling of 3D Hydrodynamic Flow Velocity in River Reaches. *Water* **2022**, *14*, 3598. [CrossRef]
32. Yossef, M. *The Effect of Groynes on Rivers: Literature Review*; Cluster Publicatienummer 03.03.04; Delft University of Technology: Delft, The Netherlands, 2002.
33. Michioku, K.; Osawa, Y.; Kanda, K. Performance of a groyne in controlling flow, sediment and morphology around a tributary confluence. *E3S Web Conf.* **2018**, *40*, 04006. [CrossRef]
34. Farzad Mohammadi, N.M. Technical Evaluation of the Performance of River Groynes Installed in Sezar and Kashkan Rivers, Lorestan, Iran. *J. Appl. Environ. Biol. Sci.* **2015**, *5*, 258–268.
35. Baker, C.; Johnson, T.; Flynn, D.; Hemida, H.; Quinn, A.; Soper, D.; Sterling, M. Chapter 4—Computational techniques. In *Train Aerodynamics*; Baker, C., Johnson, T., Flynn, D., Hemida, H., Quinn, A., Soper, D., Sterling, M., Eds.; Butterworth-Heinemann: Oxford, UK, 2019; pp. 53–71.
36. Katopodes, N.D. Chapter 8—Methods for Two-Dimensional Shallow-Water Flow. In *Free-Surface Flow*; Katopodes, N.D., Ed.; Butterworth-Heinemann: Oxford, UK, 2019; pp. 500–567.
37. U.S. Army Corps of Engineers. *HEC-RAS 2D Modeling User's Manual*; U.S. Army Corps of Engineers: Washington, DC, USA, 2021.
38. Sjölund, L. A Study of Sedimentation Problems in the Lower Reaches of the River Österdalälven. Master's Thesis, KTH Royal Institute of Technology, Stockholm, Sweden, 2018.
39. Shields, A.B. *Application of Similarity Principles and Turbulence Research to Bed-Load Movement*; California Institute of Technology: Pasadena, CA, USA, 1936.
40. Camnasio, E.; Orsi, E.; Schleiss, A.J. Experimental study of velocity fields in rectangular shallow reservoirs. *J. Hydraul. Res.* **2011**, *49*, 352–358. [CrossRef]
41. Yossef Mohamed, F.M.; de Vriend Huib, J. Sediment Exchange between a River and Its Groyne Fields: Mobile-Bed Experiment. *J. Hydraul. Eng.* **2010**, *136*, 610–625. [CrossRef]

MDPI AG
Grosspeteranlage 5
4052 Basel
Switzerland
Tel.: +41 61 683 77 34

Water Editorial Office
E-mail: water@mdpi.com
www.mdpi.com/journal/water